A voyage round the world in the years MDCCXL, I, II, III, IV

Baron George Anson Anson

Alpha Editions

This edition published in 2024

ISBN : 9789364739238

Design and Setting By
Alpha Editions
www.alphaedis.com
Email - info@alphaedis.com

As per information held with us this book is in Public Domain.
This book is a reproduction of an important historical work. Alpha Editions uses the best technology to reproduce historical work in the same manner it was first published to preserve its original nature. Any marks or number seen are left intentionally to preserve its true form.

Contents

INTRODUCTION	- 1 -
BIBLIOGRAPHY	- 11 -
AUTHOR'S INTRODUCTION	- 14 -
BOOK I	- 21 -
CHAPTER I	- 23 -
CHAPTER II	- 32 -
CHAPTER III	- 36 -
CHAPTER IV	- 46 -
CHAPTER V	- 52 -
CHAPTER VI	- 63 -
CHAPTER VII	- 72 -
CHAPTER VIII	- 76 -
CHAPTER IX	- 82 -
CHAPTER X	- 92 -
BOOK II	- 99 -
CHAPTER I	- 101 -
CHAPTER II	- 113 -
CHAPTER III	- 121 -
CHAPTER IV	- 133 -
CHAPTER V	- 143 -
CHAPTER VI	- 156 -
CHAPTER VII	- 168 -
CHAPTER VIII	- 174 -
CHAPTER IX	- 179 -
CHAPTER X	- 185 -
CHAPTER XI	- 196 -

CHAPTER XII ....- 203 -
CHAPTER XIII ....- 210 -
CHAPTER XIV ....- 218 -
BOOK III ....- 227 -
CHAPTER I ....- 229 -
CHAPTER II ....- 238 -
CHAPTER III ....- 250 -
CHAPTER IV ....- 257 -
CHAPTER V ....- 260 -
CHAPTER VI ....- 267 -
CHAPTER VII ....- 273 -
CHAPTER VIII ....- 285 -
CHAPTER IX ....- 296 -
CHAPTER X ....- 307 -

INTRODUCTION

The men-of-war in which Anson went to sea were built mostly of oak. They were painted externally yellow, with a blue stripe round the upper works. Internally, they were painted red. They carried cannon on one, two, or three decks according to their size. The biggest ships carried a hundred cannon and nearly a thousand men. The ship in which this famous voyage was made was of the middle size, then called the fourth-rate. She carried sixty cannon, and a crew of four hundred men. Her lower gun deck, a little above the level of the water, was about 140 feet long. She was of about a thousand tons burthen.

Though this seems small to us, it is not small for a wooden ship. It is not possible to build a long wooden ship. The *Centurion*, though short, was broad, bulky, and deep. She was fit for the sea. As she was built more to carry cannon than to sail, she was a slow sailer. She became slower as the barnacles gathered on her planks under the water. She carried three wooden masts, each fitted with two or three square sails, extended by wooden yards. Both yards and masts were frequently injured in bad weather.

The cannon were arranged in rows along her decks. On the lower gun deck, a little above the level of the water, she carried twenty-six twenty-four-pounders, thirteen on a side. These guns were muzzle-loading cannon which flung twenty-four-pound balls for a distance of about a mile. On the deck above this chief battery, she carried a lighter battery of twenty-six nine- or twelve-pounder guns, thirteen on a side. These guns were also muzzle-loading. They flung their balls for a distance of a little more than a mile.

On the quarter-deck, the poop, the forecastle, and aloft in the tops (the strong platforms on the masts), were lighter guns, throwing balls of from a half to six pounds' weight. Some of the lightest guns were mounted on swivels, so that they could be easily pointed in any direction. All the guns were clumsy weapons. They could not be aimed with any nicety. The iron round shot fired from them did not fit the bores of the pieces. The gun-carriages were clumsy, and difficult to move. Even when the carriage had been so moved that the gun was accurately trained, and when the gun itself had been raised or depressed till it was accurately pointed, the gunner could not tell how much the ball would wobble in the bore before it left the muzzle. For these reasons all the effective sea-fights were fought at close range, from within a quarter of a mile of the target to close alongside. At a close range, the muskets and small-arms could be used with effect.

The broadside cannon pointed through square portholes cut in the ship's sides. The ports were fitted with heavy wooden lids which could be tightly closed when necessary. In bad weather, the lower-deck gun ports could not be opened without danger of swamping the ship. Sometimes, when the lower-deck guns were fought in a gale, the men stood knee deep in water.

In action the guns were "run out" till their muzzles were well outside the port, so that the flashes might not set the ship's side on fire. The shock of the discharge made them recoil into a position in which they could be reloaded. The guns were run out by means of side tackles. They were kept from recoiling too far by strong ropes called breechings. When not in use, and not likely to be used, they were "housed," or so arranged that their muzzles could be lashed firmly to the ship's side. In a sea way, when the ship rolled very badly, there was danger of the guns breaking loose and rolling this way and that till they had knocked the ship's side out. To prevent this happening, clamps of wood were screwed behind the wheels of the gun-carriages, and extra breechings were rove, whenever bad weather threatened.

The great weight of the rows of cannon put a severe strain upon the upper works of the ship. In bad weather, during excessive rolling, this strain was often great enough to open the seams in the ship's sides. To prevent this, and other costly damage, it was the custom to keep the big men-of-war in harbour from October until the Spring. In the smaller vessels the strain was made less by striking down some of the guns into the hold.

The guns were fired by the application of a slow-match to the priming powder in the touch-holes. The slow-matches were twisted round wooden forks called linstocks. After firing, when the guns had recoiled, their bores were scraped with scrapers called "worms" to remove scraps of burning wad or cartridge. They were then sponged out with a wet sponge, and charged by the ramming home of fresh cartridges, wads, and balls. A gun's crew numbered from four to twelve men, according to the size of the piece. When a gun was trained aft or forward, to bear on an object before or abaft the beam, the gun's crew hove it about with crows and handspikes.

As this, and the other exercise of sponging, loading, and running out the guns in the heat, stench, and fury of a sea-fight was excessively hard labour, the men went into action stripped to the waist. The decks on those occasions were thickly sanded, lest the blood upon them should make them too slippery for the survivors' feet. Tubs of water were placed between the guns for the wetting of the sponges and the extinguishing of chance fires. The ship's boys carried the cartridges to the guns from the magazines below the water-line. The round-shot were placed close to hand in rope rings called garlands. Nets were spread under the masts to catch wreck from aloft. The decks were "cleared for action." All loose articles about the decks, and all movable wooden articles such as bulkheads (the partitions between cabins), mess-tables, chests, casks, etc., were flung into the hold or overboard, lest shot striking them should splinter them. Splinters were far more dangerous than shot. In this book it may be noticed that the officers hoped to have no fighting while the gun decks of the ships in the squadron were cumbered with provision casks.

The ships of war carried enormous crews. The *Centurion* carried four hundred seamen and one hundred soldiers. At sea, most of this complement was divided into two watches. Both watches were subdivided into several divisions, to each of which was allotted some special duty, as the working of the main-mast, the keeping of the main deck clean, etc., etc. Many members of the crew stood no watch, but worked at special crafts and occupations about the ship. A wooden ship of war employed and kept busy a carpenter and carpenter's mates, a sailmaker and sailmaker's mates, a cooper and a gunner, each with his mates, and many other specially skilled craftsmen and their assistants. She was a little world, carrying within herself all that she needed. Her daily business required men to sail her and steer her, men to fight her guns, men to rule her, men to drill, men to play the spy, men to teach, preach, and decorate, men to clean her, caulk her, paint her and keep her sweet, men to serve out food, water, and intoxicants, men to tinker, repair, and cook and forge, to doctor and operate, to bury and flog, to pump, fumigate and scrape, and to load and unload. She called for so many skilled craftsmen, and provided so much special employment out of the way of seamanship, that the big crew was never big enough. The special employments took away now one man, now another, till there were few left to work the ship. The soldiers and marines acted as a military guard for the prevention of mutiny. They worked about the ship, hauling ropes, etc., when not engaged in military duty.

The hundreds of men in the ship's crew lived below decks. Most of them lived on the lower gun deck in the narrow spaces (known as berths) between the guns. Here they kept their chests, mess-tables, crockery, and other gear. Here they ate and drank, made merry, danced, got drunk, and, in port, entertained their female acquaintance. Many more, including the midshipmen, surgeon, and gunner, lived below the lower gun deck, in the orlop or cable tier, where sunlight could never come and fresh air never came willingly. At night the men slept in hammocks, which they slung from the beams. They were packed together very tightly, man to man, hammock touching hammock. In the morning, the hammocks were lashed up and stowed in racks till the evening.

There was no "regulation" naval uniform until some years after the *Centurion's* return to England. The officers and men seem to have worn what clothes they pleased. The ships carried stores of clothes which were issued to the men as they needed them. The store clothes, being (perhaps) of similar patterns, may have given a sort of uniformity to the appearance of the crews after some months at sea. In some of the prints of the time the men are drawn wearing rough, buckled shoes, coarse stockings, aprons or short skirts of frieze, baize, or tarred canvas, and short jackets worn open. Anson, like most captains, took care that the men in his boat's crew all dressed alike. The marines wore their regimental uniforms.

Life at sea has always been, and may always be, a harder life than the hardest of shore lives.

Life ashore in the early and middle eighteenth century was, in the main, both hard and brutal. Society ashore was made up of a little, brilliant, artificial class, a great, dull, honest, and hardworking mass, and a brutal, dirty, and debased rabble. Society at sea was like society ashore, except that, being composed of men, and confronted with the elements, and based on a grand ceremonial tradition, it was never brilliant, and never artificial. It was, in the main, an honest and hardworking society. Much in it was brutal, dirty, and debased; but it had always behind it an order and a ceremony grand, impressive, and unfaltering. That life in that society was often barbarous and disgusting cannot be doubted. The best men in the ships were taken by force from the merchant service. The others were gathered by press-gangs and gaol-deliveries. They were knocked into shape by brutal methods and kept in hand by brutal punishments. The officers were not always gentlemen; and when they were, they were frequently incompetent. The administration was scandalously corrupt. The ships were unhealthy, the food foul, the pay small, and the treatment cruel. The attractions of the service seem to have been these: the chance of making a large sum of prize-money, and the possibility of getting drunk once a day on the enormous daily ration of intoxicating liquor. The men were crammed together into a dark, stinking, confined space, in which privacy was impossible, peace a dream, and cleanliness a memory. Here they were fed on rotten food, till they died by the score, as this book testifies.

"We sent," says Mr. Walter, chaplain in the *Centurion*, "about eighty sick from the *Centurion*; and the other ships, I believe, sent nearly as many, in proportion.... As soon as we had performed this necessary duty, we scraped our decks, and gave our ship a thorough cleaning; then smoked it between decks, and after all washed every part well with vinegar. These operations were extremely necessary for correcting the noisome stench on board, and destroying the vermin; for ... both these nuisances had increased upon us to a very loathsome degree."

"The Biscuit," says Mr. Thomas, the teacher of mathematics in the *Centurion*, "(was) so worm-eaten it was scarce anything but dust, and a little blow would reduce it to that immediately; our Beef and Pork was likewise very rusty and rotten, and the surgeon endeavoured to hinder us from eating any of it, alledging it was, tho' a slow, yet a sure Poison."

That tradition and force of will could keep life efficient, and direct it to great ends, in such circumstances, deserves our admiration and our reverence.

The traditions and unpleasantness of the sea service are suggested vividly in many pages of this book. A few glimpses of both may be obtained from the following extracts from some of the logs and papers which deal with this voyage and with Anson's entry into the Navy. The marine chapters in Smollett's *Roderick Random* give a fair picture of the way of life below decks during the years of which this book treats.

George Anson was born at Shugborough, in Staffordshire, on April 23, 1697. His first ship was the *Ruby*, Captain Peter Chamberlen, a 54-gun ship, with a scratch crew of 185 men. George Anson's name appears in her pay book between the names of John Baker, ordinary seaman, and George Hirgate, captain's servant. He joined her on February 2, 1712. The ship had lain cleaning and fitting "at Chatham and in the River Medway" since the 4th of the preceding month. Two days after the boy came aboard she weighed her anchor "at 1 afternoon," fresh gales and cloudy, and ran out to the Nore where she anchored in seven fathoms and moored.

It is not known what duties the boy performed during his first days of service. The ship fired twenty-one guns in honour of the queen's birthday on February 7. The weather was hazy, foggy, and cold, with snow and rain; lighters came off with dry provisions, and the ship's boats brought off water. On February 9, the *Centurion*, an earlier, smaller *Centurion* than the ship afterwards made famous by him, anchored close to them. On the 16th, two Dutch men-of-war, with a convoy, anchored close to them. Yards and topmasts were struck and again got up on the 17th. On the 24th, three shot were fired at a brigantine to bring her to.

On the 27th, Sir John Norris and Sir Charles Wager hoisted their flags aboard the *Cambridge* and the *Ruby* respectively, and signal was made for a court-martial. Six men of the *Dover* were tried for mutiny, theft, disorderly conduct, and desertion of their ship after she had gone ashore "near Alborough Haven." Being all found guilty they were whipped from ship to ship next morning. Each received six lashes on the bare back at the side of each ship then riding at the Nore. A week later, the *Ruby* and the *Centurion* sailed leisurely to Spithead, chasing a Danish ship on the way. On March 11, the *Ruby* anchored at Spithead and struck her topmasts. On March 18, Captain Chamberlen removed "into ye *Monmouth*" with all his "followers," Anson among them. The *Monmouth* sailed on April 13, with three other men-of-war, as a guard to the West Indian fleet, bound for Port Royal. Her master says that on June 7, in lat. 21° 36' N., long. 18° 9' W., "we duckt those men that want willing to pay for crossing the tropick." In August, off the Jamaican coast, a man fell overboard and was drowned. Later in the month, a hurricane very nearly put an end to Anson and *Monmouth* together. Both pumps were kept going, there was four feet of water on the ballast and the same between decks, the foretopmast went, the main and mizen masts were cut away, and men with buckets worked for their lives "bealing at each hatchway." Port Royal was reached on September 1. The *Monmouth* made a cruise after pirates in Blewfields Bay, and returned to Spithead in June 1713.

Anson is next heard of as a second lieutenant aboard the *Hampshire*. He was in the *Montague*, 60-gun ship, in Sir George Byng's action off Cape Passaro, in March 1718. In 1722, he commanded the *Weasel* sloop in some obscure services in the North Sea against the Dutch smugglers and French Jacobites. During this command he made several captures of brandy. From 1724 till 1735 he was employed in various commands, mostly in the American colonies, against the pirates. From 1735 till 1737 he was not employed at sea.

In 1737, he took command of the *Centurion*, and sailed in her to the Guinea Coast, to protect our gum merchants from the French. His gunner was disordered in his head during the cruise; and Sierra Leone was so unhealthy that "the merchant ships had scarce a well man on board." A man going mad and others dying were the only adventures of the voyage. He was back in the Downs to prepare for this more eventful voyage by July 21, 1739.

In November he wrote to the Admiralty that in hot climates "the Pease and Oatemeal put on board his Maj'y Ships have generally decayed and become not fitt to issue, before they have all been expended." He proposed taking instead of peas and oatmeal a proportion of "Stockfish, Grotts, Grout, and Rice." The Admiralty sanctioned the change; but the purser seems to have failed to procure the substitutes. Whether, as was the way of the pursers of that time, he pocketed money on the occasion, cannot be known. He died at sea long before the lack was discovered.

A more tragical matter took place in this November. A Mr. McKie, a naval mate, was attacked on Gosport Beach by twenty or thirty of the *Centurion's* crew, under one William Cheney, a boatswain's mate; and the said William Cheney "with a stick did cutt and bruse" the said McKie, and tore his shirt and conveyed away his "Murning ring," which was flat burglary in the said Cheney. "Mr. Cheney aledges no other reason for beating and Abusing Mr. McKie but the said McKie having got drunk at Sea, did then beat and abuse him." As Hamlet says, this was hire and salary, not revenge.

Months went by, doubtfully enlivened thus, till June 1740, when the pressing of men began. The *Centurion's* men went pressing, and got seventy-three men, a fair catch, but not enough. She despatched a tender to the Downs to press men from homeward bound merchant ships. This method of getting a crew was the best then in use, because the men obtained by it were trained seamen, which those obtained from the gaols, the gin-shops, and the slums seldom were. It was an extremely cruel method. A man within sight of his home, after a voyage of perhaps two years, might be dragged from his ship (before his wages were paid) to serve willy-nilly in the Navy, at a third of the pay, for the next half-dozen years. An impartial conscription seems noble beside such a method. Knowing how the ships were manned, it cannot seem strange that the Navy was not then a loved nor an honoured service. Nineteen of the *Centurion's* catch loved and honoured it so little that they contrived to desert (risking death at the yard-arm by doing so) during the weeks of waiting at Portsmouth.

Before the tender sailed for the Downs, Anson discovered that the dockyard men had scamped their work in the *Centurion*. They had supplied her with a defective foremast "Not fitt for Sarves." High up on the mast was "a rotten Nott eleven inches deep," a danger to spar and ship together. The dockyard officials, who had probably pocketed the money for a good spar, swore that the Nott only "wants a Plugg drove in" to be perfection. Dockyard men at this time and for many years afterwards deserved to be suspended both from their duties and by their necks. Soon after the wrangle over the spar, there was a wrangle about the *Gloucester's* beef. Forty-two out of her seventy-two puncheons of beef were found to be stinking. With some doubts as to what would happen in the leaf if such things happened in the bud, Anson got his squadron to sea. Early in the voyage his master "shoved" his boatswain while he was knotting a cable, and the boatswain complained. "The Boatswain," says the letter, "is very often Drunk and incapable of his Duty." Later in the voyage, when many hundreds had died, Mr. Cheney, who hit Mr. McKie, became boatswain in his stead.

The squadron sailed from England on September 18, 1740, with six ships of war manned by 1872 seamen and marines, twenty-four of whom were sick. At Madeira, on November 4, after less than seven weeks at sea, there were 122 sick, and fourteen had been buried. Less than eleven weeks later, at St. Catherine's in Brazil, there were 450 sick, and 160 had been buried. From this time until what was left of the squadron reached Juan Fernandez, sickness and death took continual toll. It is shocking to see the *Centurion's* muster lists slowly decreasing, by one or two a week, till she was up to the Horn, then dropping six, ten, twenty, or twenty-four a week, as the scurvy and the frost took hold. Few but the young survived. What that passage of the Horn was like may be read here at length; but perhaps nothing in this book is so eloquent of human misery as the following entries from Anson's private record:—

"1741. 8 May.—Heavy Flaws and dangerous Gusts, expecting every Moment to have my masts Carry'd away, having very little succor, from the standing rigging, every Shroud knotted, and not men able to keep the deck sufficient to take in a Topsail, all being violently afflicted with the Scurvy, and every day lessening our Number by six eight and Ten.

"1741. 1st Sept.—I mustered my Ship's Company, the number of Men I brought out of England, being Five hundred, are now reduced by Mortality to Two hundred and Thirteen, and many of them in a weak and Low condition."

Nothing in any of the records is so eloquent as the remark in Pascoe Thomas's account of the voyage:—

"I have seen 4 or 5 dead Bodies at a time, some sown up in their Hammocks and others not, washing about the Decks, for Want of Help to bury them in the Sea."

On December 7, 1741, the 1872 men had dwindled down to 201. Of the six ships of war only one, the *Centurion*, still held her course. She was leaking an inch an hour, but she showed bright to the world under a new coat of paint. On this day Anson sent home a letter to the Admiralty (from Canton in China). The letter was delivered 173 days later.

In spite of the miseries of the service, there were compensations. The entry off Payta—

"1741. 12 Nov.—I keept Possession of the Town three days and employed my Boats in plundering"—

must have been pleasant to write; and the entries for Tuesday, June 21, 1743, and following days, become almost incoherent:—

"reced 112 baggs and 6 Chests of Silver.

"11 Baggs of Virgin silver 72 Chests of Dollers and baggs of Dollers 114 Chests and 100 baggs of Dollers 4 baggs of wrought Plate and Virgin Silver."

The arrival at Portsmouth is thus described:—

"1744. Friday, 15 June.—Came to with the S Bower in 10 fath water and at 9 began to Moor."

Later interesting entries are:—

"Monday, 2nd July.—Fresh gales and Cloudy sent away the Treasure in 32 Waggons to London with 139 Officers and Seamen to guard it.

"Thursday, 19 July.—Mod and fair, found in the Fish Room three Chests of Treasure" (which had been overlooked).

The last entry of all is for:—

"Friday, 20 July.—Hard Gales with rain at 4 p.m. all the men on the spot were paid and the Pendant was Struck."

An old print represents an officer of the *Centurion* dropping booty into the apron of a lady friend. Behind him the waggons and their guard proceed, with a great display of flags. The passing of the treasure was acclaimed with much enthusiasm both upon the road and in London. It was no doubt the biggest prize ever brought to England by a single ship. Anson's share made him a rich man. The rest of the survivors profited according to their rank.

Anson's subsequent career may be told in a few words. He was created Lord Anson on June 13, 1747. From 1751 to 1756 and from 1757 till his death he was a very competent and energetic First Lord of the Admiralty. He became Admiral of the Fleet in 1761. He died on June 6, 1762. The figurehead of the *Centurion*, the lion which "was very loose" in the Cape Horn storms of 1741, was preserved at the family seat at Shugborough till it fell to pieces. A portrait of Anson, which has been frequently copied and engraved, still exists there. The face is that of a man placidly and agreeably contented. It is the face of the polite and even spirit who "always kept up his usual composure and steadiness," and only once allowed joy to "break through" "the equable and unvaried character which he had hitherto preserved." Something of that character is in this placid and agreeable story told by Mr. Walter, chaplain, from Anson's private records.

The book is one of the most popular of the English books of voyages. It is a pleasantly written work. The story is told with a grace and quietness "very grateful and refreshing." The story itself is remarkable. It bears witness to the often illustrated contrast between the excellence of Englishmen and the stupidity of their governors. The management of the squadron before it sailed gave continuous evidence of imbecility. Something fine in a couple of hundred "emaciated ship-mates" drove them on to triumph through every possible disadvantage. In the general joy over their triumph, the imbecility was forgotten. There is something pathetic in the mismanagement of the squadron. The ships were sent to sea on the longest and most dangerous of voyages with no anti-scorbutics. When scurvy broke out the only medicines available were "the pill and drop of Dr. Ward" (very violent emetic purgatives), which came not from the government, but from Anson's own stores. In the absence of proper medicines, Anson produced these things, "and first try'd them on himself." This spirit in our captains and in our common men has borne us (so far) fairly triumphantly out of the bogs into which our stupidity so often drives us.

JOHN MASEFIELD.

January 30, 1911.

BIBLIOGRAPHY

A Voyage to the South Seas, and to many other Parts of the World, performed from the Month of September in the Year 1740, to June 1744, by Commodore Anson, in his Majesty's Ship the *Centurion*. By an officer of the Fleet, 1744.

An authentic journal of the late expedition under the command of Commodore Anson. Containing a regular and exact account of the whole proceedings, etc. To which is added A Narrative of the Extraordinary Hardships suffered by the Adventurers, in this voyage. By John Philips, midshipman of the *Centurion*, 1744.

A True Journal of a Voyage to the South Seas and round the Globe, in the *Centurion*, 1745, by Pascoe Thomas.

A Voyage round the World in the Year 1740, 1, 2, 3, 4; compiled from papers and other materials of the Right Honourable George Anson, and published under his direction by Richard Walter, M.A., chaplain of His Majesty's ship *Centurion* in that expedition, 1748, and many later editions. One in 1749 was illustrated with 42 plates.

A supplement to Lord Anson's Voyage round the World, containing a discovery and description of the Island of Frivola, by the Abbé Coyer, 1752.

Anson's voyages were included in J. Harris's Navigantium atque Itinerantium Bibliotheca, etc., vol. 1., revised and enlarged edition by Campbell, 1744-8, 1764; and also in J. H. More's A New and Complete Collection of Voyages, etc., vol. i., 1780.

LIFE: By John Barrow, 1839.

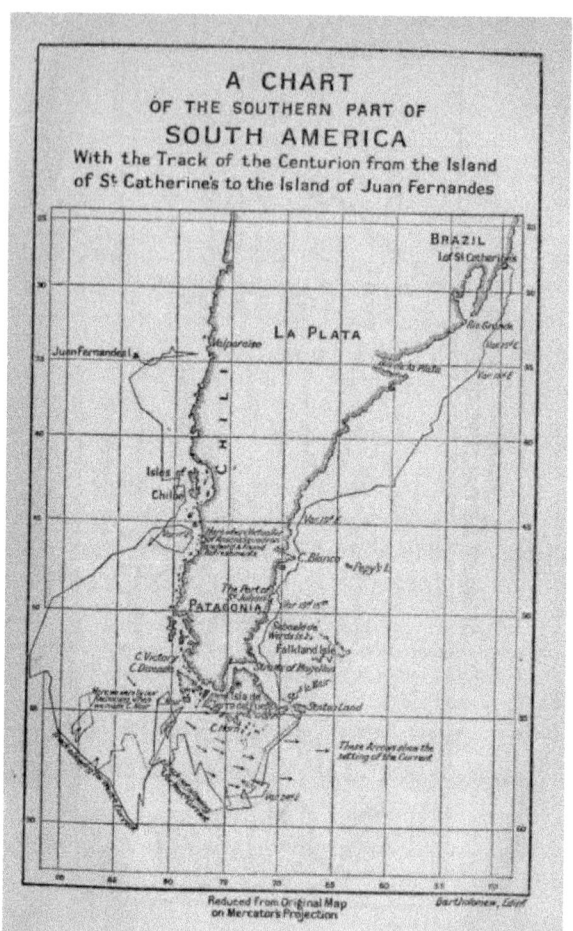

A CHART OF THE SOUTHERN PART OF SOUTH AMERICA

With the Track of the *Centurion* from the Island
of St. Catherine's to the Island of Juan Fernandas

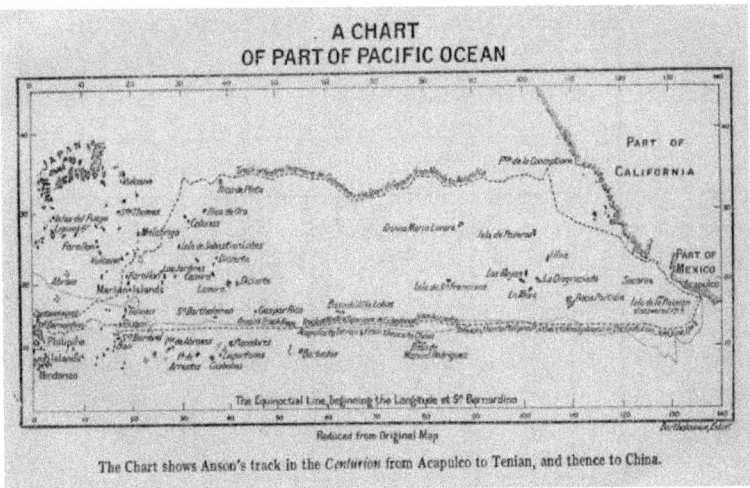

A CHART OF PART OF PACIFIC OCEAN

The Chart shows Anson's track in the *Centurion* from Acapulco to Tenian, and thence to China.

AUTHOR'S INTRODUCTION

Notwithstanding the great improvement of navigation within the last two centuries, a voyage round the world is still considered as an enterprize of so very singular a nature, that the public have never failed to be extremely inquisitive about the various accidents and turns of fortune with which this uncommon attempt is generally attended. And though the amusement expected in these narrations is doubtless one great source of that curiosity with the bulk of readers, yet the more intelligent part of mankind have always agreed that from accounts of this nature, if faithfully executed, the more important purposes of navigation, commerce, and national interest may be greatly promoted: for every authentic description of foreign coasts and countries will contribute to one or more of these great ends, in proportion to the wealth, wants, or commodities of those countries, and our ignorance of those coasts; and therefore a voyage round the world promises a species of information of all others the most desirable and interesting, since great part of it is performed in seas with which we are as yet but very imperfectly acquainted, and in the neighbourhood of a country renowned for the abundance of its wealth, though it is at the same time stigmatised for its poverty in the necessaries and conveniences of a civilized life.

These considerations have occasioned the compiling the ensuing work; which, in gratifying the inquisitive disposition of mankind, and contributing to the safety and success of future navigators, and to the extension of our commerce and power, may doubtless vie with any narration of this kind hitherto made public: since as to the first of these heads it may well be supposed that the general curiosity hath been strongly excited by the circumstances of this undertaking already known to the world; for whether we consider the force of the squadron sent on this service, or the diversified distresses that each single ship was separately involved in, or the uncommon instances of varying fortune which attended the whole enterprize, each of these articles, I conceive, must, from its rude, well-known outlines, appear worthy of a compleater and more finished delineation.

Besides these descriptions and directions relating thereto, there is inserted in the ensuing work an ample account of a particular navigation of which hitherto little more than the name has been known, except to those immediately employed in it: I mean the track described by the Manila ship, in her passage to Acapulco, through the northern part of the Pacific Ocean. This material article is collected from the draughts and journals met with on board the Manila galeon, founded on the experience of more than a hundred and fifty years' practice, and corroborated in its principal circumstances by the concurrent evidence of all the Spanish prisoners taken in that vessel. And as many of their journals, which I have examined, appear to have been not ill kept, I presume the particulars of their route may be very safely relied on by future navigators. The advantages which may be drawn from an exact knowledge of this navigation, and the beneficial projects that may be formed thereon, both in war and peace, are by no means proper to be discussed in this place, but they will easily offer themselves to the skillful in maritime affairs. However, as the Manila ships are the only ones which have ever traversed this vast ocean, except a French straggler or two which have been afterwards seized on the coast of Mexico, and as, during near two ages in which this trade has been carried on, the Spaniards have, with the greatest care, secreted all accounts of their voyages from the rest of the world, these reasons alone would authorize the insertion of those papers, and would recommend them to the inquisitive as a very great improvement in geography, and worthy of attention from the singularity of many circumstances therein recited.

Thus much it has been thought necessary to premise with regard to the ensuing work, which it is hoped the reader will, on perusal, find much ampler and more important than this slight sketch can well explain. But as there are hereafter occasionally interspersed some accounts of Spanish transactions, and many observations relating to the disposition of the American Spaniards, and to the condition of the countries bordering on the South Seas, and as herein I may appear to differ greatly from the opinions generally established, I think it behoves me particularly to recite the authorities I have been guided by in these matters, that I may not be censured as having given way either to a thoughtless credulity on one hand, or, what would be a much more criminal imputation, to a willful and deliberate misrepresentation on the other.

Mr. Anson, before he set sail upon this expedition, besides the printed journals to those parts, took care to furnish himself with the best manuscript accounts he could procure of all the Spanish settlements upon the coasts of Chili, Peru, and Mexico: these he carefully compared with the examinations of his prisoners, and the informations of several intelligent persons who fell into his hands in the South Seas. He had likewise the good fortune, in some of his captures, to possess himself of a great number of letters and papers of a public nature, many of them written by the Viceroy of Peru to the Viceroy of Santa Fee, to the Presidents of Panama and Chili, to Don Blass de Lezo, admiral of the galeons, and to divers other persons in considerable employments; and in these letters there was usually inserted a recital of those they were intended to answer; so that they contained no small part of the correspondence between those officers for some time previous to our arrival on that coast. We took, besides, many letters sent from persons entrusted by the Spanish Government to their friends and correspondents, which were frequently filled with narrations of public business, and sometimes contained undisguised animadversions on the views and conduct of their superiors.

From these materials those accounts of the Spanish affairs are drawn which may at first sight appear the most exceptionable. In particular, the history of the various casualties which befel Pizarro's squadron is for the most part composed from intercepted letters. Though indeed the relation of the insurrection of Orellana and his followers is founded on rather a less disputable authority: for it was taken from the mouth of an English gentleman then on board Pizarro, who often conversed with Orellana; and it was upon inquiry confirmed in its principal circumstances by others who were in the ship at the same time: so that the fact, however extraordinary, is, I conceive, not to be contested.

And on this occasion I cannot but mention, that though I have endeavoured with my utmost care to adhere strictly to truth in every article of the ensuing narration, yet I am apprehensive that in so complicated a work some oversights must have been committed by the inattention to which at times all mankind are liable. However, I am as yet conscious of none but literal and insignificant mistakes; and if there are others more considerable which have escaped me, I flatter myself they are not of moment enough to affect any material transaction, and therefore I hope they may justly claim the reader's indulgence.

After this general account of the ensuing work, it might be expected, perhaps, that I should proceed to the work itself, but I cannot finish this Introduction without adding a few reflections on a matter very nearly connected with the present subject, and, as I conceive, neither destitute of utility nor unworthy the attention of the public; I mean the animating my countrymen, both in their public and private stations, to the encouragement and pursuit of all kinds of geographical and nautical observations, and of every species of mechanical and commercial information. It is by a settled attachment to these seemingly minute particulars that our ambitious neighbours have established some part of that power with which we are now struggling: and as we have the means in our hands of pursuing these subjects more effectually than they can, it would be a dishonour to us longer to neglect so easy and beneficial a practice. For, as we have a navy much more numerous than theirs, great part of which is always employed in very distant nations, either in the protection of our colonies and commerce, or in assisting our allies against the common enemy, this gives us frequent opportunities of furnishing ourselves with such kind of materials as are here recommended, and such as might turn greatly to our advantage either in war or peace. Since, not to mention what might be expected from the officers of the navy, if their application to these subjects was properly encouraged, it would create no new expence to the government to establish a particular regulation for this purpose, as all that would be requisite would be constantly to embark on board some of our men-of-war which are sent on these distant cruises a person who, with the character of an engineer and the skill and talents necessary to that profession, should be employed in drawing such coasts and planning such harbours as the ship should touch at, and in making such other observations of all kinds as might either prove of advantage to future navigators, or might any ways tend to promote the public service. Persons habituated to these operations (which could not fail at the same time of improving them in their proper business) would be extremely useful in many other lights besides those already mentioned, and might tend to secure our fleets from those disgraces with which their attempts against places on shore have been often attended; and in a nation like ours, where all sciences are more eagerly and universally pursued and better understood than in any other part of the world, proper subjects for these employments could not long be wanting if due encouragement were given to them. This method here recommended is known to have been frequently practised by the French, particularly in the instance of Mons. Frazier, an engineer, who has published a celebrated voyage to the South Seas; for this person, in the year 1711, was purposely sent by the French king into that country on board a merchantman, that he might examine and describe the coast, and take plans of all the fortified places, the better to enable the French to prosecute their illicit trade, or, on a rupture between them and the court of Spain, to form their enterprizes in those seas with more readiness and certainty. Should we pursue this method, we might hope that the emulation amongst those who are commissioned for these undertakings, and the experience which even in the most peaceable intervals they would hereby acquire, might at length procure us a proper number of able engineers, and might efface the national scandal which our deficiency in that species of men has sometimes exposed us to: and surely every step to encourage and improve them is of great moment to the publick, as no persons, when they are properly instructed, make better returns in war for the distinctions and emoluments bestowed on them in time of peace. Of which the advantages the French have reaped from their dexterity (too numerous and recent to be soon forgot) are an ample confirmation.

And having mentioned engineers, or such as are skilled in drawing and the other usual practices of that profession, as the properest persons to be employed in these foreign enquiries, I cannot (as it offers itself so naturally to the subject in hand) but lament how very imperfect many of our accounts of distant countries are rendered by the relators being unskilled in drawing, and in the general principles of surveying, even where other abilities have not been wanting. Had more of our travellers been initiated in these acquirements, and had there been added thereto some little skill in the common astronomical observations (all which a person of ordinary talents might attain with a very moderate share of application), we should by this time have seen the geography of the globe much correcter than we now find it: the dangers of navigation would have been considerably lessened, and the manners, arts, and produce of foreign countries would have been better known to us than they are. Indeed, when I consider the strong incitements that all travellers have to pursue some part at least of these qualifications, especially drawing; when I consider how much it will facilitate their observations, assist and strengthen their memories, and of how tedious, and often unintelligible, a load of description it would rid them, I cannot but wonder that any person who intends to visit distant countries with a view of informing either himself or others, should be wanting in so necessary a piece of skill. And to enforce this argument still further, I must add that besides the uses of drawing, already mentioned, there is one which, tho' not so obvious, is yet perhaps of more consequence than all that has been hitherto urged; I mean the strength and distinguishing power it adds to some of our faculties. This appears from hence, that those who are used to draw objects observe them with more accuracy than others who are not habituated to that practice. For we may easily find by a little experience, that when we view any object, however simple, our attention or memory is scarcely at any time so strong as to enable us, when we have turned our eyes away from it, to recollect exactly every part it consisted of, and to recall all the circumstances of its appearance; since on examination it will be discovered that in some we were mistaken and others we had totally overlooked: but he that is accustomed to draw what he sees is at the same time accustomed to rectify this inattention; for by confronting his ideas copied on the paper with the object he intends to represent, he finds out what circumstances have deceived him in its appearance, and hence he at length acquires the habit of observing much more at one view, and retains what he sees with more correctness than he could ever have done without his practice and proficiency in drawing.

If what has been said merits the attention of travellers of all sorts, it is, I think, more particularly applicable to the gentlemen of the navy; since, without drawing and planning, neither charts nor views of land can be taken, and without these it is sufficiently evident that navigation is at a full stand. It is doubtless from a persuasion of the utility of these qualifications that his Majesty has established a drawing master at Portsmouth for the instruction of those who are presumed to be hereafter intrusted with the command of his royal navy: and tho' some have been so far misled as to suppose that the perfection of sea-officers consisted in a turn of mind and temper resembling the boisterous element they had to deal with, and have condemned all literature and science as effeminate and derogatory to that ferocity which, they would falsely persuade us, was the most unerring characteristic of courage: yet it is to be hoped that such absurdities as these have at no time been authorised by the public opinion, and that the belief of them daily diminishes. If those who adhere to these mischievous positions were capable of being influenced by reason or swayed by example, I should think it sufficient for their conviction to observe that the most valuable drawings made in the following voyage, though done with such a degree of skill that even professed artists could with difficulty imitate them, were taken by Mr. Piercy Brett, one of Mr. Anson's lieutenants, and since captain of the *Lion* man-of-war; who, in his memorable engagement with the *Elizabeth* (for the importance of the service, or the resolution with which it was conducted, inferior to none this age has seen), has given ample proof that a proficiency in the arts I have been here recommending is extremely consistent with the most exemplary bravery, and the most distinguished skill in every function belonging to the duty of a sea-officer. Indeed, when the many branches of science are attended to, of which even the common practice of navigation is composed, and the many improvements which men of skill have added to this practice within these few years, it would induce one to believe that the advantages of reflection and speculative knowledge were in no profession more eminent than in that of a sea-officer; for, not to mention some expertness in geography, geometry, and astronomy, which it would be dishonourable for him to be without (as his journal and his estimate of the daily position of the ship are founded on particular branches of these arts), it may be well supposed that the management and working of a ship, the discovery of her most eligible position in the water (usually stiled her trim), and the disposition of her sails in the most advantageous manner, are articles wherein the knowledge of mechanicks cannot but be greatly assistant. And perhaps the application of this kind of knowledge to naval subjects may produce as great improvements in sailing and working a ship as it has already done in many other matters conducive to the ease and convenience of human life. Since, when the fabric of a ship and the variety of her sails are considered, together with the artificial contrivances for adapting them to her different motions, as it cannot be doubted but these things have been brought about by more than ordinary sagacity and invention; so neither can it be doubted but that in some conjunctures a speculative and scientific turn of mind may find out the means of directing and disposing this complicated mechanism much more advantageously than can be done by mere habit, or by a servile copying of what others may perhaps have erroneously practised in similar emergencies. But it is time to finish this digression, and to leave the reader to the perusal of the ensuing work, which, with how little art soever it may be executed, will yet, from the importance of the subject and the utility and excellence of the materials, merit some share of the public attention.

BOOK I

CHAPTER I

OF THE EQUIPMENT OF THE SQUADRON: THE INCIDENTS RELATING THERETO, FROM ITS FIRST APPOINTMENT TO ITS SETTING SAIL FROM ST. HELENS

The squadron under the command of Mr. Anson (of which I here propose to recite the most material proceedings) having undergone many changes in its destination, its force, and its equipment, during the ten months between its original appointment and its final sailing from St. Helens; I conceive the history of these alterations is a detail necessary to be made public, both for the honour of those who first planned and promoted this enterprize, and for the justification of those who have been entrusted with its execution. Since it will from hence appear that the accidents the expedition was afterwards exposed to, and which prevented it from producing all the national advantages the strength of the squadron, and the expectation of the public, seemed to presage, were principally owing to a series of interruptions, which delayed the commander in the course of his preparations, and which it exceeded his utmost industry either to avoid or to get removed.

When, in the latter end of the summer of the year 1739, it was foreseen that a war with Spain was inevitable, it was the opinion of some considerable persons then trusted with the administration of affairs, that the most prudent step the nation could take, on the breaking out of the war, was attacking that crown in her distant settlements; for by this means (as at that time there was the greatest probability of success) it was supposed that we should cut off the principal resources of the enemy, and should reduce them to the necessity of sincerely desiring a peace, as they would hereby be deprived of the returns of that treasure by which alone they could be enabled to carry on a war.

In pursuance of these sentiments, several projects were examined, and several resolutions were taken by the council. And in all these deliberations it was from the first determined that George Anson, Esq., then captain of the *Centurion*, should be employed as commander-in-chief of an expedition of this kind: and he at that time being absent on a cruize, a vessel was dispatched to his station so early as the beginning of September to order him to return with his ship to Portsmouth. And soon after he came there, that is on the 10th of November following, he received a letter from Sir Charles Wager directing him to repair to London, and to attend the Board of Admiralty: where, when he arrived, he was informed by Sir Charles that two squadrons would be immediately fitted out for two secret expeditions, which, however, would have some connexion with each other; that he, Mr. Anson, was intended to command one of them, and Mr. Cornwall (who hath since lost his life gloriously in the defence of his country's honour) the other; that the squadron under Mr. Anson was to take on board three independent companies of a hundred men each, and Bland's regiment of foot; that Colonel Bland was likewise to embark with his regiment, and to command the land-forces; and that, as soon as this squadron could be fitted for the sea, they were to set sail, with express orders to touch at no place till they came to Java Head in the East Indies; that there they were only to stop to take in water, and thence to proceed directly to the city of Manila, situated on Luconia, one of the Philippine Islands; that the other squadron was to be of equal force with this commanded by Mr. Anson, and was intended to pass round Cape Horn, into the South Seas, to range along that coast; and after cruising upon the enemy in those parts, and attempting their settlements, this squadron in its return was to rendezvous at Manila, there to join the squadron under Mr. Anson, where they were to refresh their men, and refit their ships, and perhaps receive orders for other considerable enterprizes.

This scheme was doubtless extremely well projected, and could not but greatly advance the public service, and the reputation and fortune of those concerned in its execution; for had Mr. Anson proceeded for Manila at the time and in the manner proposed by Sir Charles Wager, he would, in all probability, have arrived there before they had received any advice of the war between us and Spain, and consequently before they had been in the least prepared for the reception of an enemy, or had any apprehensions of their danger. The city of Manila might be well supposed to have been at that time in the same defenceless condition with all the other Spanish settlements, just at the breaking out of the war; that is to say, their fortifications neglected, and in many places decayed; their cannon dismounted, or rendered useless by the mouldring of their carriages; their magazines, whether of military stores or provision, all empty; their garrisons unpaid, and consequently thin, ill affected, and dispirited; and the royal chests in Peru, whence alone all these disorders could receive their redress, drained to the very bottom; this, from the intercepted letters of their viceroys and governors, it is well known to have been the defenceless state of Panama, and the other Spanish places on the coast of the South Sea, for near a twelvemonth after our declaration of war. And it cannot be supposed that the city of Manila, removed still farther by almost half the circumference of the globe, should have experienced from the Spanish government a greater share of attention and concern for its security than Panama, and the other important ports in Peru and Chili, on which their possession of that immense empire depends. Indeed, it is well known that Manila was at that time incapable of making any considerable defence, and in all probability would have surrendered only on the appearance of our squadron before it. The consequence of this city, and the island it stands on, may be in some measure estimated, from the known healthiness of its air, the excellency of its port and bay, the number and wealth of its inhabitants, and the very extensive and beneficial commerce which it carries on to the principal ports in the East Indies, and China, and its exclusive trade to Acapulco, the returns for which, being made in silver, are, upon the lowest valuation, not less than three millions of dollars per annum.

On this scheme Sir Charles Wager was so intent that in a few days after this first conference, that is, on November 18, Mr. Anson received an order to take under his command the *Argyle, Severn, Pearl, Wager,* and *Tryal* sloop; and other orders were issued to him in the same month, and in the December following, relating to the victualling of this squadron. But Mr. Anson attending the Admiralty the beginning of January, he was informed by Sir Charles Wager that for reasons with which he, Sir Charles, was not acquainted, the expedition to Manila was laid aside. It may be conceived that Mr. Anson was extremely chagrined at the losing the command of so infallible, so honourable, and in every respect, so desirable an enterprize, especially too as he had already, at a very great expence, made the necessary provision for his own accommodation in this voyage, which he had reason to expect would prove a very long one. However, Sir Charles, to render his disappointment in some degree more tolerable, informed him that the expedition to the South Seas was still intended, and that he, Mr. Anson, and his squadron, as their first destination was now countermanded, should be employed in that service. And on the 10th of January he received his commission, appointing him commander-in-chief of the forementioned squadron, which (the *Argyle* being in the course of their preparation changed for the *Gloucester*) was the same he sailed with above eight months after from St. Helens. On this change of destination, the equipment of the squadron was still prosecuted with as much vigour as ever, and the victualling, and whatever depended on the commodore, was soon so far advanced that he conceived the ships might be capable of putting to sea the instant he should receive his final orders, of which he was in daily expectation. And at last, on the 28th of June 1740, the Duke of Newcastle, principal Secretary of State, delivered to him his Majesty's instructions, dated January 31, 1739, with an additional instruction from the Lords Justices, dated June 19, 1740. On the receipt of these, Mr. Anson immediately repaired to Spithead, with a resolution to sail with the first fair wind, flattering himself that all his difficulties were now at an end. For though he knew by the musters that his squadron wanted three hundred seamen of their complement (a deficiency which, with all its assiduity, he had not been able to get supplied), yet, as Sir Charles Wager informed him, that an order from the Board of Admiralty was dispatched to Sir John Norris to spare him the numbers which he wanted, he doubted not of its being complied with. But on his arrival at Portsmouth, he found himself greatly mistaken, and disappointed in this persuasion; for on his application, Sir John Norris told him he could spare him none, for he wanted men for his own fleet. This occasioned an inevitable and a very considerable delay; for it was the end of July before this deficiency was by any means supplied, and all that was then done was extremely short of his necessities and expectation. For Admiral Balchen, who succeeded to the command at Spithead, after Sir John Norris had sailed to the westward, instead of three hundred able sailors, which Mr. Anson wanted of his complement, ordered on board the squadron a hundred and seventy men only; of which thirty-two were from the hospital and sick quarter, thirty-seven from the *Salisbury*, with three officers of Colonel Lowther's regiment, and ninety-eight marines, and these were all that were ever granted to make up the forementioned deficiency.

But the commodore's mortification did not end here. It has been already observed that it was at first intended that Colonel Bland's regiment, and three independent companies of a hundred men each, should embark as land-forces on board the squadron. But this disposition was now changed, and all the land-forces that were to be allowed were five hundred invalids to be collected from the out-pensioners of Chelsea College. As these out-pensioners consist of soldiers who from their age, wounds, or other infirmities, are incapable of service in marching regiments, Mr. Anson was greatly chagrined at having such a decrepid detachment allotted him; for he was fully persuaded that the greatest part of them would perish long before they arrived at the scene of action, since the delays he had already encountered necessarily confined his passage round Cape Horn to the most rigorous season of the year. Sir Charles Wager too joined in opinion with the commodore, that invalids were no ways proper for this service, and sollicited strenuously to have them exchanged: but he was told that persons who were supposed to be better judges of soldiers than he or Mr. Anson, thought them the properest men that could be employed on this occasion. And upon this determination they were ordered on board the squadron on the 5th of August; but instead of five hundred, there came on board no more than two hundred and fifty-nine; for all those who had limbs and strength to walk out of Portsmouth deserted, leaving behind them only such as were literally invalids, most of them being sixty years of age, and some of them upwards of seventy. Indeed it is difficult to conceive a more moving scene than the embarkation of these unhappy veterans: they were themselves extremely averse to the service they were engaged in, and fully apprised of all the disasters they were afterwards exposed to; the apprehensions of which were strongly marked by the concern that appeared in their countenances, which was mixed with no small degree of indignation, to be thus hurried from their repose into a fatiguing employ, to which neither the strength of their bodies, nor the vigour of their minds, were any ways proportioned, and where, without seeing the face of an enemy, or in the least promoting the success of the enterprize, they would, in all probability, uselessly perish by lingering and painful diseases; and this too after they had spent the activity and strength of their youth in their country's service.

I cannot but observe, on this melancholy incident, how extremely unfortunate it was, both to this aged and diseased detachment, and to the expedition they were employed in, that amongst all the out-pensioners of Chelsea Hospital, which were supposed to amount to two thousand men, the most crazy and infirm only should be culled out for so laborious and perilous an undertaking. For it was well known that however unfit invalids in general might be for this service, yet by a prudent choice there might have been found amongst them five hundred men who had some remains of vigour left: and Mr. Anson fully expected that the best of them would have been allotted him; whereas the whole detachment that was sent to him seemed to be made up of the most decrepid and miserable objects that could be collected out of the whole body; and by the desertion above-mentioned, these were a second time cleared of that little health and strength which were to be found amongst them, and he was to take up with such as were much fitter for an infirmary than for any military duty.

And here it is necessary to mention another material particular in the equipment of this squadron. It was proposed to Mr. Anson, after it was resolved that he should be sent to the South Seas, to take with him two persons under the denomination of Agent Victuallers. Those who were mentioned for this employment had formerly been in the Spanish West Indies, in the South Sea Company's service, and it was supposed that by their knowledge and intelligence on that coast they might often procure provisions for him by compact with the inhabitants, when it was not to be got by force of arms: these agent victuallers were, for this purpose, to be allowed to carry to the value of £15,000 in merchandize on board the squadron; for they had represented that it would be much easier for them to procure provisions with goods than with the value of the same goods in money. Whatever colours were given to this scheme, it was difficult to persuade the generality of mankind that it was not principally intended for the enrichment of the agents, by the beneficial commerce they proposed to carry on upon that coast. Mr. Anson from the beginning objected both to the appointment of agent victuallers and the allowing them to carry a cargo on board the squadron: for he conceived that in those few amicable ports where the squadron might touch, he needed not their assistance to contract for any provisions the place afforded; and on the enemy's coast, he did not imagine that they could ever procure him the necessaries he should want, unless (which he was resolved not to comply with) the military operations of his squadron were to be regulated by the ridiculous views of their trading projects. All that he thought the government ought to have done on this occasion was to put on board to the value of £2000 or £3000 only of such goods as the Indians or the Spanish planters in the less cultivated part of the coast might be tempted with; since it was in such places only that he imagined it would be worth while to truck with the enemy for provisions: and in these places it was sufficiently evident a very small cargo would suffice.

But though the commodore objected both to the appointment of these officers and to their project, of the success of which he had no opinion; yet, as they had insinuated that their scheme, besides victualling their squadron, might contribute to settling a trade upon that coast, which might be afterwards carried on without difficulty, and might thereby prove a very considerable national advantage, they were much listened to by some considerable persons: and of the £15,000 which was to be the amount of their cargo, the government agreed to advance them 10,000 upon imprest, and the remaining 5000 they raised on bottomry bonds; and the goods purchased by this sum were all that were taken to sea by the squadron, how much soever the amount of them might be afterwards magnified by common report.

This cargo was at first shipped on board the *Wager* storeship, and one of the victuallers; no part of it being admitted on board the men-of-war. But when the commodore was at St. Catherine's, he considered that in case the squadron should be separated, it might be pretended that some of the ships were disappointed of provisions for want of a cargo to truck with, and therefore he distributed some of the least bulky commodities on board the men-of-war, leaving the remainder principally on board the *Wager*, where it was lost; and more of the goods perishing by various accidents to be recited hereafter, and no part of them being disposed of upon the coast, the few that came home to England did not produce, when sold, above a fourth part of the original price. So true was the commodore's judgment of the event of this project, which had been by many considered as infallibly productive of immense gains. But to return to the transactions at Portsmouth.

To supply the place of the two hundred and forty invalids which had deserted, as is mentioned above, there were ordered on board two hundred and ten marines detached from different regiments. These were raw and undisciplined men, for they were just raised, and had scarcely anything more of the soldier than their regimentals, none of them having been so far trained as to be permitted to fire. The last detachment of these marines came on board the 8th of August, and on the 10th the squadron sailed from Spithead to St. Helens, there to wait for a wind to proceed on the expedition.

But the delays we had already suffered had not yet spent all their influence, for we were now advanced into a season of the year when the westerly winds are usually very constant, and very violent; and it was thought proper that we should put to sea in company with the fleet commanded by Admiral Balchen, and the expedition under Lord Cathcart. As we made up in all twenty-one men-of-war, and a hundred and twenty-four sail of merchant-men and transports, we had no hopes of getting out of the Channel with so large a number of ships without the continuance of a fair wind for some considerable time. This was what we had every day less and less reason to expect, as the time of the equinox drew near; so that our golden dreams and our ideal possession of the Peruvian treasures grew each day more faint, and the difficulties and dangers of the passage round Cape Horn in the winter season filled our imaginations in their room. For it was forty days from our arrival at St. Helens, to our final departure from thence: and even then (having orders to proceed without Lord Cathcart) we tided it down the Channel with a contrary wind. But this interval of forty days was not free from the displeasing fatigue of often setting sail, and being as often obliged to return; nor exempt from dangers, greater than have been sometimes undergone in surrounding the globe. For the wind coming fair for the first time, on the 23d of August, we got under sail, and Mr. Balchen shewed himself truly solicitous to have proceeded to sea, but the wind soon returning to its old quarter obliged us to put back to St. Helens, not without considerable hazard, and some damage received by two of the transports, who, in tacking, ran foul of each other. Besides this, we made two or three more attempts to sail, but without any better success. And, on the 6th of September, being returned to an anchor at St. Helens, after one of these fruitless efforts, the wind blew so fresh that the whole fleet struck their yards and topmasts to prevent driving: yet, notwithstanding this precaution, the *Centurion* drove the next evening, and brought both cables ahead, and we were in no small danger of driving foul of the *Prince Frederick*, a seventy gun ship, moored at a small distance under our stern; though we happily escaped, by her driving at the same time, and so preserving her distance: but we did not think ourselves secure till we at last let go the sheet-anchor, which fortunately brought us up. However, on the 9th of September, we were in some degree relieved from this lingring vexatious situation, by an order which Mr. Anson received from the Lords Justices, to put to sea the first opportunity with his own squadron only, if Lord Cathcart should not be ready. Being thus freed from the troublesome company of so large a fleet, our commodore resolved to weigh and tide it down the Channel, as soon as the weather should become sufficiently moderate; and this might easily have been done with our own squadron alone full two months sooner, had the orders of the Admiralty, for supplying us with seamen, been punctually complied with, and had we met with none of those other delays mentioned in this narration. It is true, our hopes of a speedy departure were even now somewhat damped by a subsequent order which Mr. Anson received on the 12th of September; for by that he was required to take under his convoy the

St. Albans, with the Turky fleet, and to join the *Dragon* and the *Winchester*, with the Streights and the American trade, at Torbay or Plymouth, and to proceed with them to sea as far as their way and ours lay together. This incumbrance of a convoy gave us some uneasiness, as we feared it might prove the means of lengthening our passage to the Maderas. However, Mr. Anson, now having the command himself, resolved to adhere to his former determination, and to tide it down the Channel with the first moderate weather; and that the junction of his convoy might occasion as little loss of time as possible, he immediately sent directions to Torbay that the fleets he was there to take under his care might be in a readiness to join him instantly on his approach. And at last, on the 18th of September, he weighed from St. Helens; and though the wind was at first contrary, had the good fortune to get clear of the Channel in four days, as will be more particularly related in the ensuing chapter.

Having thus gone through the respective steps taken in the equipment of this squadron, it is sufficiently obvious how different an aspect this expedition bore at its first appointment in the beginning of January, from what it had in the latter end of September, when it left the Channel; and how much its numbers, its strength, and the probability of its success were diminished by the various incidents which took place in that interval. For instead of having all our old and ordinary seamen exchanged for such as were young and able (which the commodore was at first promised) and having our numbers compleated to their full complement, we were obliged to retain our first crews, which were very indifferent; and a deficiency of three hundred men in our numbers was no otherwise made up to us than by sending us on board a hundred and seventy men, the greatest part composed of such as were discharged from hospitals, or new-raised marines who had never been at sea before. And in the land-forces allotted to us, the change was still more disadvantageous; for there, instead of three independent companies of a hundred men each, and Bland's regiment of foot, which was an old one, we had only four hundred and seventy invalids and marines, one part of them incapable of action by their age and infirmities, and the other part useless by their ignorance of their duty. But the diminishing the strength of the squadron was not the greatest inconveniency which attended these alterations; for the contests, representations, and difficulties which they continually produced (as we have above seen, that in these cases the authority of the Admiralty was not always submitted to) occasioned a delay and waste of time, which in its consequences was the source of all the disasters to which this enterprise was afterwards exposed: for by this means we were obliged to make our passage round Cape Horn in the most tempestuous season of the year; whence proceeded the separation of our squadron, the loss of numbers of our men, and the imminent hazard of our total destruction. By this delay, too, the enemy had been so well informed of our designs, that a person who had been employed in the South Sea Company's service, and arrived from Panama three or four days before we left Portsmouth, was able to relate to Mr. Anson most of the particulars of the destination and strength of our squadron, from what he had learnt amongst the Spaniards before he left them. And this was afterwards confirmed by a more extraordinary circumstance: For we shall find that, when the Spaniards (fully satisfied that our expedition was intended for the South Seas) had fitted out a squadron to oppose us, which had so far got the start of us, as to arrive before us off the island of Madera, the commander of this squadron was so well instructed in the form and make of Mr. Anson's broad pendant, and had imitated it so exactly, that he thereby decoyed the *Pearl*, one of our squadron, within gun-shot of him, before the captain of the *Pearl* was able to discover his mistake.

CHAPTER II

THE PASSAGE FROM ST. HELENS TO THE ISLAND OF MADERA; WITH A SHORT ACCOUNT OF THAT ISLAND, AND OF OUR STAY THERE

On the 18th of September 1740, the squadron, as we have observed in the preceding chapter, weighed from St. Helens with a contrary wind, the commodore proposing to tide it down the Channel, as he dreaded less the inconveniences he should thereby have to struggle with, than the risk he should run of ruining the enterprise by an uncertain, and, in all probability, a tedious attendance for a fair wind.

The squadron allotted to this service consisted of five men-of-war, a sloop of war and two victualling ships. They were the *Centurion* of sixty guns, four hundred men, George Anson, Esq., commander; the *Gloucester* of fifty guns, three hundred men, Richard Norris, commander; the *Severn* of fifty guns, three hundred men, the Honourable Edward Legg, commander; the *Pearl* of forty guns, two hundred and fifty men, Matthew Mitchel, commander; the *Wager* of twenty-eight guns, one hundred and sixty men, Dandy Kidd, commander; and the *Tryal* sloop of eight guns, one hundred men, the Honourable John Murray, commander; the two victuallers were pinks, the largest of about four hundred, and the other of about two hundred tons burthen. These were to attend us till the provisions we had taken on board were so far consumed as to make room for the additional quantity they carried with them, which, when we had taken into our ships, they were to be discharged. Besides the complement of men borne by the above-mentioned ships as their crews, there were embarked on board the squadron about four hundred and seventy invalids and marines, under the denomination of land forces (as has been particularly mentioned in the preceding chapter) which were commanded by Lieutenant Colonel Cracherode. With this squadron, together with the *St. Albans* and the *Lark*, and the trade under their convoy, Mr. Anson, after weighing from St. Helen's, tided it down the Channel for the first forty-eight hours; and, on the 20th, in the morning, we discovered off the Ram Head the *Dragon*, *Winchester*, *South Sea Castle*, and *Rye*, with a number of merchantmen under their convoy. These we joined about noon the same day, our commodore having orders to see them (together with the *St. Albans* and *Lark*) as far into the sea as their course and ours lay together. When we came in sight of this last mentioned fleet, Mr. Anson first hoisted his broad pennant, and was saluted by all the men-of-war in company.

When we had joined this last convoy, we made up eleven men-of-war, and about one hundred and fifty sail of merchantmen, consisting of the Turky, the Streights, and the American trade. Mr. Anson, the same day, made a signal for all the captains of the men-of-war to come on board him, where he delivered them their fighting and sailing instructions, and then, with a fair wind, we all stood towards the south-west; and the next day at noon, being the 21st, we had run forty leagues from the Ram Head. Being now clear of the land, our commodore, to render our view more extensive, ordered Captain Mitchel, in the *Pearl*, to make sail two leagues ahead of the fleet every morning, and to repair to his station every evening. Thus we proceeded till the 25th, when the *Winchester* and the American convoy made the concerted signal for leave to separate, which being answered by the commodore, they left us: as the *St. Albans* and the *Dragon*, with the Turkey and Streights convoy, did on the 29th. After which separation, there remained in company only our own squadron and our two victuallers, with which we kept on our course for the island of Madera. But the winds were so contrary that we had the mortification to be forty days in our passage thither from St. Helens, though it is known to be often done in ten or twelve. This delay was a most unpleasing circumstance, productive of much discontent and ill-humour amongst our people, of which those only can have a tolerable idea who have had the experience of a like situation. For besides the peevishness and despondency which foul and contrary winds and a lingering voyage never fail to create on all occasions, we, in particular, had very substantial reasons to be greatly alarmed at this unexpected impediment. Since as we had departed from England much later than we ought to have done, we had placed almost all our hopes of success in the chance of retrieving in some measure at sea the time we had so unhappily wasted at Spithead and St. Helens. However, at last, on Monday, October the 25th, at five in the morning, we, to our great joy, made the land, and in the afternoon came to an anchor in Madera Road, in forty fathom water; the Brazen Head bearing from us E. by S. the Loo N.N.W. and great church N.N.E. We had hardly let go our anchor when an English privateer sloop ran under our stern and saluted the commodore with nine guns, which we returned with five. And, the next day, the consul of the island visiting the commodore, we saluted him with nine guns on his coming on board.

This island of Madera, where we are now arrived, is famous through all our American settlements for its excellent wines, which seem to be designed by Providence for the refreshment of the inhabitants of the torrid zone. It is situated in a fine climate, in the latitude of 32° 27' north; and in the longitude from London (by our different reckonings) of 18-½° to 19-½° west, though laid down in the charts in 17°. It is composed of one continued hill, of a considerable height, extending itself from east to west: the declivity of which, on the south side, is cultivated and interspersed with vineyards: and in the midst of this slope the merchants have fixed their country seats, which help to form a very agreeable prospect. There is but one considerable town in the whole island; it is named Fonchiale, and is seated on the south part of the island, at the bottom of a large bay. Towards the sea, it is defended by a high wall, with a battery of cannon, besides a castle on the Loo, which is a rock standing in the water at a small distance from the shore. Fonchiale is the only place of trade, and indeed the only place where it is possible for a boat to land. And even here the beach is covered with large stones, and a violent surf continually beats upon it; so that the commodore did not care to venture the ships' long-boats to fetch the water off, there was so much danger of their being lost; and therefore ordered the captains of the squadron to employ Portuguese boats on that service.

We continued about a week at this island, watering our ships, and providing the squadron with wine and other refreshments. Here on the 3d of November, Captain Richard Norris signified by a letter to the commodore, his desire to quit his command on board the *Gloucester* in order to return to England for the recovery of his health. This request the commodore complied with; and thereupon was pleased to appoint Captain Matthew Mitchel to command the *Gloucester* in his room, and to remove Captain Kidd from the *Wager* to the *Pearl*, and Captain Murray from the *Tryal* sloop to the *Wager*, giving command of the *Tryal* to Lieutenant Cheap. These promotions being settled, with other changes in the lieutenancies, the commodore, on the following day, gave to the captains their orders, appointing St. Jago, one of the Cape de Verd Islands, to be the first place of rendezvous in case of separation; and directing them, if they did not meet the *Centurion* there, to make the best of their way to the island of St. Catherine's, on the coast of Brazil. The water for the squadron being the same day compleated, and each ship supplied with as much wine and other refreshments as they could take in, we weighed anchor in the afternoon, and took our leave of the island of Madera. But before I go on with the narration of our own transactions, I think it necessary to give some account of the proceedings of the enemy, and of the measures they had taken to render all our designs abortive.

When Mr. Anson visited the governor of Madera, he received information from him that for three or four days, in the latter end of October, there had appeared to the westward of that island, seven or eight ships of the line, and a patache, which last was sent every day close in to make the land. The governor assured the commodore, upon his honour, that none upon the island had either given them intelligence, or had in any sort communicated with them, but that he believed them to be either French or Spanish, but was rather inclined to think them Spanish. On this intelligence Mr. Anson sent an officer in a clean sloop, eight leagues to the westward, to reconnoitre them, and, if possible, to discover what they were: but the officer returned without being able to get a sight of them, so that we still remained in uncertainty. However, we could not but conjecture that this fleet was intended to put a stop to our expedition, which, had they cruised to the eastward of the island instead of the westward, they could not but have executed with great facility. For as, in that case, they must have certainly fallen in with us, we should have been obliged to throw overboard vast quantities of provision to clear our ships for an engagement; and this alone, without any regard to the event of the action, would have effectually prevented our progress. This was so obvious a measure that we could not help imagining reasons which might have prevented them from pursuing it. And we therefore supposed that this French or Spanish squadron was sent out, upon advice of our sailing in company with Admiral Balchen and Lord Cathcart's expedition: and thence, from an apprehension of being overmatched, they might not think it adviseable to meet with us till we had parted company, which they might judge would not happen before our arrival at this island. These were our speculations at that time, and from hence we had reason to suppose that we might still fall in with them in our way to the Cape de Verd Islands. We afterwards, in the course of our expedition, were persuaded that this was the Spanish squadron, commanded by Don Joseph Pizarro, which was sent out purposely to traverse the views and enterprizes of our squadron, to which in strength they were greatly superior. As this Spanish armament then was so nearly connected with our expedition, and as the catastrophe it underwent, though not effected by our force, was yet a considerable advantage to this nation, produced in consequence of our equipment, I have, in the following chapter, given a summary account of their proceedings, from their first setting out from Spain in the year 1740, till the *Asia*, the only ship of the whole squadron which returned to Europe, arrived at the Groyne in the beginning of the year 1746.

CHAPTER III

THE HISTORY OF THE SPANISH SQUADRON COMMANDED BY DON JOSEPH PIZARRO

The squadron fitted out by the court of Spain to attend our motions, and traverse our projects, we supposed to have been the ships seen off Madera, as mentioned in the preceding chapter. As this force was sent out particularly against our expedition, I cannot but imagine that the following history of the casualties it met with, as far as by intercepted letters and other information the same has come to my knowledge, is a very essential part of the present work. For hence it will appear that we were the occasion that a considerable part of the naval power of Spain was diverted from the prosecution of the ambitious views of that court in Europe. And whatever men and ships were lost by the enemy in this undertaking, were lost in consequence of the precautions they took to secure themselves against our enterprizes.

This squadron (besides two ships intended for the West Indies, which did not part company till after they had left the Maderas) was composed of the following men-of-war, commanded by Don Joseph Pizarro:—

The *Asia* of sixty-six guns, and seven hundred men; this was the admiral's ship.

The *Guipuscoa* of seventy-four guns, and seven hundred men.

The *Hermiona* of fifty-four guns, and five hundred men.

The *Esperanza* of fifty guns, and four hundred and fifty men.

The *St. Estevan* of forty guns, and three hundred and fifty men.

And a patache of twenty guns.

These ships, over and above their complement of sailors and marines, had on board an old Spanish regiment of foot, intended to reinforce the garisons on the coast of the South Seas. When this fleet had cruised for some days to the leeward of the Maderas, as is mentioned in the preceding chapter, they left that station in the beginning of November, and steered for the river of Plate, where they arrived the 5th of January, O.S., and coming to an anchor in the bay of Maldonado, at the mouth of that river, their admiral Pizarro sent immediately to Buenos Ayres for a supply of provisions; for they had departed from Spain with only four months' provisions on board. While they lay here expecting this supply, they received intelligence, by the treachery of the Portuguese governor of St. Catherine's, of Mr. Anson's having arrived at that island on the 21st of December preceding, and of his preparing to put to sea again with the utmost expedition. Pizarro, notwithstanding his superior force, had his reasons (and as some say, his orders, likewise) for avoiding our squadron anywhere short of the South Seas. He was besides extremely desirous of getting round Cape Horn before us, as he imagined that step alone would effectually baffle all our designs; and therefore, on hearing that we were in his neighbourhood, and that we should soon be ready to proceed for Cape Horn, he weighed anchor with the five large ships (the patache being disabled and condemned, and the men taken out of her), after a stay of seventeen days only, and got under sail without his provisions, which arrived at Maldonado within a day or two after his departure. But notwithstanding the precipitation with which he departed, we put to sea from St. Catherine's four days before him, and in some part of our passage to Cape Horn the two squadrons were so near together that the *Pearl*, one of our ships, being separated from the rest, fell in with the Spanish fleet, and mistaking the *Asia* for the *Centurion*, had got within gun-shot of Pizarro before she discovered her error, and narrowly escaped being taken.

It being the 22d of January when the Spaniards weighed from Maldonado (as has been already mentioned), they could not expect to get into the latitude of Cape Horn before the equinox; and as they had reason to apprehend very tempestuous weather in doubling it at that season, and as the Spanish sailors, being for the most part accustomed to a fair weather country, might be expected to be very averse to so dangerous and fatiguing a navigation, the better to encourage them, some part of their pay was advanced to them in European goods, which they were to be permitted to dispose of in the South Seas, that so the hopes of the great profit each man was to make on his venture might animate him in his duty, and render him less disposed to repine at the labour, the hardships, and the perils he would in all probability meet with before his arrival on the coast of Peru.

Pizarro with his squadron having, towards the latter end of February, run the length of Cape Horn, he then stood to the westward, in order to double it; but in the night of the last day of February, O.S., while with this view they were turning to windward, the *Guipuscoa*, the *Hermiona*, and the *Esperanza* were separated from the admiral; and, on the 6th of March following, the *Guipuscoa* was separated from the other two; and, on the 7th (being the day after we had passed Streights le Maire), there came on a most furious storm at N.W. which, in despight of all their efforts, drove the whole squadron to the eastward, and after several fruitless attempts, obliged them to bear away for the river of Plate, where Pizarro in the *Asia* arrived about the middle of May, and a few days after him the *Esperanza* and the *Estevan*. The *Hermiona* was supposed to founder at sea, for she was never heard of more; and the *Guipuscoa* was run on shore and sunk on the coast of Brazil. The calamities of all kinds which this squadron underwent in this unsuccessful navigation can only be paralleled by what we ourselves experienced in the same climate, when buffeted by the same storms. There was indeed some diversity in our distresses, which rendered it difficult to decide whose situation was most worthy of commiseration. For to all the misfortunes we had in common with each other, as shattered rigging, leaky ships, and the fatigues and despondency which necessarily attend these disasters, there was superadded on board our squadron the ravage of a most destructive and incurable disease, and on board the Spanish squadron the devastation of famine.

For this squadron, either from the hurry of their outset, their presumption of a supply at Buenos Ayres, or from other less obvious motives, departed from Spain, as has been already observed, with no more than four months' provision on board, and even that, as it is said, at short allowance only; so that, when by the storms they met with off Cape Horn their continuance at sea was prolonged a month or more beyond their expectation, they were reduced to such infinite distress, that rats, when they could be caught, were sold for four dollars apiece; and a sailor who died on board had his death concealed for some days by his brother, who during that time lay in the same hammock with the corpse, only to receive the dead man's allowance of provisions. In this dreadful situation they were alarmed (if their horrors were capable of augmentation) by the discovery of a conspiracy among the marines on board the *Asia*, the admiral's ship. This had taken its rise chiefly from the miseries they endured: for though no less was proposed by the conspirators than the massacring the officers and the whole crew, yet their motive for this bloody resolution seemed to be no more than their desire of relieving their hunger by appropriating the whole ship's provisions to themselves. But their designs were prevented, when just upon the point of execution, by means of one of their confessors; and three of their ringleaders were immediately put to death. However, though the conspiracy was suppressed, their other calamities admitted of no alleviation, but grew each day more and more destructive. So that by the complicated distress of fatigue, sickness, and hunger, the three ships which escaped lost the greatest part of their men. The *Asia*, their admiral's ship, arrived at Monte Vedio, in the river of Plate, with half her crew only; the *St. Estevan* had lost, in like manner, half her hands when she anchored in the bay of Barragan; the *Esperanza*, a fifty-gun ship, was still more unfortunate; for of four hundred and fifty hands which she brought from Spain, only fifty-eight remained alive, and the whole regiment of foot perished except sixty men. But to give the reader a more distinct and particular idea of what they underwent upon this occasion, I shall lay before him a short account of the fate of the *Guipuscoa*, extracted from a letter written by Don Joseph Mendinuetta, her captain, to a person of distinction at Lima, a copy of which fell into our hands afterwards in the South Seas.

He mentions that he separated from the *Hermiona* and the *Esperanza* in a fog on the 6th of March, being then, as I suppose, to the S.E. of Staten-land, and plying to the westward; that in the night after it blew a furious storm at N.W. which, at half an hour after ten, split his main-sail, and obliged him to bear away with his fore-sail; that the ship went ten knots an hour with a prodigious sea, and often run her gangway under water; that he likewise sprung his mainmast; and the ship made so much water, that with four pumps and bailing he could not free her. That on the 9th it was calm, but the sea continued so high that the ship in rolling opened all her upper works and seams, and started the butt ends of her planking, and the greatest part of her top timbers, the bolts being drawn by the violence of her roll: that in this condition, with other additional disasters to the hull and rigging, they continued beating to the westward till the 12th: that they were then in sixty degrees of south latitude, in great want of provisions, numbers every day perishing by the fatigue of pumping, and those who survived being quite dispirited by labour, hunger, and the severity of the weather, they having two spans of snow upon the decks: that then finding the wind fixed in the western quarter, and blowing strong, and consequently their passage to the westward impossible, they resolved to bear away for the river of Plate: that on the 22d they were obliged to throw overboard all the upper-deck guns and an anchor, and to take six turns of the cable round the ship to prevent her opening: that on the 4th of April, it being calm, but a very high sea, the ship rolled so much that the main-mast came by the board, and in a few hours after she lost, in like manner, her fore-mast and her mizen-mast: and that, to accumulate their misfortunes, they were soon obliged to cut away their bowsprit, to diminish, if possible, the leakage at her head; that by this time he had lost two hundred and fifty men by hunger and fatigue; for those who were capable of working at the pumps (at which every officer without exception took his turn) were allowed only an ounce and half of biscuit per diem; and those who were so sick or so weak that they could not assist in this necessary labour, had no more than an ounce of wheat; so that it was common for the men to fall down dead at the pumps: that, including the officers, they could only muster from eighty to a hundred persons capable of duty: that the south-west winds blew so fresh after they had lost their masts, that they could not immediately set up jury-masts, but were obliged to drive like a wreck, between the latitudes of 32 and 28, till the 24th of April, when they made the coast of Brazil at Rio de Patas, ten leagues to the southward of the island of St. Catherine's; that here they came to an anchor, and that the captain was very desirous of proceeding to St. Catherine's, if possible, in order to save the hull of the ship, and the guns and stores on board her; but the crew instantly left off pumping, and being enraged at the hardships they had suffered, and the numbers they had lost (there being at that time no less than thirty dead bodies lying on the deck), they all with one voice cried out, "On shore, on shore!" and obliged the captain to run the ship in directly for the land, where, the 5th day after, she sunk with her stores and all her furniture on board her; but the remainder of the crew, whom hunger and fatigue had spared, to the number of four hundred, got safe on shore.

From this account of the adventures and catastrophe of the *Guipuscoa* we may form some conjecture of the manner in which the *Hermiona* was lost, and of the distresses endured by the three remaining ships of the squadron, which got into the river of Plate. These last being in great want of masts, yards, rigging, and all kinds of naval stores, and having no supply at Buenos Ayres, nor in any of their neighbouring settlements, Pizarro dispatched an advice-boat with a letter of credit to Rio Janeiro, to purchase what was wanting from the Portuguese. He, at the same time, sent an express across the continent to St. Jago in Chili, to be thence forwarded to the Viceroy of Peru, informing him of the disasters that had befallen his squadron, and desiring a remittance of 200,000 dollars from the royal chests at Lima, to enable him to victual and refit his remaining ships, that he might be again in a condition to attempt the passage to the South Seas, as soon as the season of the year should be more favourable. It is mentioned by the Spaniards as a most extraordinary circumstance that the Indian charged with this express (though it was then the depth of winter, when the Cordilleras are esteemed impassable on account of the snow) was only thirteen days in his journey from Buenos Ayres to St. Jago in Chili, though these places are distant three hundred Spanish leagues, near forty of which are amongst the snows and precipices of the Cordilleras.

The return to this dispatch of Pizarro's from the Viceroy of Peru was no ways favourable; instead of 200,000 dollars, the sum demanded, the viceroy remitted him only 100,000, telling him that it was with great difficulty he was able to procure him even that: though the inhabitants of Lima, who considered the presence of Pizarro as absolutely necessary to their security, were much discontented at this procedure, and did not fail to assert that it was not the want of money, but the interested views of some of the viceroy's confidents, that prevented Pizarro from having the whole sum he had asked for.

The advice-boat sent to Rio Janeiro also executed her commission but imperfectly; for though she brought back a considerable quantity of pitch, tar, and cordage, yet she could not procure either masts or yards: and, as an additional misfortune, Pizarro was disappointed of some masts he expected from Paragua; for a carpenter, whom he had entrusted with a large sum of money, and had sent there to cut masts, instead of prosecuting the business he was employed in, had married in the country, and refused to return. However, by removing the masts of the *Esperanza* into the *Asia*, and making use of what spare masts and yards they had on board, they made a shift to refit the *Asia* and the *St. Estevan*. And in the October following, Pizarro was preparing to put to sea with these two ships, in order to attempt the passage round Cape Horn a second time, but the *St. Estevan*, in coming down the river Plate, ran on a shoal, and beat off her rudder, on which and other damages she received she was condemned and broke up, and Pizarro in the *Asia* proceeded to sea without her. Having now the summer before him, and the winds favourable, no doubt was made of his having a fortunate and speedy passage; but being off Cape Horn, and going right before the wind in very moderate weather, though in a swelling sea, by some misconduct of the officer of the watch, the ship rolled away her masts, and was a second time obliged to put back to the river of Plate in great distress.

The *Asia* having considerably suffered in this second unfortunate expedition, the *Esperanza*, which had been left behind at Monte Vedio, was ordered to be refitted, and the command of her being given to Mindinuetta, who was captain of the *Guipuscoa* when she was lost; he, in the November of the succeeding year, that is, in November 1742, sailed from the river of Plate for the South Seas, and arrived safe on the coast of Chili, where his commodore, Pizarro, passing overland from Buenos Ayres, met him. There were great animosities and contests between these two gentlemen at their meeting, occasioned principally by the claim of Pizarro to command the *Esperanza*, which Mindinuetta had brought round; for Mindinuetta refused to deliver her up to him, insisting that, as he came into the South Seas alone and under no superior, it was not now in the power of Pizarro to resume that authority which he had once parted with. However, the President of Chili interposing, and declaring for Pizarro, Mindinuetta, after a long and obstinate struggle, was obliged to submit.

But Pizarro had not yet compleated the series of his adventures, for when he and Mindinuetta came back by land from Chili to Buenos Ayres, in the year 1745, they found at Monte Vedio the *Asia*, which near three years before they had left there.

This ship they resolved, if possible, to carry to Europe; and with this view they refitted her in the best manner they could; but their great difficulty was to procure a sufficient number of hands to navigate her, for all the remaining sailors of the squadron to be met with in the neighbourhood of Buenos Ayres did not amount to a hundred men. They endeavoured to supply this defect by pressing many of the inhabitants of Buenos Ayres, and putting on board besides all the English prisoners then in their custody, together with a number of Portuguese smugglers which they had taken at different times, and some of the Indians of the country. Among these last there was a chief and ten of his followers which had been surprised by a party of Spanish soldiers about three months before. The name of this chief was Orellana; he belonged to a very powerful tribe which had committed great ravages in the neighbourhood of Buenos Ayres. With this motley crew (all of them, except the European Spaniards, extremely averse to the voyage) Pizarro set sail from Monte Vedio in the river of Plate, about the beginning of November 1745; and the native Spaniards, being no strangers to the dissatisfaction of their forced men, treated both those, the English prisoners, and the Indians, with great insolence and barbarity, but more particularly the Indians, for it was common for the meanest officers in the ship to beat them most cruelly on the slightest pretences, and oftentimes only to exert their superiority. Orellana and his followers, though in appearance sufficiently patient and submissive, meditated a severe revenge for all these inhumanities. As he conversed very well in Spanish (these Indians having, in time of peace, a great intercourse with Buenos Ayres) he affected to talk with such of the English as understood that language, and seemed very desirous of being informed how many Englishmen there were on board, and which they were. As he knew that the English were as much enemies to the Spaniards as himself, he had doubtless an intention of disclosing his purposes to them, and making them partners in the scheme he had projected for revenging his wrongs, and recovering his liberty; but having sounded them at a distance, and not finding them so precipitate and vindictive as he expected, he proceeded no further with them, but resolved to trust alone to the resolution of his ten faithful followers. These, it should seem, readily engaged to observe his directions, and to execute whatever commands he gave them; and having agreed on the measures necessary to be taken, they first furnished themselves with Dutch knives sharp at the point, which being common knives used in the ship, they found no difficulty in procuring: besides this, they employed their leisure in secretly cutting out thongs from raw hides, of which there were great numbers on board, and in fixing to each end of these thongs the double-headed shot of the small quarter-deck guns; this, when swung round their heads, according to the practice of their country, was a most mischievous weapon, in the use of which the Indians about Buenos Ayres are trained from their infancy, and consequently are extremely expert. These particulars being in good forwardness, the execution of their scheme was perhaps precipitated by a particular outrage committed on Orellana himself. For one of the officers, who was a very brutal fellow, ordered Orellana aloft, which being what he was incapable of performing, the officer, under pretence of his disobedience, beat him with such violence, that he left him bleeding on the deck, and stupified for some time with his bruises and wounds. This usage undoubtedly heightened his thirst for revenge, and made him eager and impatient, till the means of executing it were in his power; so that within a day or two after this incident, he and his followers opened their desperate resolves in the ensuing manner.

It was about nine in the evening, when many of the principal officers were on the quarter-deck, indulging in the freshness of the night air; the waste of the ship was filled with live cattle, and the forecastle was manned with its customary watch. Orellana and his companions, under cover of the night, having prepared their weapons, and thrown off their trousers and the more cumbrous part of their dress, came all together on the quarter-deck, and drew towards the door of the great cabin. The boatswain immediately reprimanded them and ordered them to be gone. On this Orellana spoke to his followers in his native language, when four of them drew off, two towards each gangway, and the chief and the six remaining Indians seemed to be slowly quitting the quarter-deck. When the detached Indians had taken possession of the gangway, Orellana placed his hands hollow to his mouth, and bellowed out the war-cry used by those savages, which is said to be the harshest and most terrifying sound known in nature. This hideous yell was the signal for beginning the massacre: For on this they all drew their knives, and brandished their prepared double-headed shot; and the six with their chief, which remained on the quarter-deck, immediately fell on the Spaniards who were intermingled with them, and laid near forty of them at their feet, of which above twenty were killed on the spot, and the rest disabled. Many of the officers, in the beginning of the tumult, pushed into the great cabin, where they put out the lights, and barricadoed the door; whilst of the others, who had avoided the first fury of the Indians, some endeavoured to escape along the gangways into the forecastle, where the Indians, placed on purpose, stabbed the greatest part of them, as they attempted to pass by, or forced them off the gangways into the waste: some threw themselves voluntarily over the barricadoes into the waste, and thought themselves fortunate to lie concealed amongst the cattle. But the greatest part escaped up the main shrouds, and sheltered themselves either in the tops or rigging. And though the Indians attacked only the quarter-deck, yet the watch in the forecastle finding their communication cut off, and being terrified by the wounds of the few, who, not being killed on the spot, had strength sufficient to force their passage, and not knowing either who their enemies were, or what were their numbers, they likewise gave all over for lost, and in great confusion ran up into the rigging of the foremast and bowsprit.

Thus these eleven Indians, with a resolution perhaps without example, possessed themselves almost in an instant of the quarter-deck of a ship mounting sixty-six guns, and mann'd with near five hundred hands, and continued in peaceable possession of this post a considerable time. For the officers in the great cabbin (amongst whom were Pizarro and Mindinuetta), the crew between decks, and those who had escaped into the tops and rigging, were only anxious for their own safety, and were for a long time incapable of forming any project for suppressing the insurrection, and recovering the possession of the ship. It is true, the yells of the Indians, the groans of the wounded, and the confused clamours of the crew, all heightened by the obscurity of the night, had at first greatly magnified their danger, and had filled them with the imaginary terrors, which darkness, disorder, and an ignorance of the real strength of an enemy never fail to produce. For as the Spaniards were sensible of the disaffection of their prest hands, and were also conscious of their barbarity to their prisoners, they imagined the conspiracy was general, and considered their own destruction as infallible; so that, it is said, some of them had once taken the resolution of leaping into the sea, but were prevented by their companions.

However, when the Indians had entirely cleared the quarter-deck, the tumult in a great measure subsided, for those who had escaped were kept silent by their fears, and the Indians were incapable of pursuing them to renew the disorder. Orellana, when he saw himself master of the quarter-deck, broke open the arm chest, which, on a slight suspicion of mutiny, had been ordered there a few days before, as to a place of the greatest security. Here he took it for granted he should find cutlasses sufficient for himself and his companions, in the use of which weapon they were all extremely skilful, and with these, it was imagined, they proposed to have forced the great cabbin. But on opening the chest, there appeared nothing but fire-arms, which to them were of no use. There were indeed cutlasses in the chest, but they were hid by the fire-arms being laid over them. This was a sensible disappointment to them, and by this time Pizarro and his companions in the great cabbin were capable of conversing aloud through the cabbin windows and port-holes with those in the gun-room and between decks, and from thence they learnt that the English (whom they principally suspected) were all safe below, and had not intermeddled in this mutiny; and by other particulars they at last discovered that none were concerned in it but Orellana and his people. On this Pizarro and the officers resolved to attack them on the quarter-deck, before any of the discontented on board should so far recover their first surprize, as to reflect on the facility and certainty of seizing the ship by a junction with the Indians in the present emergency. With this view Pizarro got together what arms were in the cabbin, and distributed them to those who were with him. But there were no other fire-arms to be met with but pistols, and for these they had neither powder nor ball. However, having now settled a correspondence with the gun-room, they lowered down a bucket out of the cabbin window, into which the gunner, out of one of the gun-room ports, put a quantity of pistol-cartridges. When they had thus procured ammunition, and had loaded their pistols, they set the cabbin door partly open, and fired several shots amongst the Indians on the quarter-deck, though at first without effect. But at last Mindinuetta, whom we have often mentioned, had the good fortune to shoot Orellona dead on the spot; on which his faithful companions, abandoning all thoughts of farther resistance, instantly leaped into the sea, where they every man perished. Thus was this insurrection quelled, and the possession of the quarter-deck regained, after it had been full two hours in the power of this great and daring chief, and his gallant unhappy countrymen.

Pizarro having escaped this imminent peril, steered for Europe, and arrived safe on the coast of Gallicia, in the beginning of the year 1746, after having been absent between four and five years, and having, by his attendance on our expedition, diminished the naval power of Spain by above three thousand hands (the flower of their sailors), and by four considerable ships of war and a patache. For we have seen that the *Hermiona* foundered at sea; the *Guipuscoa* was stranded, and sunk on the coast of Brazil; the *St. Estevan* was condemned, and broke up in the river of Plate; and the *Esperanza*, being left in the South Seas, is doubtless by this time incapable of returning to Spain. So that the *Asia*, only, with less than one hundred hands, may be regarded as all the remains of that squadron with which Pizarro first put to sea. And whoever considers the very large proportion, which this squadron bore to the whole navy of Spain, will, I believe, confess, that, had our undertaking been attended with no other advantages than that of ruining so great a part of the sea force of so dangerous an enemy, this alone would be a sufficient equivalent for our equipment, and an incontestable proof of the service which the nation has thence received. Having thus concluded this summary of Pizarro's adventures, I shall now return again to the narration of our own transactions.

CHAPTER IV

FROM MADERA TO ST. CATHERINE'S

I have already mentioned, that on the 3d of November we weighed from Madera, after orders had been given to the captains to rendezvous at St. Jago, one of the Cape de Verd Islands, in case the squadron was separated. But the next day, when we were got to sea, the commodore considering that the season was far advanced, and that touching at St. Jago would create a new delay, he for this reason thought proper to alter his rendezvous, and to appoint the island of St. Catherine's, on the coast of Brazil, to be the first place to which the ships of the squadron were to repair in case of separation.

In our passage to the island of St. Catherine's, we found the direction of the trade-winds to differ considerably from what we had reason to expect, both from the general histories given of these winds, and the experience of former navigators. For the learned Dr. Halley, in his account of the trade-winds which take place in the Ethiopia and Atlantic Ocean, tells us that from the latitude of 28° N., to the latitude of 10° N., there is generally a fresh gale of N.E. wind, which towards the African side rarely comes to the eastward of E.N.E., or passes to the northward of N.N.E.; but on the American side the wind is somewhat more easterly, though most commonly even there it is a point or two to the northward of the east. That from 10° N. to 4° N., the calms and tornadoes take place; and from 4° N. to 30° S., the winds are generally and perpetually between the south and the east. This account we expected to have verified by our own experience; but we found considerable variations from it, both in respect to the steadiness of the winds, and the quarter from whence they blew. For though we met with a N.E. wind about the latitude of 28° N., yet from the latitude of 25° to the latitude of 18° N., the wind was never once to the northward of the east, but, on the contrary, almost constantly to the southward of it. However, from thence to the latitude of 6° 20' N., we had it usually to the northward of the east, though not entirely, it having for a short time changed to E.S.E. From hence, to about 4° 46' N., the weather was very unsettled; sometimes the wind was N.E. then changed to S.E., and sometimes we had a dead calm attended with small rain and lightning. After this, the wind continued almost invariably between the S. and E., to the latitude of 7° 30' S.; and then again as invariably between the N. and E., to the latitude of 15° 30' S.; then E. and S.E., to 21° 37' S. But after this, even to the latitude of 27° 44' S., the wind was never once between the S. and the E., though we had it at all times in all the other quarters of the compass. But this last circumstance may be in some measure accounted for from our approach to the main continent of the Brazils. I mention not these particulars with a view of cavilling at the received accounts of these trade-winds, which I doubt not are in general sufficiently accurate; but I thought it a matter worthy of public notice that such deviations from the established rules do sometimes take place. Besides, this observation may not only be of service to navigators, by putting them on their guard against these hitherto unexpected irregularities, but is a circumstance necessary to be attended to in the solution of that great question about the causes of trade-winds, and monsoons, a question which, in my opinion, has not been hitherto discussed with that clearness and accuracy which its importance (whether it be considered as a naval or philosophical inquiry) seems to demand.

On the 16th of November, one of our victuallers made a signal to speak with the commodore, and we shortened sail for her to come up with us. The master came on board, and acquainted Mr. Anson that he had complied with the terms of his charter-party, and desired to be unloaded and dismissed. Mr. Anson, on consulting the captains of the squadron, found all the ships had still such quantities of provision between their decks, and were withal so deep, that they could not, without great difficulty, take in their several proportions of brandy from the *Industry* pink, one of the victuallers only: consequently he was obliged to continue the other of them, the *Anna* pink, in the service of attending the squadron. This being resolved on, the commodore the next day made a signal for the ships to bring to, and to take on board their shares of the brandy from the *Industry* pink; and in this the long boats of the squadron were employed the three following days, that is, till the 19th in the evening, when the pink being unloaded, she parted company with us, being bound for Barbadoes, there to take in a freight for England. Most of the officers of the squadron took the opportunity of writing to their friends at home by this ship; but she was afterwards, as I have been since informed, unhappily taken by the Spaniards.

On the 20th of November, the captains of the squadron represented to the commodore that their ships' companies were very sickly, and that it was their own opinion, as well as their surgeons', that it would tend to the preservation of the men to let in more air between decks, but that their ships were so deep they could not possibly open their lower ports. On this representation, the commodore ordered six air scuttles to be cut in each ship, in such places where they would least weaken it.

And on this occasion I cannot but observe, how much it is the duty of all those who either by office or authority have any influence in the direction of our naval affairs, to attend to this important article, the preservation of the lives and health of our seamen. If it could be supposed that the motives of humanity were insufficient for this purpose, yet policy, and a regard to the success of our arms, and the interest and honour of each particular commander, should naturally lead us to a careful and impartial examination of every probable method proposed for maintaining a ship's crew in health and vigour. But hath this been always done? Have the late invented plain and obvious methods of keeping our ships sweet and clean by a constant supply of fresh air been considered with that candour and temper which the great benefits promised hereby ought naturally to have inspired? On the contrary, have not these salutary schemes been often treated with neglect and contempt? And have not some of those who have been entrusted with experimenting their effects, been guilty of the most indefensible partiality in the accounts they have given of these trials? Indeed, it must be confessed that many distinguished persons, both in the direction and command of our fleets, have exerted themselves on these occasions with a judicious and dispassionate examination becoming the interesting nature of the inquiry; but the wonder is, that any could be found irrational enough to act a contrary part in despight of the strongest dictates of prudence and humanity. I must, however, own that I do not believe this conduct to have arisen from motive so savage as the first reflection thereon does naturally suggest: but I rather impute it to an obstinate, and in some degree, superstitious attachment to such practices as have been long established, and to a settled contempt and hatred of all kinds of innovations, especial such as are projected by landmen and persons residing on shore. But let us return from this, I hope not, impertinent digression.

We crossed the equinoxial with a fine fresh gale at S.E. on Friday the 28th of November, at four in the morning, being then in the longitude of 27° 59' west from London. And on the 2d of December, in the morning, we saw a sail in the N.W. quarter, and made the *Gloucester's* and *Tryal's* signals to chace, and half an hour after we let out our reefs and chased with the squadron; and about noon a signal was made for the *Wager* to take our remaining victualler, the *Anna* pink, in tow. But at seven in the evening, finding we did not near the chace, and that the *Wager* was very far astern, we shortened sail, and made a signal for the cruizers to join the squadron. The next day but one we again discovered a sail, which on a nearer approach we judged to be the same vessel. We chased her the whole day, and though we rather gained upon her, yet night came on before we could overtake her, which obliged us to give over the chace, to collect our scattered squadron. We were much chagrined at the escape of this vessel, as we then apprehended her to be an advice-boat sent from Old Spain to Buenos Ayres with notice of our expedition. But we have since learnt that we were deceived in this conjecture, and that it was our East India Company's packet bound to St. Helena.

On the 10th of December, being by our accounts in the latitude of 20° S., and 36° 30' longitude west from London, the *Tryal* fired a gun to denote soundings. We immediately sounded, and found sixty fathom water, the bottom coarse ground with broken shells. The *Tryal* being ahead of us, had at one time thirty-seven fathom, which afterwards increased to ninety: and then she found no bottom, which happened to us too at our second trial, though we sounded with a hundred and fifty fathom of line. This is the shoal which is laid down in most charts by the name of the Abrollos; and it appeared we were upon the very edge of it; perhaps farther in it may be extremely dangerous. We were then, by our different accounts, from ninety to sixty leagues east of the coast of Brazil. The next day but one we spoke with a Portuguese brigantine from Rio Janeiro, bound to Bahia del todos Santos, who informed us that we were thirty-four leagues from Cape St. Thomas, and forty leagues from Cape Frio, which last bore from us W.S.W. By our accounts we were near eighty leagues from Cape Frio; and though on the information of this brigantine we altered our course and stood more to the southward, yet by our coming in with the land afterwards, we were fully convinced that our reckoning was much correcter than our Portuguese intelligence. We found a considerable current setting to the southward after we had passed the latitude of 16° S. And the same took place all along the coast of Brazil, and even to the southward of the river of Plate, it amounting sometimes to thirty miles in twenty-four hours, and once to above forty miles.

If this current is occasioned (as it is most probable) by the running off of the water accumulated on the coast of Brazil by the constant sweeping of the eastern trade-wind over the Ethiopic Ocean, then it is most natural to suppose that its general course is determined by the bearings of the adjacent shore. Perhaps too, in almost every other instance of currents, the same may hold true, as I believe no examples occur of considerable currents being observed at any great distance from land. If this then could be laid down for a general principle, it would be always easy to correct the reckoning by the observed latitude. But it were much to be wished, for the general interests of navigation, that the actual settings of the different currents which are known to take place in various parts of the world were examined more frequently and accurately than hitherto appears to have been done.

We now began to grow impatient for a sight of land, both for the recovery of our sick, and for the refreshment and security of those who as yet continued healthy. When we departed from St. Helens, we were in so good a condition that we lost but two men on board the *Centurion* in our long passage to Madera. But in this present run between Madera and St. Catherine's we were remarkably sickly, so that many died, and great numbers were confined to their hammocks, both in our own ship, and in the rest of the squadron, and several of these past all hopes of recovery. The disorders they in general laboured under were such as are common to the hot climates, and what most ships bound to the southward experience in a greater or less degree. These are those kind of fevers which they usually call calentures: a disease which was not only terrible in its first instance, but even the remains of it often proved fatal to those who considered themselves as recovered from it; for it always left them in a very weak and helpless condition, and usually afflicted either with fluxes or tenesmus's. By our continuance at sea all these complaints were every day increasing, so that it was with great joy we discovered the coast of Brazil on the 16th of December, at seven in the morning.

The coast of Brazil appeared high and mountainous land, extending from W. to W.S.W., and when we first saw it, it was about seventeen leagues distant. At noon we perceived a low double land bearing W.S.W. about ten leagues distant, which we took to be the island of St. Catherine's. That afternoon and the next morning, the wind being N.N.W., we gained very little to windward, and were apprehensive of being driven to the leeward of the island; but a little before noon the next day the wind came about to the southward, and enabled us to steer in between the north point of St. Catherine's and the neighbouring island of Alvoredo. As we stood in for the land, we had regular soundings, gradually decreasing from thirty-six to twelve fathom, all muddy ground. In this last depth of water we let go our anchor at five o'clock in the evening of the 18th, the north-west point of the island of St. Catherine's bearing S.S.W., distant three miles; and the island Alvoredo N.N.E., distant two leagues. Here we found the tide to set S.S.E. and N.N.W., at the rate of two knots, the tide of flood coming from the southward. We could from our ships observe two fortifications at a considerable distance within us, which seemed designed to prevent the passage of an enemy between the island of St. Catherine's and the main. And we could soon perceive that our squadron had alarmed the coast, for we saw the two forts hoist their colours and fire several guns, which we supposed were signals for assembling the inhabitants. To prevent any confusion, the commodore immediately sent a boat with an officer on shore to compliment the governor, and to desire a pilot to carry us into the road. The governor returned a very civil answer, and ordered us a pilot. On the morning of the 20th we weighed and stood in, and towards noon the pilot came on board us, who, the same afternoon, brought us to an anchor in five fathom and an half, in a large commodious bay on the continent side, called by the French Bon Port. In standing from our last anchorage to this place we everywhere found an ouzy bottom, with a depth of water first regularly decreasing to five fathom, and then increasing to seven, after which we had six and five fathom alternately. The next morning we weighed again with the squadron, in order to run above the two fortifications we have mentioned, which are called the castles of Santa Cruz and St. Juan. Our soundings now between the island and the main were four, five, and six fathom, with muddy ground. As we passed by the castle of Santa Cruz, we saluted it with eleven guns, and were answered by an equal number;

and at one in the afternoon the squadron came to an anchor in five fathom and a half, the governor's island bearing N.N.W., St. Juan's Castle N.E.½E., and the island of St. Antonio south. In this position we moored at the island of St. Catherine's on Sunday the 21st of December, the whole squadron being, as I have already mentioned, sickly, and in great want of refreshments; both which inconveniencies we hoped to have soon removed at this settlement, celebrated by former navigators for its healthiness, and the plenty of its provisions, and for the freedom, indulgence, and friendly assistance there given to the ships of all European nations in amity with the crown of Portugal.

CHAPTER V

PROCEEDINGS AT ST. CATHERINE'S, AND A DESCRIPTION OF THE PLACE, WITH A SHORT ACCOUNT OF BRAZIL

Our first care, after having moored our ships, was to get our sick men on shore, preparatory to which, each ship was ordered by the commodore to erect two tents: one of them for the reception of the diseased, and the other for the accommodation of the surgeon and his assistants. We sent about eighty sick from the *Centurion*; and the other ships, I believe, sent nearly as many, in proportion to the number of their hands. As soon as we had performed this necessary duty, we scraped our decks, and gave our ship a thorough cleansing: then smoked it between decks, and after all washed every part well with vinegar. These operations were extremely necessary for correcting the noisome stench on board, and destroying the vermin; for from the number of our men, and the heat of the climate, both these nusances had increased upon us to a very loathsome degree; and besides being most intolerably offensive, they were doubtless in some sort productive of the sickness we had laboured under for a considerable time before our arrival at this island.

Our next employment was wooding and watering our squadron, caulking our ships' sides and decks, overhauling our rigging, and securing our masts against the tempestuous weather we were, in all probability, to meet with in our passage round Cape Horn in so advanced and inconvenient a season. But before I engage in the particulars of these transactions, it will not be improper to give some account of the present state of this island of St. Catherine's, and of the neighbouring country, both as the circumstances of this place are now greatly changed from what they were in the time of former writers, and as these changes laid us under many more difficulties and perplexities than we had reason to expect, or than other British ships, hereafter bound to the South Seas, may perhaps think it prudent to struggle with.

This island is esteemed by the natives to be nowhere above two leagues in breadth, though about nine in length; it lies in 49° 45' of west longitude of London, and extends from the south latitude of 27° 35', to that of 28°. Although it be of a considerable height, yet it is scarce discernible at the distance of ten leagues, being then obscured under the continent of Brazil, whose mountains are exceeding high; but on a nearer approach it is easy to be distinguished, and may be readily known by a number of small islands lying at each end, and scattered along the east side of it. Frezier has given a draught of the island of St. Catherine's, and of the neighbouring coast, and the minuter isles adjacent; but he has, by mistake, called the island of

Alvoredo the Isle de Gal, whereas the true Isle de Gal lies seven or eight miles to the north-westward of it, and is much smaller. He has also called an island to the southward of St. Catherine's Alvoredo, and has omitted the island Masaqura; in other respects his plan is sufficiently exact.

The north entrance of the harbour is in breadth about five miles, and the distance from thence to the island of St. Antonio is eight miles, and the course from the entrance to St. Antonio is S.S.W.½W. About the middle of the island the harbour is contracted by two points of land to a narrow channel no more than a quarter of a mile broad; and to defend this passage, a battery was erecting on the point of land on the island side. But this seems to be a very useless work, as the channel has no more than two fathom water, and consequently is navigable only for barks and boats, and therefore seems to be a passage that an enemy could have no inducement to attempt, especially as the common passage at the north end of the island is so broad and safe that no squadron can be prevented from coming in by any of their fortifications when the sea breeze is made. However, the Brigadier Don Jose Sylva de Paz, the governor of this settlement, is deemed an expert engineer, and he doubtless understands one branch of his business very well, which is the advantages which new works bring to those who are entrusted with the care of erecting them; for besides the battery mentioned above, there are three other forts carrying on for the defence of the harbour, none of which are yet compleated. The first of these, called St. Juan, is built on a point of St. Catherine's, near Parrot Island; the second, in form of a half-moon, is on the island of St. Antonio; and the third, which seems to be the chief, and has some appearance of a regular fortification, is on an island near the continent, where the governor resides.

The soil of the island is truly luxuriant, producing fruits of many kinds spontaneously; and the ground is covered over with one continued forest of trees of a perpetual verdure, which, from the exuberance of the soil, are so entangled with briars, thorns, and underwood, as to form a thicket absolutely impenetrable, except by some narrow pathways which the inhabitants have made for their own convenience. These, with a few spots cleared for plantations along the shore facing the continent, are the only uncovered parts of the island. The woods are extremely fragrant, from the many aromatic trees and shrubs with which they abound, and the fruits and vegetables of all climates thrive here, almost without culture, and are to be procured in great plenty, so that here is no want of pine-apples, peaches, grapes, oranges, lemons, citrons, melons, apricots, nor plantains. There are besides great abundance of two other productions of no small consideration for a sea-store, I mean onions and potatoes. The flesh provisions are, however, much inferior to the vegetables. There are indeed small wild cattle to be purchased, somewhat like buffaloes, but these are very indifferent food, their flesh being of a loose contexture, and generally of a disagreeable flavour, which is probably owing to the wild calabash on which they feed. There are likewise great plenty of pheasants, but they are not to be compared in taste to those we have in England. The other provisions of the place are monkeys, parrots, and, above all, fish of various sorts; these abound in the harbour, are exceeding good, and are easily catched, for there are a great number of small sandy bays very convenient for haling the Seyne.

The water both on the island and the opposite continent is excellent, and preserves at sea as well as that of the Thames. For after it has been in the cask a day or two it begins to purge itself, and stinks most intolerably, and is soon covered over with a green scum. But this, in a few days, subsides to the bottom, and leaves the water as clear as chrystal, and perfectly sweet. The French (who, during their South Sea trade in Queen Anne's reign, first brought this place into repute) usually wooded and watered in Bon Port, on the continent side, where they anchored with great safety in six fathom water; and this is doubtless the most commodious road for such ships as intend to make only a short stay. But we watered on the St. Catherine's side, at a plantation opposite the island of St. Antonio.

These are the advantages of this island of St. Catherine's; but there are many inconveniencies attending it, partly from its climate, but more from its new regulations, and the late form of government established there. With regard to the climate, it must be remembered that the woods and hills which surround the harbour prevent a free circulation of the air. And the vigorous vegetation which constantly takes place there, furnishes such a prodigious quantity of vapour that all the night, and great part of the morning, a thick fog covers the whole country, and continues till either the sun gathers strength to dissipate it, or it is dispersed by a brisk sea-breeze. This renders the place close and humid, and probably occasioned the many fevers and fluxes we were there afflicted with. To these exceptions I must not omit to add, that all the day we were pestered with great numbers of muscatos, which are not much unlike the gnats in England, but more venomous in their stings. And at sunset, when the muscatos retired, they were succeeded by an infinity of sand-flies, which, though scarce discernible to the naked eye, make a mighty buzzing, and, wherever they bite, raise a small bump in the flesh which is soon attended with a painful itching, like that arising from the bite of an English harvest bug. But as the only light in which this place deserves our consideration is its favourable situation for supplying and refreshing our cruisers intended for the South Seas, in this view its greatest inconveniences remain still to be related. And to do this more distinctly, it will not be amiss to consider the changes which it has lately undergone, both in its inhabitants, its police, and its governor.

In the time of Frezier and Shelvocke, this place served only as a retreat to vagabonds and outlaws, who fled thither from all parts of Brazil. They did indeed acknowledge a subjection to the crown of Portugal, and had a person among them whom they called their captain, who was considered in some sort as their governor; but both their allegiance to their king, and their obedience to their captain, seemed to be little more than verbal. For as they had plenty of provisions, but no money, they were in a condition to support themselves without the assistance of any neighbouring settlements, and had not amongst them the means of tempting any adjacent governor to busy his authority about them. In this situation they were extremely hospitable and friendly to such foreign ships as came amongst them. For these ships wanting only provisions, of which the natives had great store; and the natives wanting clothes (for they often despised money, and refused to take it), which the ships furnished them with in exchange for their provisions, both sides found their account in this traffick; and their captain or governor had neither power nor interest to restrain it, or to tax it. But of late (for reasons which shall be hereafter mentioned) these honest vagabonds have been obliged to receive amongst them a new colony, and to submit to new laws and new forms of government. Instead of their former ragged bare-legged captain (whom, however, they took care to keep innocent), they have now the honour to be governed by Don Jose Sylva de Paz, a brigadier of the armies of Portugal. This gentleman has with him a garrison of soldiers, and has consequently a more extensive and a better supported power than any of his predecessors; and as he wears better clothes, and lives more splendidly, and has besides a much better knowledge of the importance of money than they could ever pretend to, so he puts in practice certain methods of procuring it, with which they were utterly unacquainted. But it may be much doubted if the inhabitants consider these methods as tending to promote either their interests or that of their sovereign the King of Portugal. This is certain, that his behaviour cannot but be extremely embarrassing to such British ships as touch there in their way to the South Seas. For one of his practices was placing centinels at all the avenues, to prevent the people from selling us any refreshments, except at such exorbitant rates as we could not afford to give. His pretence for this extraordinary stretch of power was, that he was obliged to preserve their provisions for upwards of an hundred families, which they daily expected to reinforce their colony. Hence he appears to be no novice in his profession, by his readiness at inventing a plausible pretence for his interested management. However, this, though sufficiently provoking, was far from being the most exceptionable part of his conduct. For by the neighbourhood of the river Plate, a considerable smuggling traffick is carried on between the Portuguese and the Spaniards, especially in the exchanging gold for silver, by which both princes are defrauded of their fifths; and in this prohibited commerce, Don Jose was so deeply engaged, that in order to ingratiate himself with his Spanish correspondents (for no other reason can be given for his procedure) he treacherously dispatched an express to Buenos Ayres, in the river of Plate, where Pizarro then lay, with an account of our arrival, and of the strength of our squadron; particularly mentioning the number of ships, guns, and men, and every circumstance which he could suppose our enemy desirous of being acquainted with. And the same perfidy every British cruiser may expect who touches at St. Catherine's while it is under the government of Don Jose Sylva de Paz.

Thus much, with what we shall be necessitated to relate in the course of our own proceedings, may suffice as to the present state of St. Catherine's, and the character of its governor. But as the reader may be desirous of knowing to what causes the late new modelling of this settlement is owing; to satisfy him in this particular, it will be necessary to give a short account of the adjacent continent of Brazil, and of the wonderful discoveries which have been made there within these last forty years, which, from a country of but mean estimation, has rendered it now perhaps the most considerable colony on the face of the globe.

This country was first discovered by Americus Vesputio, a Florentine, who had the good fortune to be honoured with giving his name to the immense continent, some time before found out by Columbus. Vesputio being in the service of the Portuguese, it was settled and planted by that nation, and, with the other dominions of Portugal, devolved to the crown of Spain when that kingdom became subject to it. During the long war between Spain and the States of Holland, the Dutch possessed themselves of the northermost part of Brazil, and were masters of it for some years. But when the Portuguese revolted from the Spanish Government, this country took part in the revolt, and soon re-possessed themselves of the places the Dutch had taken; since which time it has continued without interruption under the crown of Portugal, being, till the beginning of the present century, only productive of sugar and tobacco, and a few other commodities of very little account.

But this country, which for many years was only considered for the produce of its plantations, has been lately discovered to abound with the two minerals which mankind hold in the greatest esteem, and which they exert their utmost art and industry in acquiring, I mean gold and diamonds. Gold was first found in the mountains which lie adjacent to the city of Rio Janeiro. The occasion of its discovery is variously related, but the most common account is, that the Indians lying on the back of the Portuguese settlements, were observed by the soldiers employed in an expedition against them, to make use of this metal for their fish-hooks, and their manner of procuring it being enquired into, it appeared that great quantities of it were annually washed from the hills, and left amongst the sand and gravel, which remained in the vallies after the running off, or evaporation of the water. It is now little more than forty years since any quantities of gold worth notice have been imported to Europe from Brazil, but since that time the annual imports from thence have been continually augmented by the discovery of places in other provinces, where it is to be met with as plentifully as at first about Rio Janeiro. And it is now said that there is a small slender vein of it spread through all the country at about twenty-four feet from the surface, but that this vein is too thin and poor to answer the expence of digging; however, where the rivers or rains have had any course for a considerable time, there gold is always to be collected, the water having separated the metal from the earth, and deposited it b the sands, thereby saving the expences of digging: so that it is esteemed an infallible gain to be able to divert a stream from its channel, and to ransack its bed. From this account of gathering this metal, it should follow that there are properly no gold mines in Brazil; and this the governor of Rio Grande (who being at St. Catherine's, frequently visited Mr. Anson) did most confidently affirm, assuring us that the gold was all collected either from rivers, or from the beds of torrents after floods. It is indeed asserted, that in the mountains large rocks are found abounding with this metal, and I myself have seen the fragment of one of these rocks, with a considerable lump of gold intangled in it; but even in this case the workmen break off the rocks, and do not properly mine into them, and the great expence in subsisting among these mountains, and afterwards in separating the metal from the stone, makes this method of procuring gold to be but rarely put in practice.

The examining the bottoms of rivers and the gullies of torrents, and the washing the gold found therein from the sand and dirt with which it is always mixed, are works performed by slaves, who are principally negroes, kept in great numbers by the Portuguese for these purposes. The regulation of the duty of these slaves is singular, for they are each of them obliged to furnish their master with the eighth part of an ounce of gold per diem; and if they are either so fortunate or industrious as to collect a greater quantity, the surplus is considered as their own property, and they have the liberty of disposing of it as they think fit. So that it is said, some negroes who have accidentally fallen upon rich washing places have themselves purchased slaves, and have lived afterwards in great splendour, their original master having no other demand on them than the daily supply of the forementioned eighth, which, as the Portuguese ounce is somewhat lighter than our troy ounce, may amount to about nine shillings sterling.

The quantity of gold thus collected in the Brazils, and returned annually to Lisbon, may be in some degree estimated from the amount of the king's fifth. This hath of late been esteemed, one year with another, to be one hundred and fifty arroves of 32 lb. Portuguese weight each, which at £4 the troy ounce, makes very near £300,000 sterling, and consequently the capital, of which this is the fifth, is about a million and a half sterling. It is obvious that the annual return of gold to Lisbon cannot be less than this, though it be difficult to determine how much it exceeds it; perhaps we may not be very much mistaken in our conjecture, if we suppose the gold exchanged for silver with the Spaniards at Buenos Ayres, and what is brought privily to Europe and escapes the duty, amounts to near half a million more, which will make the whole annual produce of the Brazilian gold near two millions sterling, a prodigious sum to be found in a country which a few years since was not known to furnish a single grain.

I have already mentioned that besides gold this country does likewise produce diamonds. The discovery of these valuable stones is much more recent than that of gold, it being as yet scarce 20 years since the first were brought to Europe. They are found in the same manner as the gold, in the gullies of torrents and beds of rivers, but only in particular places, and not so universally spread through the country. They were often found in washing the gold before they were known to be diamonds, and were consequently thrown away with the sand and gravel separated from it; and it is very well remembered that numbers of very large stones, which would have made the fortunes of the possessors, have passed unregarded through the hands of those who now with impatience support the mortifying reflection. However, about twenty years since a person acquainted with the appearance of rough diamonds conceived that these pebbles, as they were then esteemed, were of the same kind: but it is said that there was a considerable interval between the first starting of this opinion and the confirmation of it by proper trials and examination, it proving difficult to persuade the inhabitants that what they had been long accustomed to despise could be of the importance represented by this discovery, and I have been informed that in this interval a governor of one of their places procured a good number of these stones, which he pretended to make use of at cards to mark with instead of counters. But it was at last confirmed by skilful jewellers in Europe, consulted on this occasion, that the stones thus found in Brazil were truly diamonds, many of which were not inferior, either in lustre, or any other quality, to those of the East Indies. On this determination, the Portuguese, in the neighbourhood of those places where they had first been observed, set themselves to search for them with great assiduity. And they were not without great hopes of discovering considerable masses of them, as they found large rocks of chrystal in many of the mountains, from whence the streams came which washed down the diamonds.

But it was soon represented to the King of Portugal that if such plenty of diamonds should be met with, as their sanguine conjectures seemed to indicate, this would so debase their value and diminish their estimation, that besides ruining all the Europeans who had any quantity of Indian diamonds in their possession, it would render the discovery itself of no importance, and would prevent his majesty from receiving any advantages from it, and on these considerations his majesty has thought proper to restrain the general search of diamonds, and has erected a diamond company for that purpose, with an exclusive charter. This company, in consideration of a sum paid to the king, is vested with the property of all diamonds found in Brazil: but to hinder their collecting too large quantities, and thereby reducing their value, they are prohibited from employing above eight hundred slaves in searching after them. And to prevent any of his other subjects from acting the same part, and likewise to secure the company from being defrauded by the interfering of interlopers in their trade and property, he has depopulated a large town and a considerable district round it, and has obliged the inhabitants, who are said to amount to six thousand, to remove to another part of the country; for this town being in the neighbourhood of the diamonds, it was thought impossible to prevent such a number of people who were on the spot from frequently smuggling.

In consequence of these important discoveries in Brazil, new laws, new governments, and new regulations have been established in many parts of the country. For not long since a considerable tract, possessed by a set of inhabitants who from their principal settlement were called Paulists, was almost independent of the crown of Portugal, to which it scarcely acknowledged more than a nominal allegiance. These Paulists are said to be descendants of those Portuguese who retired from the northern part of Brazil when it was invaded and possessed by the Dutch. As from the confusion of the times they were long neglected by their superiors, and were obliged to provide for their own security and defence, the necessity of their affairs produced a kind of government amongst them which they found sufficient for the confined manner of life to which they were inured. And being thus habituated to their own regulations, they at length grew fond of their independency: so that rejecting and despising the mandates of the court of Lisbon, they were often engaged in a state of downright rebellion, and the mountains surrounding their country, and the difficulty of clearing the few passages that open into it, generally put it in their power to make their own terms before they submitted. But as gold was found to abound in this country of the Paulists, the present King of Portugal (during whose reign almost the whole discoveries I have mentioned were begun and compleated) thought it incumbent on him to reduce this province, which now became of great consequence, to the same dependency and obedience with the rest of the country, which, I am told, he has at last, though with great difficulty, happily effected. And the same motives which induced his Majesty to undertake the reduction of the Paulists has also occasioned the changes I have mentioned to have taken place at the island of St. Catherine's. For the governor of Rio Grande, of whom I have already spoken, assured us that in the neighbourhood of this island there were considerable rivers which were found to be extremely rich, and that this was the reason that a garrison, a military governor, and a new colony was settled there. And as the harbour at this island is by much the securest and the most capacious of any on the coast, it is not improbable, if the riches of the neighbourhood answer their expectation, but it may become in time the principal settlement in Brazil, and the most considerable port in all South America.

Thus much I have thought necessary to insert in relation to the present state of Brazil, and of the island of St. Catherine's. For as this last place has been generally recommended as the most eligible port for our cruisers to refresh at, which are bound to the South Seas, I believed it to be my duty to instruct my countrymen in the hitherto unsuspected inconveniencies which attend that place; and as the Brazilian gold and diamonds are subjects about which, from their novelty, very few particulars have been hitherto published, I conceived this account I had collected of them would appear to the reader to be neither a trifling nor a useless digression. These subjects being thus dispatched, I shall now return to the series of our own proceedings.

When we first arrived at St. Catherine's we were employed in refreshing our sick on shore, in wooding and watering the squadron, cleansing our ships, and examining and securing our masts and rigging, as I have already observed in the foregoing chapter. At the same time Mr. Anson gave directions that the ships' companies should be supplied with fresh meat, and that they should be victualled with whole allowance of all the kinds of provision. In consequence of these orders, we had fresh beef sent on board us continually for our daily expence, and what was wanting to make up our allowance we received from our victualler the *Anna* pink, in order to preserve the provisions on board our squadron entire for our future service. The season of the year growing each day less favourable for our passage round Cape Horn, Mr. Anson was very desirous of leaving this place as soon as possible, and we were at first in hopes that our whole business would be done and we should be in a readiness to sail in about a fortnight from our arrival; but, on examining the *Tryal's* masts, we, to our no small vexation, found inevitable employment for twice that time. For, on a survey, it was found that the main-mast was sprung at the upper woulding, though it was thought capable of being secured by a couple of fishes; but the fore-mast was reported to be unfit for service, and thereupon the carpenters were sent into the woods to endeavour to find a stick proper for a fore-mast. But after a search of four days they returned without having been able to meet with any tree fit for the purpose. This obliged them to come to a second consultation about the old fore-mast, when it was agreed to endeavour to secure it by casing it with three fishes, and in this work the carpenters were employed till within a day or two of our sailing. In the meantime, the commodore, thinking it necessary to have a clean vessel on our arrival in the South Seas, ordered the *Tryal* to be hove down, as this would not occasion any loss of time, but might be compleated while the carpenters were refitting her masts, which was done on shore.

On the 27th of December we discovered a sail in the offing, and not knowing but she might be a Spaniard, the eighteen-oared boat was manned and armed, and sent under the command of our second lieutenant, to examine her before she arrived within the protection of the forts. She proved to be a Portuguese brigantine from Rio Grande; and though our officer, as it appeared on inquiry, had behaved with the utmost civility to the master, and had refused to accept a calf which the master would have forced on him as a present, yet the governor took great offence at our sending our boat, and talked of it in a high strain as a violation of the peace subsisting between the crowns of Great Britain and Portugal. We at first imputed this ridiculous blustering to no deeper a cause than Don Jose's insolence; but as we found he proceeded so far as to charge our officer with behaving rudely, and opening letters, and particularly with an attempt to take out of the vessel by violence the very calf which we knew he had refused to receive as a present (a circumstance which we were satisfied the governor was well acquainted with), we had thence reason to suspect that he purposely sought this quarrel, and had more important motives for engaging in it than the mere captious bias of his temper. What these motives were it was not so easy for us to determine at that time; but as we afterwards found by letters which fell into our hands in the South Seas that he had dispatched an express to Buenos Ayres, where Pizarro then lay, with an account of our squadron's arrival at St. Catherine's, together with the most ample and circumstantial intelligence of our force and condition, we thence conjectured that Don Jose had raised this groundless clamour only to prevent our visiting the brigantine when she should put to sea again, lest we might there find proofs of his perfidious behaviour, and perhaps, at the same time, discover the secret of his smuggling correspondence with his neighbouring governors, and the Spaniards at Buenos Ayres. But to proceed.

It was near a month before the *Tryal* was refitted; for not only her lower masts were defective, as hath been already mentioned, but her main top-mast and fore-yard were likewise decayed and rotten. While this work was carrying on, the other ships of the squadron fixed new standing rigging, and set up a sufficient number of preventer shrouds to each mast, to secure them in the most effectual manner. And in order to render the ships stiffer and to enable them to carry more sail abroad, and to prevent their straining their upper works in hard gales of wind, each captain had orders given him to strike down some of their great guns into the hold. These precautions being complied with, and each ship having taken in as much wood and water as there was room for, the *Tryal* was at last compleated, and the whole squadron was ready for the sea: on which the tents on shore were struck, and all the sick were received on board. And here we had a melancholy proof how much the healthiness of this place had been over-rated by former writers, for we found that though the *Centurion* alone had buried no less than twenty-eight men since our arrival, yet the number of her sick was in the same interval increased from eighty to ninety-six. When our crews were embarked, and everything was prepared for our departure, the commodore made a signal for all captains, and delivered them their orders, containing the successive place of rendezvous from hence to the coast of China. And then, on the next day, being the 18th of January, the signal was made for weighing, and the squadron put to sea, leaving without regret this island of St. Catherine's, where we had been so extremely disappointed in our refreshments, in our accommodations, and in the humane and friendly offices which we had been taught to expect in a place which hath been so much celebrated for its hospitality, freedom, and conveniency.

CHAPTER VI

THE RUN FROM ST. CATHERINE'S TO PORT ST. JULIAN, WITH SOME ACCOUNT OF THAT PORT, AND OF THE COUNTRY TO THE SOUTHWARD OF THE RIVER OF PLATE

In leaving St. Catherine's, we left the last amicable port we proposed to touch at, and were now proceeding to an hostile, or at best, a desart and inhospitable coast. And as we were to expect a more boisterous climate to the southward than any we had yet experienced, not only our danger of separation would by this means be much greater than it had been hitherto, but other accidents of a more mischievous nature were likewise to be apprehended, and as much as possible to be provided against. Mr. Anson, therefore, in appointing the various stations at which the ships of the squadron were to rendezvous, had considered that it was possible his own ship might be disabled from getting round Cape Horn, or might be lost, and had given proper direction, that even in that case the expedition should not be abandoned. For the orders delivered to the captains, the day before we sailed from St. Catherine's, were, that in case of separation, which they were with the utmost care to endeavour to avoid, the first place of rendezvous should be the bay of port St. Julian; describing the place from Sir John Narborough's account of it. There they were to supply themselves with as much salt as they could take in, both for their own use and for the use of the squadron; and if, after a stay of ten days, they were not joined by the commodore, they were then to proceed through Streights le Maire round Cape Horn into the South Seas, where the next place of rendezvous was to be the island of Nostra Senora del Socoro, in the latitude of 45° south, and longitude from the Lizard 71° 12' west. They were to bring this island to bear E.N.E. and to cruize from five to twelve leagues distance from it as long as their store of wood and water would permit, both which they were to expend with the utmost frugality. And when they were under an absolute necessity of a fresh supply, they were to stand in, and endeavour to find out an anchoring place; and in case they could not, and the weather made it dangerous to supply their ships by standing off and on, they were then to make the best of their way to the island of Juan Fernandes, in the latitude of 33° 37' south. At this island, as soon as they had recruited their wood and water, they were to continue cruizing off the anchoring place for fifty-six days; in which time, if they were not joined by the commodore, they might conclude that some accident had befallen him, and they were forthwith to put themselves under the command of the senior officer, who was to use his utmost endeavours to annoy the enemy both by sea and land. With these views their new commodore was to continue in those seas as long as his provisions lasted, or as long as they were recruited by what he should take from the enemy, reserving only a sufficient quantity to carry him and the ships under his command to Macao, at the entrance of the river of Canton on the coast of China, where, having supplied himself with a new stock of provisions, he was thence, without delay, to make the best of his way to England. And as it was found impossible as yet to unload our victualler, the *Anna* pink, the commodore gave the master of her the same rendezvous, and the same orders to put himself under the command of the remaining senior officer.

Under these orders the squadron sailed from St. Catherine's on Sunday the 18th of January, as hath been already mentioned in the preceding chapter. The next day we had very squally weather, attended with rain, lightning, and thunder, but it soon became fair again with light breezes, and continued thus till Wednesday evening, when it blew fresh again; and increasing all night, by eight the next morning it became a most violent storm, and we had with it so thick a fog, that it was impossible to see the distance of two ships' length, so that the whole squadron disappeared. On this a signal was made, by firing guns, to bring to with the larboard tacks, the wind being then due east. We ourselves immediately handed the top-sails, bunted the main-sail, and lay to under a reefed mizen till noon, when the fog dispersed, and we soon discovered all the ships of the squadron, except the *Pearl*, who did not join us till near a month afterwards. Indeed the *Tryal* sloop was a great way to the leeward, having lost her main-mast in the squall, and having been obliged, for fear of bilging, to cut away the raft. We therefore bore down with the squadron to her relief, and the *Gloucester* was ordered to take her in tow, for the weather did not entirely abate till the day after, and even then a great swell continued from the eastward in consequence of the preceding storm.

After this accident we stood to the southward with little interruption, and here we experienced the same setting of the current which we had observed before our arrival at St. Catherine's; that is, we generally found ourselves to the southward of our reckoning by about twenty miles each day. This deviation, with a little inequality, lasted till we had passed the latitude of the river of Plate; and even then we discovered that the same current, however difficult to be accounted for, did yet undoubtedly take place; for we were not satisfied in deducing it from the error in our reckoning, but we actually tried it more than once when a calm made it practicable.

As soon as we had passed the latitude of the river of Plate, we had soundings which continued all along the coast of Patagonia. These soundings, when well ascertained, being of great use in determining the position of the ships, and we having tried them more frequently, and in greater depths, and with more attention than, I believe, hath been done before us, I shall recite our observations as succinctly as I can. In the latitude of 36° 52' we had sixty fathom of water, with a bottom of fine black and grey sand; from thence to 39° 55' we varied our depths from fifty to eighty fathom, though we had constantly the same bottom as before; between the last mentioned latitude and 43° 16' we had only fine grey sand, with the same variation of depths, except that we once or twice lessened our water to forty fathom. After this, we continued in forty fathom for about half a degree, having a bottom of coarse sand and broken shells, at which time we were in sight of land, and not above seven leagues from it. As we edged from the land, we met with variety of soundings; first black sand, then muddy, and soon after rough ground with stones; but when we had increased our water to forty-eight fathom we had a muddy bottom to the latitude of 46° 10'. Hence, drawing towards the shore, we had first thirty-six fathom, and still kept shoaling our water, till at length we came into twelve fathom, having constantly small stones and pebbles at the bottom. Part of this time we had a view of Cape Blanco, which lies in about the latitude of 47° 10', and longitude west from London 69°. This is the most remarkable land upon the coast. Steering from hence S. by E. nearly, we, in a run of about thirty leagues, deepned our water to fifty fathom without once altering the bottom: and then drawing towards the shore with a S.W. course, varying rather to the westward, we had constantly a sandy bottom till our coming into thirty fathom, where we had again a sight of land distant from us about eight leagues lying in the latitudes of 48° 31'. We made this land on the 17th of February, and at five that afternoon we came to an anchor, having the same soundings as before, in the latitude of 48° 58', the souther-most land then in view bearing S.S.W., the northermost N.½E., a small island N.W., and the westermost hummock W.S.W. In this station we found the tide to set S. by W.; and weighing again at five the next morning, we an hour afterwards discovered a sail, upon which the *Severn* and *Gloucester* were both directed to give chase; but we soon perceived it to be the *Pearl*, which separated from us a few days after we left St. Catherine's, and on this we made a signal for the *Severn* to rejoin the squadron, leaving the *Gloucester* alone in the pursuit. And now we were surprised to see that on the *Gloucester's* approach the people on board the *Pearl* increased their sail, and stood from her. However, the *Gloucester* came up with them, but found them with their hammocks in their nettings, and every thing ready for an engagement. At two in the afternoon the *Pearl* joined us, and running up under our stern, Lieutenant Salt haled the commodore, and acquainted him that Captain Kidd died on the 31st of January. He likewise informed us that he had seen five large ships the 10th instant, which he for some time imagined to be our squadron: so that he suffered the commanding ship, which wore a red broad pendant exactly resembling that of the commodore at the main top-mast head, to come within gun-shot of him before he discovered his mistake; but then, finding it not to be the *Centurion*, he haled close upon the wind, and crowded from them with all his sail, and standing cross a ripling, where they hesitated to follow him, he happily escaped. He made them to be five Spanish men-of-war, one of them exceedingly like the *Gloucester*, which was the occasion of his apprehensions when the *Gloucester* chaced him. By their appearance he thought they consisted of two ships of seventy guns, two of fifty, and one of forty guns. It seems the whole squadron continued in chace of him all that day, but at night, finding they could not get near him, they gave over the chace, and directed their course to the southward.

Had it not been for the necessity we were under of refitting the *Tryal*, this piece of intelligence would have prevented our making any stay at St. Julian's; but as it was impossible for that sloop to proceed round the Cape in her present condition, some stay there was inevitable, and therefore the same evening we came to an anchor again in twenty-five fathom water; the bottom a mixture of mud and sand, and the high hummock bearing S.W. by W. And weighing at nine in the morning, we sent the two cutters belonging to the *Centurion* and *Severn* in shore to discover the harbour of St. Julian, while the ships kept standing along the coast, about the distance of a league from the land. At six o'clock we anchored in the bay of St. Julian in nineteen fathom, the bottom muddy ground with sand, the northermost land in sight bearing N. and by E., the southermost S.-½E., and the high hummock, to which Sir John Narborough formerly gave the name of Wood's Mount, W.S.W. Soon after, the cutter returned on board, having discovered the harbour, which did not appear to us in our situation, the northermost point shutting in upon the southermost, and in appearance closing the entrance.

Being come to an anchor in this bay of St. Julian, principally with a view of refitting the *Tryal*, the carpenters were immediately employed in that business, and continued so during our whole stay at the place. The *Tryal's* main-mast having been carried away about twelve feet below the cap, they contrived to make the remaining part of the mast serve again; and the *Wager* was ordered to supply her with a spare main top-mast, which the carpenters converted into a new fore-mast. And I cannot help observing that this accident to the *Tryal's* mast, which gave us so much uneasiness at that time, on account of the delay it occasioned, was, in all probability, the means of preserving the sloop, and all her crew. For before this, her masts, how well soever proportioned to a better climate, were much too lofty for these high southern latitudes, so that had they weathered the preceding storm, it would have been impossible for them to have stood against those seas and tempests we afterwards encountered in passing round Cape Horn, and the loss of masts in that boisterous climate would scarcely have been attended with less than the loss of the vessel and of every man on board her, since it would have been impracticable for the other ships to have given them any relief during the continuance of those impetuous storms.

Whilst we stayed at this place, the commodore appointed the Honourable Captain Murray to succeed to the *Pearl*, and Captain Cheap to the *Wager*, and he promoted Mr. Charles Saunders, his first lieutenant, to the command of the *Tryal* sloop. But Captain Saunders lying dangerously ill of a fever on board the *Centurion*, and it being the opinion of the surgeons that the removing him on board his own ship in his present condition might tend to the hazard of his life, Mr. Anson gave an order to Mr. Saumarez, first lieutenant of the *Centurion*, to act as master and commander of the *Tryal* during the illness of Captain Saunders.

Here the commodore, too, in order to ease the expedition of all unnecessary expence, held a farther consultation with his captains about unloading and discharging the *Anna* pink; but they represented to him that they were so far from being in a condition of taking any part of her loading on board, that they had still great quantities of provisions in the way of their guns between decks, and that their ships were withal so very deep that they were not fit for action without being cleared. This put the commodore under a necessity of retaining the pink in the service; and as it was apprehended we should certainly meet with the Spanish squadron in passing the Cape, Mr. Anson thought it adviseable to give orders to the captains to put all their provisions which were in the way of their guns on board the *Anna* pink, and to remount such of their guns as had formerly, for the ease of their ships, been ordered into the hold.

This bay of St. Julian, where we are now at anchor, being a convenient rendezvous, in case of separation, for all cruizers bound to the southward, and the whole coast of Patagonia, from the river of Plate to the Streights of Magellan, lying nearly parallel to their usual route, a short account of the singularity of this country, with a particular description of port St. Julian, may, perhaps, be neither unacceptable to the curious, nor unworthy the attention of future navigators, as some of them, by unforeseen accidents, may be obliged to run in with the land, and to make some stay on this coast, in which case the knowledge of the country, its produce and inhabitants, cannot but be of the utmost consequence to them.

To begin then with the tract of country usually styled Patagonia. This is the name often given to the southermost part of South America, which is unpossessed by the Spaniards, extending from their settlements to the Streights of Magellan. This country on the east side is extremely remarkable for a peculiarity not to be paralleled in any other known part of the globe, for though the whole territory to the northward of the river of Plate is full of wood, and stored with immense quantities of large timber trees, yet to the southward of the river no trees of any kind are to be met with except a few peach trees, first planted and cultivated by the Spaniards in the neighbourhood of Buenos Ayres. So that on the whole eastern coast of

Patagonia, extending near four hundred leagues in length, and reaching as far back as any discoveries have yet been made, no other wood has been found than a few insignificant shrubs. Sir John Narborough in particular, who was sent out by King Charles the Second expressly to examine this country and the Streights of Magellan, and who, in pursuance of his orders, wintered upon this coast in Port St. Julian and Port Desire, in the year 1670; Sir John Narborough, I say, tells us that he never saw a stick of wood in the country large enough to make the handle of an hatchet.

But though the country be so destitute of wood, it abounds with pasture. For the land appears in general to be made up of downs, of a light dry gravelly soil, and produces great quantities of long coarse grass, which grows in tufts, interspersed with large barren spots of gravel between them. This grass, in many places, feeds immense herds of cattle, for the Spaniards at Buenos Ayres having, soon after their first settling there, brought over a few black cattle from Europe, they have thriven prodigiously by the plenty of herbage which they everywhere met with, and are now increased to that degree, and are extended so far into different parts of Patagonia, that they are not considered as private property; but many thousands at a time are slaughtered every year by the hunters, only for their hides and tallow. The manner of killing these cattle, being a practice peculiar to that part of the world, merits a more circumstantial description. The hunters employed on this occasion, being all of them mounted on horseback (and both the Spaniards and Indians in that part of the world are usually most excellent horsemen), they arm themselves with a kind of a spear, which, at its end, instead of a blade fixed in the same line with the wood in the usual manner, has its blade fixed across. With this instrument they ride at a beast, and surround him, when the hunter that comes behind him hamstrings him, and as after this operation the beast soon tumbles, without being able to raise himself again, they leave him on the ground and pursue others, whom they serve in the same manner. Sometimes there is a second party, who attend the hunters to skin the cattle as they fall, but it is said that at other times the hunters chuse to let them languish in torment till the next day, from an opinion that the anguish, which the animal in the meantime endures, may burst the lymphaticks, and thereby facilitate the separation of the skin from the carcase; and though their priests have loudly condemned this most barbarous practice, and have gone so far, if my memory does not fail me, as to excommunicate those who follow it, yet all their efforts to put an entire stop to it have hitherto proved ineffectual.

Besides the numbers of cattle which are every year slaughtered for their hides and tallow, in the manner already described, it is often necessary for the uses of agriculture, and for other purposes, to take them alive without wounding them. This is performed with a most wonderful and almost incredible dexterity, and principally by the use of a machine which the English who have resided at Buenos Ayres generally denominate a lash. It is made of a thong of several fathoms in length, and very strong, with a running noose at one end of it. This the hunters (who in this case are also mounted on horseback) take in their right hands, it being first properly coiled up, and having its end opposite to the noose fastened to the saddle; and thus prepared they ride at a herd of cattle. When they arrive within a certain distance of a beast, they throw their thong at him with such exactness that they never fail of fixing the noose about his horns. The beast, when he finds himself entangled, generally runs, but the horse, being swifter, attends him, and prevents the thong from being too much strained, till a second hunter, who follows the game, throws another noose about one of its hind legs; and this being done, both horses (for they are trained to this practice) instantly turn different ways, in order to strain the two thongs in contrary directions, on which the beast, by their opposite pulls, is presently overthrown, and then the horses stop, keeping the thongs still upon the stretch. Being thus on the ground, and incapable of resistance (for he is extended between the two horses), the hunters alight, and secure him in such a manner that they afterwards easily convey him to whatever place they please. They in like manner noose horses, and, as it is said, even tigers; and however strange this last circumstance may appear, there are not wanting persons of credit who assert it. Indeed, it must be owned, that the address both of the Spaniards and Indians in that part of the world in the use of this lash or noose, and the certainty with which they throw it, and fix it on any intended part of the beast at a considerable distance, are matters only to be believed from the repeated and concurrent testimony of all who have frequented that country, and might reasonably be questioned, did it rely on a single report, or had it been ever contradicted or denied by any one who had resided at Buenos Ayres.

The cattle which are killed in the manner I have already observed are slaughtered only for their hides and tallow, to which sometimes are added their tongues, but the rest of their flesh is left to putrify, or to be devoured by the birds and wild beasts. The greatest part of this carrion falls to the share of the wild dogs, of which there are immense numbers to be found in that country.

These are supposed to have been originally produced by Spanish dogs from Buenos Ayres, who, allured by the great quantity of carrion, and the facility they had by that means of subsisting, left their masters, and ran wild amongst the cattle; for they are plainly of the breed of European dogs, an animal not originally found in America. But though these dogs are said to be some thousands in a company, they hitherto neither diminish nor prevent the increase of the cattle, not daring to attack the herds, by reason of the numbers, which constantly feed together; but contenting themselves with the carrion left them by the hunters, and perhaps now and then with a few stragglers, who, by accidents, are separated from the main body they belong to.

Besides the wild cattle which have spread themselves in such vast herds from Buenos Ayres towards the southward, the same country is in like manner furnished with horses. These too were first brought from Spain, and are also prodigiously increased, and run wild to a much greater distance than the black cattle; and though many of them are excellent, yet their number makes them of very little value, the best of them being often sold in the neighbouring settlements, where money is plenty and commodities very dear, for not more than a dollar a-piece. It is not as yet certain how far to the southward these herds of wild cattle and horses have extended themselves, but there is some reason to conjecture that stragglers of both kinds are to be met with very near the Streights of Magellan; and they will in time doubtless fill all the southern part of this continent with their breed, which cannot fail of proving of considerable advantage to such ships as may touch upon the coast, for the horses themselves are said to be very good eating, and, as such, are preferred by some of the Indians even before the black cattle. But whatever plenty of flesh provisions may be hereafter found here, there is one material refreshment which this eastern side of Patagonia seems to be very defective in, and that is fresh water, for the land being generally of a nitrous and saline nature, the ponds and streams are frequently brackish. However, as good water has been found there, though in small quantities, it is not improbable, but on a further search, this inconvenience may be removed.

To the account already given, I must add that there are in all parts of this country a good number of vicunnas or Peruvian sheep; but these, by reason of their shyness and swiftness, are killed with difficulty. On the eastern coast, too, there are found immense quantities of seals, and a vast variety of sea-fowl, amongst which the most remarkable are the penguins. They are in size and shape like a goose, but instead of wings, they have short stumps like fins, which are of no use to them, except in the water. Their bills are narrow, like that of an albitross, and they stand and walk in an erect posture. From this and their white bellies, Sir John Narborough has whimsically likened them to little children standing up in white aprons.

The inhabitants of this eastern coast (to which I have all along hitherto confined my relation) appear to be but few, and have rarely been seen more than two or three at a time by any ships that have touched here. We, during our stay at the port of St. Julian, saw none. However, towards Buenos Ayres they are sufficiently numerous, and oftentimes very troublesome to the Spaniards, but there the greater breadth and variety of the country, and a milder climate, yield them a better protection, for in that place the continent is between three and four hundred leagues in breadth, whereas at Port St. Julian it is little more than a hundred; so that I conceive the same Indians who frequent the western coast of Patagonia and the Streights of Magellan often ramble to this side. As the Indians near Buenos Ayres exceed these southern Indians in number, so they greatly surpass them in activity and spirit, and seem in their manners to be nearly allied to those gallant Chilian Indians who have long set the whole Spanish power at defiance, have often ravaged their country, and remain to this hour independent. For the Indians about Buenos Ayres have learnt to be excellent horsemen, and are extremely expert in the management of all cutting weapons, though ignorant of the use of firearms, which the Spaniards are very solicitous to keep out of their hands. And of the vigour and resolution of these Indians, the behaviour of Orellana and his followers, whom we have formerly mentioned, is a memorable instance. Indeed were we disposed to aim at the utter subversion of the Spanish power in America, no means seem more probable to effect it than due encouragement and assistance given to these Indians and those of Chili.

Thus much may suffice in relation to the eastern coast of Patagonia. The western coast is of less extent, and by reason of the Andes which skirt it, and stretch quite down to the water, is a very rocky and dangerous shore. However, I shall be hereafter necessitated to make further mention of it, and therefore shall not enlarge thereon at this time, but shall conclude this account with a short description of the harbour of St. Julian.

We, on our first arrival here, sent an officer on shore to the salt pond in order to procure a quantity of salt for the use of the squadron, Sir John Narborough having observed when he was here that the salt produced in that place was very white and good, and that in February there was enough of it to fill a thousand ships; but our officer returned with a sample which was very bad, and he told us that even of this there was but little to be got. I suppose the weather had been more rainy than ordinary, and had destroyed it.

CHAPTER VII

DEPARTURE FROM THE BAY OF ST. JULIAN, AND THE PASSAGE
FROM THENCE TO STREIGHTS LE MAIRE

The *Tryal* being nearly refitted, which was our principal occupation at this bay of St. Julian, and the sole occasion of our stay, the commodore thought it necessary, as we were now directly bound for the South Seas and the enemy's coasts, to fix the plan of his first operations; and, therefore, on the 24th of February, a signal was made for all captains, and a council of war was held on board the *Centurion*, at which were present the Honourable Edward Legg, Captain Matthew Mitchell, the Honourable George Murray, Captain David Cheap, together with Colonel Mordaunt Cracherode, commander of the land-forces. At this council Mr. Anson proposed that their first attempt, after their arrival in the South Seas, should be the attack of the town and harbour of Baldivia, the principal frontier of the district of Chili; Mr. Anson informing them, at the same time, that it was an article contained in his Majesty's instructions to him to endeavour to secure some port in the South Seas where the ships of the squadron might be careened and refitted. To this proposition made by the commodore, the council unanimously and readily agreed, and, in consequence of this resolution, new instructions were given to the captains of the squadron, by which, though they were still directed, in case of separation, to make the best of their way to the island of Nuestra Senora del Socoro (yet notwithstanding the orders they had formerly given them at St. Catherine's) they were to cruise off that island only ten days; from whence, if not joined by the commodore, they were to proceed and cruise off the harbour of Baldivia, making the land between the latitudes of 40° and 40° 30', and taking care to keep to the southward of the port; and if in fourteen days they were not joined by the rest of the squadron, they were then to quit this station, and to direct their course to the island of Juan Fernandes, after which they were to regulate their further proceedings by their former orders. The same directions were also given to the master of the *Anna* pink, who was not to fail in answering the signals made by any ship of the squadron, and was to be very careful to destroy his papers and orders if he should be so unfortunate as to fall into the hands of the enemy. And as the separation of the squadron might prove of the utmost prejudice to his Majesty's service, each captain was ordered to give it in charge to the respective officers of the watch not to keep their ship at a greater distance from the *Centurion* than two miles, as they would answer it at their peril; and if any captain should find his ship beyond the distance specified, he was to acquaint the commodore with the name of the officer who had thus neglected his duty.

These necessary regulations being established, and the *Tryal* sloop compleated, the squadron weighed on Friday the 27th of February, at seven in the morning, and stood to the sea; the *Gloucester* indeed found a difficulty in purchasing her anchor, and was left a considerable way a-stern, so that in the night we fired several guns as a signal to her captain to make sail, but he did not come up to us till the next morning, when we found that they had been obliged to cut their cable, and leave their best bower behind them. At ten in the morning, the day after our departure, Wood's Mount, the high land over St. Julian, bore from us N. by W. distant ten leagues, and we had fifty-two fathom of water. And now standing to the southward, we had great expectation of falling in with Pizarro's squadron; for, during our stay at Port St. Julian, there had generally been hard gales between the W.N.W. and S.W., so that we had reason to conclude the Spaniards had gained no ground upon us in that interval. Indeed it was the prospect of meeting with them that had occasioned our commodore to be so very solicitous to prevent the separation of our ships, for had we been solely intent upon getting round Cape Horn in the shortest time, the properest method for this purpose would have been to have ordered each ship to have made the best of her way to the rendezvous without waiting for the rest.

From our departure from St. Julian to the 4th of March we had little wind, with thick hazy weather and some rain; and our soundings were generally from forty to fifty fathom, with a bottom of black and grey sand, sometimes intermixed with pebble stones. On the 4th of March we were in sight of Cape Virgin Mary, and not more than six or seven leagues distant from it. This cape is the northern boundary of the entrance of the Straights of Magellan; it lies in the latitude of 52° 21' south, and longitude from London 71° 44' west, and seems to be a low flat land, ending in a point. Off this cape our depth of water was from thirty-five to forty-eight fathom. The afternoon of this day was very bright and clear, with small breezes of wind, inclinable to a calm, and most of the captains took the opportunity of this favourable weather to pay a visit to the commodore; but while they were in company together, they were all greatly alarmed by a sudden flame which burst out on board the *Gloucester*, and which was succeeded by a cloud of smoke. However, they were soon relieved from their apprehensions by receiving information that the blast was occasioned by a spark of fire from the forge lighting on some gunpowder and other combustibles which an officer on board was preparing for use, in case we should fall in with the Spanish fleet, and that it had been extinguished without any damage to the ship.

We here found what was constantly verified by all our observations in these high latitudes, that fair weather was always of an exceeding short duration, and that when it was remarkably fine it was a certain presage of a succeeding storm, for the calm and sunshine of our afternoon ended in a most turbulent night, the wind freshning from the S.W. as the night came on, and increasing its violence continually till nine in the morning the next day, when it blew so hard that we were obliged to bring-to with the squadron, and to continue under a reefed mizen till eleven at night, having in that time from forty-three to fifty-seven fathom water, with black sand and gravel; and by an observation we had at noon, we concluded a current had set us twelve miles to the southward of our reckoning. Towards midnight, the wind abating, we made sail again, and steering south, we discovered in the morning, for the first time, the land called Terra del Fuego, stretching from the S. by W., to the S.E.-½E. This indeed afforded us but a very uncomfortable prospect, it appearing of a stupendous height, covered everywhere with snow. The dreariness of this scene can be but imperfectly represented by any drawing. We steered along this shore all day, having soundings from forty to fifty fathom, with stones and gravel. And as we intended to pass through Streights Le Maire next day, we lay-to at night, that we might not overshoot them, and took this opportunity to prepare ourselves for the tempestuous climate we were soon to be engaged in; with which view we employed ourselves good part of the night in bending an entire new suit of sails to the yards. At four the next morning, being the 7th of March, we made sail, and at eight saw the land, and soon after we began to open the streights, at which time Cape St. James bore from us E.S.E., Cape St. Vincent S.E.½E., the middlemost of the Three Brothers S. and by W., Montegorda south, and Cape St. Bartholomew, which is the souther-most point of Staten-land, E.S.E. And here I must observe, that though Frezier has given us a very correct prospect of the part of Terra del Fuego which borders on the streights, yet he has omitted that of Staten-land, which forms the opposite shore: hence we found it difficult to determine exactly where the streights lay, till they began to open to our view; and for want of this, if we had not happened to have coasted a considerable way along shore, we might have missed the streights, and have got to the eastward of Staten-land before we knew it. This is an accident that has happened to many ships, particularly, as Frezier mentions, to the *Incarnation* and *Concord*, who intending to pass through Streights Le Maire, were deceived by three hills on Staten-land like the Three Brothers, and some creeks resembling those of Terra del Fuego, and thereby overshot the streights.

And on occasion of the prospect of Staten-land, I cannot but remark, that though Terra del Fuego had an aspect extremely barren and desolate, yet this island of Staten-land far surpasses it in the wildness and horror of its appearance, it seeming to be entirely composed of inaccessible rocks, without the least mixture of earth or mold between them. These rocks terminate in a vast number of ragged points, which spire up to a prodigious height, and are all of them covered with everlasting snow; the points themselves are on every side surrounded with frightful precipices, and often overhang in a most astonishing manner, and the hills which bear them are generally separated from each other by narrow clefts which appear as if the country had been frequently rent by earthquakes; for these chasms are nearly perpendicular, and extend through the substance of the main rocks, almost to their very bottoms: so that nothing can be imagined more savage and gloomy than the whole aspect of this coast. But to proceed.

I have above mentioned, that on the 7th of March, in the morning, we opened Streights Le Maire, and soon after, or about ten o'clock, the *Pearl* and the *Tryal* being ordered to keep ahead of the squadron, we entered them with fair weather and a brisk gale, and were hurried through by the rapidity of the tide in about two hours, though they are between seven and eight leagues in length. As these Streights are often esteemed to be the boundary between the Atlantic and Pacific Oceans, and as we presumed we had nothing before us from hence but an open sea, till we arrived on those opulent coasts where all our hopes and wishes centered, we could not help perswading ourselves that the greatest difficulty of our voyage was now at an end, and that our most sanguine dreams were upon the point of being realised; and hence we indulged our imaginations in those romantic schemes which the fancied possession of the Chilian gold and Peruvian silver might be conceived to inspire. These joyous ideas were considerably heightened by the brightness of the sky and serenity of the weather, which was indeed most remarkably pleasing; for though the winter was now advancing apace, yet the morning of this day, in its brilliancy and mildness, gave place to none we had seen since our departure from England. Thus animated by these flattering delusions, we passed those memorable Streights, ignorant of the dreadful calamities which were then impending, and just ready to break upon us; ignorant that the time drew near when the squadron would be separated never to unite again, and that this day of our passage was the last chearful day that the greatest part of us would ever live to enjoy.

CHAPTER VIII

FROM STREIGHTS LE MAIRE TO CAPE NOIR

We had scarcely reached the southern extremity of the Streights Le Maire, when our flattering hopes were instantly lost in the apprehensions of immediate destruction: for before the sternmost ships of the squadron were clear of the streights, the serenity of the sky was suddenly obscured, and we observed all the presages of an impending storm; and presently the wind shifted to the southward, and blew in such violent squalls that we were obliged to hand our topsails, and reef our main-sail; whilst the tide, too, which had hitherto favoured us, at once turned furiously against us, and drove us to the eastward with prodigious rapidity, so that we were in great anxiety for the *Wager* and the *Anna* pink, the two sternmost vessels, fearing they would be dashed to pieces against the shore of Staten-land: nor were our apprehensions without foundation, for it was with the utmost difficulty they escaped. And now the whole squadron, instead of pursuing their intended course to the S.W., were driven to the eastward by the united force of the storm and of the currents; so that next day in the morning we found ourselves near seven leagues to the eastward of Streights Le Maire, which then bore from us N.W. The violence of the current, which had set us with so much precipitation to the eastward, together with the fierceness and constancy of the westerly winds, soon taught us to consider the doubling of Cape Horn as an enterprize that might prove too mighty for our efforts, though some amongst us had lately treated the difficulties which former voyagers were said to have met with in this undertaking as little better than chimerical, and had supposed them to arise rather from timidity and unskilfulness than from the real embarrassments of the winds and seas: but we were now severely convinced that these censures were rash and ill-grounded, for the distresses with which we struggled, during the three succeeding months, will not easily be paralleled in the relation of any former naval expedition. This will, I doubt not, be readily allowed by those who shall carefully peruse the ensuing narration.

From the storm which came on before we had well got clear of Streights Le Maire, we had a continual succession of such tempestuous weather as surprized the oldest and most experienced mariners on board, and obliged them to confess that what they had hitherto called storms were inconsiderable gales compared with the violence of these winds, which raised such short, and at the same time such mountainous waves, as greatly surpassed in danger all seas known in any other part of the globe: and it was not without great reason that this unusual appearance filled us with continual terror; for, had any one of these waves broke fairly over us, it must, in all probability, have sent us to the bottom. Nor did we escape with terror only; for the ship rolling incessantly gunwale to, gave us such quick and violent motions that the men were in perpetual danger of being dashed to pieces against the decks or sides of the ship. And though we were extremely careful to secure ourselves from these shocks by grasping some fixed body, yet many of our people were forced from their hold, some of whom were killed, and others greatly injured; in particular, one of our best seamen was canted overboard and drowned, another dislocated his neck, a third was thrown into the main hold and broke his thigh, and one of our boatswain's mates broke his collar-bone twice; not to mention many other accidents of the same kind. These tempests, so dreadful in themselves, though unattended by any other unfavourable circumstance, were yet rendered more mischievous to us by their inequality, and the deceitful intervals which they at sometimes afforded; for though we were oftentimes obliged to lie-to, for days together, under a reefed mizen, and were frequently reduced to lie at the mercy of the waves under our bare poles, yet now and then we ventured to make sail with our courses double reefed; and the weather proving more tolerable, would perhaps encourage us to set our top-sails; after which, the wind, without any previous notice, would return upon us with redoubled force, and would in an instant tear our sails from the yards. And that no circumstance might be wanting which could aggrandize our distress, these blasts generally brought with them a great quantity of snow and sleet, which cased our rigging, and froze our sails, thereby rendering them and our cordage brittle, and apt to snap upon the slightest strain, adding great difficulty and labour to the working of the ship, benumbing the limbs of our people, and making them incapable of exerting themselves with their usual activity, and even disabling many of them by mortifying their toes and fingers. It were indeed endless to enumerate the various disasters of different kinds which befel us; and I shall only mention the most material, which will sufficiently evince the calamitous condition of the whole squadron during the course of this navigation.

It was on the 7th of March, as hath been already observed, that we passed Streights Le Maire, and were immediately afterwards driven to the eastward by a violent storm, and the force of the current which set that way. For the four or five succeeding days we had hard gales of wind from the same quarter, with a most prodigious swell; so that though we stood, during all that time, towards the S.W., yet we had no reason to imagine we had made any way to the westward. In this interval we had frequent squalls of rain and snow, and shipped great quantities of water; after which, for three or four days, though the seas ran mountains high, yet the weather was rather more moderate: but on the 18th, we had again strong gales of wind, with extreme cold, and at midnight the main top-sail split, and one of the straps of the main dead-eyes broke. From hence to the 23d, the weather was more favourable, though often intermixed with rain and sleet, and some hard gales; but as the waves did not subside, the ship, by labouring in this lofty sea, was now grown so loose in her upper works that she let in the water at every seam, so that every part within board was constantly exposed to the seawater, and scarcely any of the officers ever lay in dry beds. Indeed it was very rare that two nights ever passed without many of them being driven from their beds by the deluge of water that came in upon them.

On the 23d, we had a most violent storm of wind, hail, and rain, with a very great sea; and though we handed the main top-sail before the height of the squall, yet we found the yard sprung; and soon after the foot rope of the main-sail breaking, the main-sail itself split instantly to rags, and, in spite of our endeavours to save it, much the greater part of it was blown overboard. On this, the commodore made the signal for the squadron to bring-to; and the storm at length flattening to a calm, we had an opportunity of getting down our main top-sail yard to put the carpenters to work upon it, and of repairing our rigging; after which, having bent a new main-sail, we got under sail again with a moderate breeze; but in less than twenty-four hours we were attacked by another storm still more furious than the former; for it proved a perfect hurricane, and reduced us to the necessity of lying-to under our bare poles. As our ship kept the wind better than any of the rest, we were obliged, in the afternoon, to wear ship, in order to join the squadron to the leeward, which otherwise we should have been in danger of losing in the night: and as we dared not venture any sail abroad, we were obliged to make use of an expedient which answered our purpose; this was putting the helm a weather, and manning the fore-shrouds: but though this method proved successful for the end intended, yet in the execution of it one of our ablest seamen was canted overboard; we perceived that, notwithstanding the prodigious agitation of the waves, he swam very strong, and it was with the utmost concern that we found ourselves incapable of assisting him; indeed we were the more grieved at his unhappy fate, as we lost sight of him struggling with the waves, and conceived from the manner in which he swam that he might continue sensible, for a considerable time longer, of the horror attending his irretrievable situation.

Before this last-mentioned storm was quite abated, we found two of our main-shrouds and one mizen-shroud broke, all which we knotted, and set up immediately. From hence we had an interval of three or four days less tempestuous than usual, but accompanied with a thick fog, in which we were obliged to fire guns almost every half-hour, to keep our squadron together. On the 31st, we were alarmed by a gun fired from the *Gloucester*, and a signal made by her to speak with the commodore; we immediately bore down to her, and were prepared to hear of some terrible disaster; but we were apprized of it before we joined her, for we saw that her main-yard was broke in the slings. This was a grievous misfortune to us all at this juncture, as it was obvious it would prove an hindrance to our sailing, and would detain us the longer in these inhospitable latitudes. But our future success and safety was not to be promoted by repining, but by resolution and activity; and therefore that this unhappy incident might delay us as little as possible, the commodore ordered several carpenters to be put on board the *Gloucester* from the other ships of the squadron, in order to repair her damage with the utmost expedition. And the captain of the *Tryal* complaining at the same time that his pumps were so bad, and the sloop made so great a quantity of water, that he was scarcely able to keep her free, the commodore ordered him a pump ready fitted from his own ship. It was very fortunate for the *Gloucester* and the *Tryal* that the weather proved more favourable this day than for many days, both before and after; since by this means they were enabled to receive the assistance which seemed essential to their preservation, and which they could scarcely have had at any other time, as it would have been extremely hazardous to have ventured a boat on board.

The next day, that is, on the 1st of April, the weather returned again to its customary bias, the sky looked dark and gloomy, and the wind began to freshen and to blow in squalls; however, it was not yet so boisterous as to prevent our carrying our top-sails close reefed; but its appearance was such as plainly prognosticated that a still severer tempest was at hand: and accordingly, on the 3d of April, there came on a storm which both in its violence and continuation (for it lasted three days) exceeded all that we had hitherto encountered. In its first onset we received a furious shock from a sea which broke upon our larboard quarter, where it stove in the quarter gallery, and rushed into the ship like a deluge; our rigging too suffered extremely from the blow; amongst the rest, one of the straps of the main dead-eyes was broke, as was also a main-shroud and puttock-shroud, so that to ease the stress upon the masts and shrouds, we lowered both our main and fore-yards, and furled all our sails, and in this posture we lay-to for three days, when the storm somewhat abating, we ventured to make sail under our courses only; but even this we could not do long; for the next day, which was the 7th, we had another hard gale of wind, with lightening and rain, which obliged us to lie-to again till night. It was wonderful that notwithstanding the hard weather we had endured, no extraordinary accident had happened to any of the squadron since the breaking of the *Gloucester's* main-yard: but this good fortune now no longer attended us; for at three the next morning, several guns were fired to leeward as signals of distress: and the commodore making a signal for the squadron to bring-to, we, at daybreak, saw the *Wager* a considerable way to leeward of any of the other ships; and we soon perceived that she had lost her mizen-mast and main top-sail yard. We immediately bore down to her, and found this disaster had arisen from the badness of her iron work; for all the chain-plates to windward had given way, upon the ship's fetching a deep roll. This proved the more unfortunate to the *Wager*, as her carpenter had been on board the *Gloucester* ever since the 31st of March, and the weather was now too severe to permit him to return. Nor was the *Wager* the only ship of the squadron that suffered in this tempest; for the next day a signal of distress was made by the *Anna* pink, and, upon speaking with the master, we learnt that they had broke their fore-stay and the gammon of the bowsprit, and were in no small danger of having all their masts come by the board; so that we were obliged to bear away until they had made all fast, after which we haled upon a wind again.

And now after all our solicitude, and the numerous ills of every kind to which we had been incessantly exposed for near forty days, we had great consolation in the flattering hopes we entertained that our fatigues were drawing to a period, and that we should soon arrive in a more hospitable climate, where we should be amply repayed for all our past sufferings. For, towards the latter end of March, we were advanced, by our reckoning, near 10° to the westward of the westernmost point of Terra del Fuego, and this allowance being double what former navigators have thought necessary to be taken, in order to compensate the drift of the western current, we esteemed ourselves to be well advanced within the limits of the southern ocean, and had therefore been ever since standing to the northward with as much expedition as the turbulence of the weather and our frequent disasters permitted. And, on the 13th of April, we were but a degree in latitude to the southward of the west entrance of the Streights of Magellan; so that we fully expected, in a very few days, to have experienced the celebrated tranquillity of the Pacifick Ocean.

But these were delusions which only served to render our disappointment more terrible; for the next morning, between one and two, as we were standing to the northward, and the weather, which had till then been hazy, accidentally cleared up, the pink made a signal for seeing land right ahead; and it being but two miles distant, we were all under the most dreadful apprehensions of running on shore, which, had either the wind blown from its usual quarter with its wonted vigour, or had not the moon suddenly shone out, not a ship amongst us could possibly have avoided: but the wind, which some few hours before blew in squalls from the S.W., having fortunately shifted to W.N.W., we were enabled to stand to the southward, and to clear ourselves of this unexpected danger, and were fortunate enough by noon to have gained an offing of near twenty leagues.

By the latitude of this land we fell in with, it was agreed to be a part of Terra del Fuego, near the southern outlet described in Frezier's chart of the Streights of Magellan, and was supposed to be that point called by him Cape Noir. It was indeed most wonderful that the currents should have driven us to the eastward with such strength; for the whole squadron esteemed themselves upwards of ten degrees more westerly than this land, so that in running down, by our account, about nineteen degrees of longitude, we had not really advanced half that distance. And now, instead of having our labours and anxieties relieved by approaching a warmer climate and more tranquil seas, we were to steer again to the southward, and were again to combat those western blasts which had so often terrified us; and this too when we were greatly enfeebled by our men falling sick, and dying apace, and when our spirits, dejected by a long continuance at sea, and by our late disappointment, were much less capable of supporting us in the various difficulties which we could not but expect in this new undertaking. Add to all this, too, the discouragement we received by the diminution of the strength of the squadron; for, three days before this, we lost sight of the *Severn* and the *Pearl* in the morning, and though we spread our ships, and beat about for them some time, yet we never saw them more; whence we had apprehensions that they too might have fallen in with this land in the night, and by being less favoured by the wind and the moon than we were, might have run on shore and have perished. Full of these desponding thoughts and gloomy presages, we stood away to the S.W., prepared by our late disaster to suspect that how large soever an allowance we made in our westing for the drift of the western current, we might still, upon a second trial, perhaps find it insufficient.

CHAPTER IX

OBSERVATIONS AND DIRECTIONS FOR FACILITATING THE PASSAGE OF OUR FUTURE CRUISERS ROUND CAPE HORN

The improper season of the year in which we attempted to double Cape Horn, and to which is to be imputed the disappointment (recited in the foregoing chapter) of falling in with Terra del Fuego, when we reckoned ourselves above a hundred leagues to the westward of that whole coast, and consequently well advanced into the Pacifick Ocean; this unseasonable navigation, I say, to which we were necessitated by our too late departure from England, was the fatal source of all the misfortunes we afterwards encountered. For from hence proceeded the separation of our ships, the destruction of our people, the ruin of our project on Baldivia, and of all our other views on the Spanish places, and the reduction of our squadron from the formidable condition in which it passed Streights Le Maire to a couple of shattered half-manned cruisers and a sloop, so far disabled that in many climates they scarcely durst have put to sea. To prevent, therefore, as much as in me lies, all ships hereafter bound to the South Seas from suffering the same calamities, I think it my duty to insert in this place such directions and observations as either my own experience and reflection, or the conversation of the most skilful navigators on board the squadron, could furnish me with, in relation to the most eligible manner of doubling Cape Horn, whether in regard to the season of the year, the course proper to be steered, or the places of refreshment both on the east and west side of South America.

And first, with regard to the proper place for refreshment on the east side of South America. For this purpose the island of St. Catherine's has been usually recommended by former writers, and on their faith we put in there, as has been formerly mentioned. But the treatment we met with, and the small store of refreshments we could procure there, are sufficient reasons to render all ships for the future cautious how they trust themselves in the government of Don Jose Sylva de Paz, for they may certainly depend on having their strength, condition, and designs betrayed to the Spaniards, as far as the knowledge the governor can procure of these particulars will give him leave. And as this treacherous conduct is inspired by the views of private gain in the illicit commerce carried on to the river of Plate, rather than by any national affection which the Portuguese bear the Spaniards, the same perfidy may perhaps be expected from most of the governors of the Brazil coast, since these smuggling engagements are doubtless very extensive and general. And though the governors should themselves detest so faithless a procedure, yet as ships are perpetually passing from some or other of the Brazil ports to the river of Plate, the Spaniards could scarcely fail of receiving, by this means, casual intelligence of any British ships upon the coast, which, however imperfect such intelligence might be, would prove of dangerous import to the views and interests of those cruisers who were thus discovered.

For the Spanish trade in the South Seas running all in one track from north to south, with very little deviation to the eastward or westward, it is in the power of two or three cruisers, properly stationed in different parts of this track, to possess themselves of every ship that puts to sea; but this is only so long as they can continue concealed from the neighbouring coast, for the instant an enemy is known to be in those seas, all navigation is prohibited, and consequently all captures are at an end, since the Spaniards, well apprized of these advantages of the enemy, send expresses along the coast and lay a general embargo on all their trade; a measure which they prudentially foresee will not only prevent their vessels being taken, but will soon lay any cruisers who have not strength sufficient to attempt their places under necessity of returning home. Hence then appears the great importance of concealing all expeditions of this kind, and hence, too, it follows how extremely prejudicial that intelligence may prove which is given by the Portuguese governors to the Spaniards in relation to the designs of ships touching at the ports of Brazil.

However, notwithstanding the inconveniences we have mentioned of touching on the coast of Brazil, it will often-times happen that ships bound round Cape Horn will be obliged to call there for a supply of wood and water, and other refreshments. In this case, St. Catherine's is the last place I would recommend, both as the proper animals for a live stock at sea, as hogs, sheep, and fowls, cannot be procured there (for want of which we found ourselves greatly distressed, by being reduced to live almost entirely on salt provisions), and also because, from its being nearer the river of Plate than many of their other settlements, the inducements and conveniences of betraying us are much stronger. The place I would recommend is Rio Janeiro, where two of our squadron put in after they were separated from us in passing Cape Horn, for here, as I have been informed by one of the gentlemen on board those ships, any quantity of hogs and poultry may be procured; and this place being more distant from the river of Plate, the difficulty of intelligence is somewhat inhanced, and consequently the chance of continuing there undiscovered in some degree augmented. Other measures, which may effectually obviate all these embarrassments, shall be considered more at large hereafter.

I next proceed to the consideration of the proper course to be steered for doubling Cape Horn. And here, I think, I am sufficiently authorised by our own fatal experience, and by a careful comparison and examination of the journals of former navigators, to give this piece of advice, which in prudence

I think ought never to be departed from: that is, that all ships bound to the South Seas, instead of passing through Streights Le Maire, should constantly pass to the eastward of Staten-land, and should be invariably bent on running to the southward, as far as the latitude of 61 or 62 degrees, before they endeavour to stand to the westward; and that when they are got into that latitude they should then make sure of sufficient westing before they once think of steering to the northward.

But as directions diametrically opposite to these have been formerly given by other writers, it is incumbent on me to produce my reasons for each part of this maxim. And first, as to the passing to the eastward of Staten-land. Those who have attended to the risque we ran in passing Streights Le Maire, the danger we were in of being driven upon Staten-land by the current, when, though we happily escaped being put on shore, we were yet carried to the eastward of that island; those who reflect on this, and the like accidents which have happened to other ships, will surely not esteem it prudent to pass through Streights Le Maire, and run the risque of shipwreck, and after all find themselves no farther to the westward (the only reason hitherto given for this practice) than they might have been in the same time by a secure navigation in an open sea.

And next as to the directions I have given for running into the latitude of 61 or 62 south, before any endeavour is made to stand to the westward. The reasons for this precept are, that in all probability the violence of the currents will be hereby avoided, and the weather will prove less tempestuous and uncertain. This last circumstance we ourselves experienced most remarkably, for after we had unexpectedly fallen in with the land, as has been mentioned in the preceding chapter, we stood away to the southward to run clear of it, and were no sooner advanced into sixty degrees or upwards but we met with much better weather and smoother water than in any other part of the whole passage. The air indeed was very cold and sharp, and we had strong gales, but they were steady and uniform, and we had at the same time sunshine and a clear sky; whereas in the lower latitudes the winds every now and then intermitted, as it were, to recover new strength, and then returned suddenly in the most violent gusts, threatening at each blast the loss of our masts, which must have ended in our certain destruction. And that the currents in this high latitude would be of much less efficacy than nearer the land seems to be evinced from these considerations, that all currents run with greater violence near the shore than at sea, and that at great distances from shore they are scarcely perceptible. Indeed the reason of this seems sufficiently obvious, if we consider that constant currents are, in all probability, produced by constant winds, the wind driving before it, though with a slow and imperceptible motion, a large body of water, which being accumulated upon any coast that it meets with, must escape along the shore by the endeavours of its surface to reduce itself to the same level with the rest of the ocean. And it is reasonable to suppose that those violent gusts of wind which we experienced near the shore, so very different from what we found in the latitude of sixty degrees and upwards, may be owing to a similar cause, for a westerly wind almost perpetually prevails in the southern part of the Pacific Ocean. And this current of air being interrupted by those immense hills called the Andes, and by the mountains on Terra del Fuego, which together bar up the whole country to the southward as far as Cape Horn, a part of it only can force its way over the tops of those prodigious precipices, whilst the rest must naturally follow the direction of the coast, and must range down the land to the southward, and sweep with an impetuous and irregular blast round Cape Horn and the southermost part of Terra del Fuego. However, not to rely on these speculations, we may, I believe, establish as incontestible these matters of fact, that both the rapidity of the currents and the violence of the western gales are less sensible in the latitude of 61 or 62 degrees than nearer the shore of Terra del Fuego.

But though I am satisfied from both our own experience, and the relations of other navigators, of the importance of the precept I here insist on, that of running into the latitude of 61 or 62 degrees, before any endeavours are made to stand to the westward, yet I would advise no ships hereafter to trust so far to this management as to neglect another most essential maxim, which is the making this passage in the height of summer—that is, in the months of December and January; and the more distant the time of passing is taken from this season, the more disastrous it may be reasonably expected to prove. Indeed, if the mere violence of the western winds be considered, the time of our passage, which was about the equinox, was perhaps the most unfavourable of the whole year; but then it must be remembered that independent of the winds there are in the depth of winter many other inconveniences to be apprehended which are almost insuperable, for the severity of the cold and the shortness of the days would render it impracticable at that season to run so far to the southward as is here recommended; and the same reasons would greatly augment the alarms of sailing in the neighbourhood of an unknown shore, dreadful in its appearance in the midst of summer, and would make a winter navigation on this coast to be, of all others, the most dismaying and terrible. As I would therefore advise all ships to make their passage in December and January, if possible, so I would warn them never to attempt the doubling Cape Horn, from the eastward, after the month of March.

And now, as to the remaining consideration, that is, the properest port for cruisers to refresh at on their first arrival in the South Seas. On this head there is scarcely any choice, the island of Juan Fernandes being the only place that can be prudently recommended for this purpose. For though there are many ports on the western side of Patagonia, between the Streights of Magellan and the Spanish settlements, where ships might ride in great safety, might recruit their wood and water, and might procure some few refreshments, yet that coast is in itself so dangerous from its numerous rocks and breakers, and from the violence of the western winds which blow constantly full upon it, that it is by no means adviseable to fall in with that land, at least till the roads, channels, and anchorage, in each part of it are accurately surveyed, and both the perils and shelter it abounds with are more distinctly known.

Thus having given the best directions in my power for the success of our cruisers who may be hereafter bound to the South Seas, it might be expected that I should again resume the thread of my narration. Yet as both in the preceding and subsequent parts of this work I have thought it my duty not only to recite all such facts, and to inculcate such maxims as had the least appearance of proving beneficial to future navigators, but also occasionally to recommend such measures to the public as I conceive are adapted to promote the same laudable purpose, I cannot desist from the present subject without beseeching those to whom the conduct of our naval affairs is committed to endeavour to remove the many perplexities and embarrassments with which the navigation to the South Seas is at present necessarily encumbered. An effort of this kind could not fail of proving highly honourable to themselves, and extremely beneficial to their country. For it seems to be sufficiently evident, that whatever improvements navigation shall receive, either by the invention of methods that shall render its practice less hazardous, or by the more accurate delineation of the coasts, roads, and ports already known, or by the discovery of new nations, or new species of commerce; it seems, I say, sufficiently evident, that by whatever means navigation is promoted, the conveniences hence arising must ultimately redound to the emolument of Great Britain. Since as our fleets are at present superior to those of the whole world united, it must be a matchless degree of supineness or mean-spiritedness if we permitted any of the advantages which new discoveries, or a more extended navigation, may produce to mankind to be ravished from us.

As, therefore, it appears that all our future expeditions to the South Seas must run a considerable risque of proving abortive whilst in our passage thither, we are under the necessity of touching at Brazil, the discovery of some place more to the southward, where ships might refresh and supply themselves with the necessary sea-stock for their voyage round Cape Horn, would be an expedient which would relieve us from this embarrassment, and would surely be a matter worthy of the attention of the public. Nor does this seem difficult to be effected. For we have already the imperfect knowledge of two places which might perhaps, on examination, prove extremely convenient for this purpose. One of them is Pepys's Island, in the latitude of 47° south, and laid down by Dr. Halley about eighty leagues to the eastward of Cape Blanco, on the coast of Patagonia; the other is Falkland's Isles, in the latitude of 51-½°, lying nearly south of Pepys's Island. The first of these was discovered by Captain Cowley in his voyage round the world in the year 1686, who represents it as a commodious place for ships to wood and water at, and says it is provided with a very good and capacious harbour, where a thousand sail of ships might ride at anchor in great safety; that it abounds with fowls, and that as the shore is either rocks or sands, it seems to promise great plenty of fish. The second place, or Falkland's Isles, has been seen by many ships, both French and English, being the land laid down by Frezier, in his chart of the extremity of South America, under the title of the New Islands. Woodes Rogers, who run along the N.E. coast of these isles in the year 1708, tells us that they extended about two degrees in length, and appeared with gentle descents from hill to hill, and seemed to be good ground, interspersed with woods, and not destitute of harbours. Either of these places, as they are islands at a considerable distance from the continent, may be supposed, from their latitude, to lie in a climate sufficiently temperate. It is true, they are too little known to be at present recommended as the most eligible places of refreshment for ships bound to the southward, but if the Admiralty should think it adviseable to order them to be surveyed, which may be done at a very small expence by a vessel fitted out on purpose, and if, on this examination, one or both of these places should appear proper for the purpose intended, it is scarcely to be conceived of what prodigious import a convenient station might prove, situated so far to the southward, and so near Cape Horn. The Duke and Duchess of Bristol were but thirty-five days from their losing sight of Falkland's Isles to their arrival at Juan Fernandes in the South Seas: and as the returning back is much facilitated by the western winds, I doubt not but a voyage might be made from Falkland's Isles to Juan Fernandes and back again in little more than two months. This, even in time of peace, might be of great consequence to this nation; and, in time of war, would make us masters of those seas.

And as all discoveries of this kind, though extremely honourable to those who direct and promote them, may yet be carried on at an inconsiderable expence, since small vessels are much the properest to be employed in this service: it were to be wished that the whole coast of Patagonia, Terra del Fuego, and Staten-land were carefully surveyed, and the numerous channels, roads, and harbours with which they abound were accurately examined. This might open to us facilities of passing into the Pacifick Ocean, which as yet we may be unacquainted with, and would render all that southern navigation infinitely securer than at present; particularly an exact draught of the west coast of Patagonia, from the Streights of Magellan to the Spanish settlements, might perhaps furnish us with better and more convenient ports for refreshment, and better situated for the purposes either of war or commerce, and above a fortnight's sail nearer to Falkland's Island than the island of Juan Fernandes. The discovery of this coast hath formerly been thought of such consequence by reason of its neighbourhood to the Araucos and other Chilian Indians, who are generally at war, or at least on ill terms, with their Spanish neighbours, that Sir John Narborough was purposely fitted out in the reign of King Charles II. to survey the Streights of Magellan, the neighbouring coast of Patagonia, and the Spanish ports on that frontier, with directions, if possible, to procure some intercourse with the Chilian Indians, and to establish a commerce and a lasting correspondence with them. His Majesty's views in employing Sir John Narborough in this expedition were not solely the advantage he might hope to receive from the alliance of those savages, in restraining and intimidating the crown of Spain; but he conceived that, independent of those motives, the immediate traffick with these Indians might prove extremely advantageous to the English nation. For it is well known that at the first discovery of Chili by the Spaniards, it abounded with vast quantities of gold, much beyond what it has at any time produced since it has been in their possession. And hence it has been generally believed that the richest mines are carefully concealed by the Indians, as well knowing that the discovery of them would only excite in the Spaniards a greater thirst for conquest and tyranny, and would render their own independence more precarious. But with respect to their commerce with the English, these reasons would no longer influence them; since it would be in our power to furnish them with arms and ammunition of all kinds, of which they are extremely desirous, together with many other conveniencies which their intercourse with the Spaniards has taught them to relish. They would then, in all probability, open their mines, and gladly embrace a traffick of such mutual convenience to both nations; for then their gold, instead of proving an incitement to enslave them, would procure them weapons to assert their liberty, to chastise their tyrants, and to secure themselves for ever from the Spanish yoke; whilst with our assistance, and under our protection, they might become a considerable people, and might secure to us that wealth which formerly by the House of Austria, and lately by the House of Bourbon, has been most mischievously lavished in the pursuit of universal monarchy.

It is true, Sir John Narborough did not succeed in opening this commerce, which, in appearance, promised so many advantages to this nation. However, his disappointment was merely accidental, and his transactions upon that coast (besides the many valuable improvements he furnished to geography and navigation) are rather an encouragement for future trials of this kind than any objection against them; his principal misfortune being the losing company of a small bark which attended him, and having some of his people trepanned at Baldivia. However, it appeared, by the precautions and fears of the Spaniards, that they were fully convinced of the practicability of the scheme he was sent to execute, and extremely alarmed with the apprehension of its consequences. It is said that his Majesty King Charles the Second was so far prepossessed with the belief of the emoluments which might redound to the publick from this expedition, and was so eager to be informed of the event of it, that having intelligence of Sir John Narborough's passing through the Downs on his return, he had not patience to attend his arrival at court, but went himself in his barge to Gravesend to meet him.

To facilitate as much as possible any attempts of this kind which may be hereafter undertaken, I prepared a chart of that part of the world, as far as it is hitherto known, which I flatter myself is, in some respects, much correcter than any which has been yet published. To evince which, it may be necessary to mention what materials I have principally made use of, and what changes I have introduced different from other authors.

The two most celebrated charts hitherto published of the southermost part of South America, are those of Dr. Halley, in his general chart of the magnetic variation, and of Frezier in his voyage to the South Seas. But besides these, there is a chart of the Streights of Magellan, and of some part of the adjacent coast, by Sir John Narborough above-mentioned, which is doubtless infinitely exacter in that part than Frezier's, and in some respects superior to Halley's, particularly in what relates to the longitudes of the different parts of those streights. The coast from Cape Blanco to Terra del Fuego, and thence to Streights Le Maire, we were in some measure capable of correcting by our own observations, as we ranged that shore generally in sight of land. The position of the land to the northward of the Streights of Magellan, on the west side, is doubtless laid down but very imperfectly; and yet I believe it to be much nearer the truth than what has hitherto been done, as it is drawn from the information of some of the *Wager's* crew who were shipwrecked on that shore and afterwards coasted it down, and as it agrees pretty nearly with the description of some Spanish manuscripts I have seen. The channel dividing Terra del Fuego is drawn from Frezier; but Sir Francis Drake, who first discovered Cape Horn, and the S.W. part of Terra del Fuego, observed that whole coast to be divided by a great number of inlets, all which he conceived did communicate with the Streights of Magellan. And I doubt not that whenever this country is thoroughly examined this circumstance will be verified, and Terra del Fuego will be found to consist of several islands.

And having mentioned Frezier so often, I must not omit warning all future navigators against relying on the longitude of Streights Le Maire, or of any part of that coast, laid down in his chart, the whole being from 8 to 10 degrees too far to the eastward, if any faith can be given to the concurrent evidences of a great number of journals, verified in some particulars by astronomical observation. For instance, Sir John Narborough places Cape Virgin Mary in 65° 42' of west longitude from the Lizard, that is in about 71-½° from London. And the ships of our squadron, who took their departure from St. Catherine's (where the longitude was rectified by an observation of the eclipse of the moon), found Cape Virgin Mary to be from 70-½° to 72-½° from London, according to their different reckonings; and since there were no circumstances in our run that could render it considerably erroneous, it cannot be esteemed in less than 71 degrees of west longitude; whereas Frezier lays it down in less than 66 degrees from Paris, that is, little more than 63 degrees from London, which is doubtless 8 degrees short of its true quantity. Again, our squadron found Cape Virgin Mary and Streights Le Maire to be not more than 2-½° different in longitude, which in Frezier are distant near 4 degrees, so that not only the longitude of Cape St. Bartholomew is laid down in him near 10 degrees too little, but the coast from the Streights of Magellan to Streights Le Maire is enlarged to near double its real extent.

But to have done with Frezier, whose errors, the importance of the subject, and not a fondness for cavilling, has obliged me to remark (though his treatment of Dr. Halley might, on the present occasion, authorise much severer usage), I must, in the next place, relate wherein I differ from that of our learned countryman last mentioned.

It is well known that this gentleman was sent abroad by the public to make such geographical and astronomical observations as might facilitate the future practice of navigation, and particularly to determine the variation of the compass in such places as he should touch at, and, if possible, to ascertain its general laws and affections. These things Dr. Halley, to his immortal reputation and the honour of our nation, in good measure accomplished, especially with regard to the variation of the compass, a subject, of all others, the most interesting to those employed in the art of navigation. He likewise corrected the position of the coast of Brazil, which had been very erroneously laid down by all former hydrographers; and from a judicious comparison of the observations of others, he happily succeeded in settling the geography of many considerable places where he had not himself been. So that the chart he composed, with the variation of the needle marked thereon, being the result of his labours on this subject, was allowed by all Europe to be far compleater in its geography than any that had till then been published, whilst it was at the same time most surprisingly exact in the quantity of variation assigned to the different parts of the globe; a subject so very intricate and perplexing, that all general determinations about it had been usually deemed impossible.

But as the only means he had of correcting the situation of those coasts, where he did not touch himself, were the observations of others, when those observations were wanting, or were inaccurate, it was no imputation on his skill that his decisions were defective. And this, upon the best comparison I have been able to make, is the case with regard to that part of his chart, which contains the south coast of South America. For though the coast of Brazil, and the opposite coast of Peru on the South Seas, are laid down, I presume, with the greatest accuracy, yet from about the river of Plate on the east side, and its opposite point on the west, the coast gradually declines too much to the westward, so as at the Streights of Magellan to be, as I conceive, about fifty leagues removed from its true position; at least, this is the result of the observations of our squadron, which agree extremely well with those of Sir John Narborough. I must add that Dr. Halley has, in the philosophical transactions, given the foundation on which he has proceeded in fixing Port St. Julian in 76-½° of west longitude, which the concurrent journals of our squadron place from 70-¾° to 71-½°. This, he tells us, was an observation of an eclipse of the moon made at that place by Mr. Wood, then Sir John Narborough's lieutenant, and which is said to have happened there at eight in the evening, on the 18th of September 1670. But Captain Wood's journal of this whole voyage under Sir John Narborough is since published together with this observation, in which he determines the longitude of Port St. Julian to be 73 degrees from London, and the time of the eclipse to have been different from Dr. Halley's account. But the numbers he has given are so faultily printed that nothing can be determined from them.

CHAPTER X

FROM CAPE NOIR TO THE ISLAND OF JUAN FERNANDES

After the mortifying disappointment of falling in with the coast of Terra del Fuego, when we esteemed ourselves ten degrees to the westward of it, as hath been at large recited in the eighth chapter, we stood away to the S.W. till the 22d of April, when we were in upwards of 60° of south latitude, and by our account near 6° to the westward of Cape Noir. In this run, we had a series of as favourable weather as could well be expected in that part of the world, even in a better season, so that this interval, setting the inquietude of our thoughts aside, was by far the most eligible of any we enjoyed from Streights Le Maire to the west coast of America. This moderate weather continued with little variation till the 24th, but on the 24th, in the evening, the wind began to blow fresh, and soon increased to a prodigious storm, and the weather being extremely thick, about midnight we lost sight of the other four ships of the squadron, which, notwithstanding the violence of the preceding storms, had hitherto kept in company with us. Nor was this our sole misfortune, for the next morning, endeavouring to hand the top-sails, the clew-lines and bunt-lines broke, and the sheets being half-flown, every seam in the top-sails was soon split from top to bottom, and the main top-sail shook so strongly in the wind that it carried away the top lanthorn, and endangered the head of the mast; however, at length some of the most daring of our men ventured upon the yard and cut the sail away close to the reefs, though with the utmost hazard of their lives. Whilst at the same time the fore top-sail beat about the yard with so much fury that it was soon blown to pieces; nor was our attention to our top-sails our sole employment, for the main-sail blew loose, which obliged us to lower down the yard to secure the sail, and the fore-yard being likewise lowered, we lay-to under a mizen. In this storm, besides the loss of our top-sails, we had much of our rigging broke, and lost a main-studding sail-boom out of the chains.

On the 25th, about noon, the weather became more moderate, which enabled us to sway up our yards, and to repair, in the best manner we could, our shattered rigging, but still we had no sight of the rest of our squadron, nor indeed were we joined by any of them again till after our arrival at Juan Fernandes, nor did any two of them, as we have since learned, continue in company together. This total and almost instantaneous separation was the more wonderful as we had hitherto kept together for seven weeks, through all the reiterated tempests of this turbulent climate. It must indeed be owned that we had hence room to expect that we might make our passage in a shorter time than if we had continued together, because we could now make the best of our way without being retarded by the misfortunes of the other ships; but then we had the melancholy reflection that we ourselves were hereby deprived of the assistance of others, and our safety would depend upon our single ship, so that if a plank started, or any other accident of the same nature should take place, we must all irrecoverably perish; or should we be driven on shore, we had the uncomfortable prospect of ending our days on some desolate coast, without any reasonable hope of ever getting off again, whereas, with another ship in company, all these calamities are much less formidable, since in every kind of danger there would be some probability that one ship at least might escape, and might be capable of preserving or relieving the crew of the other.

The remaining part of this month of April we had generally hard gales, although we had been every day, since the 22d, edging to the northward; however, on the last day of the month, we flattered ourselves with the expectation of soon terminating all our sufferings, for we that day found ourselves in the latitude of 52° 13', which being to the northward of the Streights of Magellan, we were assured that we had compleated our passage, and had arrived in the confines of the southern ocean; and this ocean being denominated Pacifick, from the equability of the seasons which are said to prevail there, and the facility and security with which navigation is there carried on, we doubted not but we should be speedily cheared with the moderate gales, the smooth water, and the temperate air for which that track of the globe has been so renowned. And under the influence of these pleasing circumstances we hoped to experience some kind of compensation for the complicated miseries which had so constantly attended us for the last eight weeks. But here we were again disappointed, for in the succeeding month of May our sufferings rose to a much higher pitch than they had ever yet done, whether we consider the violence of the storms, the shattering of our sails and rigging, or the diminishing and weakening of our crew by deaths and sickness, and the probable prospect of our total destruction. All this will be sufficiently evident from the following circumstantial account of our diversified misfortunes.

Soon after our passing Streights Le Maire, the scurvy began to make its appearance amongst us, and our long continuance at sea, the fatigue we underwent, and the various disappointments we met with, had occasioned its spreading to such a degree that at the latter end of April there were but few on board who were not in some degree afflicted with it, and in that month no less than forty-three died of it on board the *Centurion*. But though we thought that the distemper had then risen to an extraordinary height, and were willing to hope that as we advanced to the northward its malignity would abate, yet we found, on the contrary, that in the month of May we lost near double that number; and as we did not get to land till the middle of June, the mortality went on increasing, and the disease extended itself so prodigiously that, after the loss of above two hundred men, we could not at last muster more than six fore-mast men in a watch capable of duty.

This disease, so frequently attending long voyages, and so particularly destructive to us, is surely the most singular and unaccountable of any that affects the human body. Its symptoms are inconstant and innumerable, and its progress and effects extremely irregular; for scarcely any two persons have complaints exactly resembling each other, and where there hath been found some conformity in the symptoms, the order of their appearance has been totally different. However, though it frequently puts on the form of many other diseases, and is therefore not to be described by any exclusive and infallible criterions, yet there are some symptoms which are more general than the rest, and, occurring the oftenest, deserve a more particular enumeration. These common appearances are large discoloured spots dispersed over the whole surface of the body, swelled legs, putrid gums, and, above all, an extraordinary lassitude of the whole body, especially after any exercise, however inconsiderable; and this lassitude at last degenerates into a proneness to swoon, and even die, on the least exertion of strength, or even on the least motion.

This disease is likewise usually attended with a strange dejection of the spirits, and with shiverings, tremblings, and a disposition to be seized with the most dreadful terrors on the slightest accident. Indeed it was most remarkable, in all our reiterated experience of this malady, that whatever discouraged our people, or at any time damped their hopes, never failed to add new vigour to the distemper; for it usually killed those who were in the last stages of it, and confined those to their hammocks who were before capable of some kind of duty; so that it seemed as if alacrity of mind, and sanguine thoughts, were no contemptible preservatives from its fatal malignity.

But it is not easy to compleat the long roll of the various concomitants of this disease; for it often produced putrid fevers, pleurisies, the jaundice, and violent rheumatic pains, and sometimes it occasioned an obstinate costiveness, which was generally attended with a difficulty of breathing, and this was esteemed the most deadly of all the scorbutick symptoms; at other times the whole body, but more especially the legs, were subject to ulcers of the worst kind, attended with rotten bones, and such a luxuriancy of fungous flesh as yielded to no remedy. But a most extraordinary circumstance, and what would be scarcely credible upon any single evidence, is, that the scars of wounds which had been for many years healed were forced open again by this virulent distemper. Of this there was a remarkable instance in one of the invalids on board the *Centurion*, who had been wounded above fifty years before at the battle of the Boyne, for though he was cured soon after, and had continued well for a great number of years past, yet on his being attacked by the scurvy, his wounds, in the progress of his disease, broke out afresh, and appeared as if they had never been healed: nay, what is still more astonishing, the callus of a broken bone, which had been compleatly formed for a long time, was found to be hereby dissolved, and the fracture seemed as if it had never been consolidated. Indeed, the effects of this disease were in almost every instance wonderful; for many of our people, though confined to their hammocks, appeared to have no inconsiderable share of health, for they eat and drank heartily, were chearful, and talked with much seeming vigour, and with a loud strong tone of voice; and yet, on their being the least moved, though it was from only one part of the ship to the other, and that too in their hammocks, they have immediately expired; and others, who have confided in their seeming strength, and have resolved to get out of their hammocks, have died before they could well reach the deck; nor was it an uncommon thing for those who were able to walk the deck, and to do some kind of duty, to drop down dead in an instant, on any endeavours to act with their utmost effort, many of our people having perished in this manner during the course of this voyage.

With this terrible disease we struggled the greatest part of the time of our beating round Cape Horn; and though it did not then rage with its utmost violence, yet we buried no less than forty-three men on board the *Centurion* in the month of April, as hath been already observed; however, we still entertained hopes that when we should have once secured our passage round the Cape, we should put a period to this, and all the other evils which had so constantly pursued us. But it was our misfortune to find that the Pacifick Ocean was to us less hospitable than the turbulent neighbourhood of Terra del Fuego and Cape Horn. For being arrived, on the 8th of May, off the island of Socoro, which was the first rendezvous appointed for the squadron, and where we hoped to have met with some of our companions, we cruised for them in that station several days. But here we were not only disappointed in our expectations of being joined by our friends, and were thereby induced to favour the gloomy suggestions of their having all perished; but we were likewise perpetually alarmed with the fears of being driven on shore upon this coast, which appeared too craggy and irregular to give us the least prospect that in such a case any of us could possibly escape immediate destruction. For the land had indeed a most tremendous aspect: the most distant part of it, and which appeared far within the country, being the mountains usually called the Andes or Cordilleras, was extremely high and covered with snow; and the coast itself seemed quite rocky and barren, and the water's edge skirted with precipices. In some places indeed we discerned several deep bays running into the land, but the entrance into them was generally blocked up by numbers of little islands; and though it was not improbable but there might be convenient shelter in some of those bays, and proper channels leading thereto, yet, as we were utterly ignorant of the coast, had we been driven ashore by the western winds which blew almost constantly there, we did not expect to have avoided the loss of our ship, and of our lives.

This continued peril, which lasted for above a fortnight, was greatly aggravated by the difficulties we found in working the ship, as the scurvy had by this time destroyed so great a part of our hands, and had in some degree affected almost the whole crew. Nor did we, as we hoped, find the winds less violent as we advanced to the northward; for we had often prodigious squalls which split our sails, greatly damaged our rigging, and endangered our masts. Indeed, during the greatest part of the time we were upon this coast, the wind blew so hard, that in another situation, where we had sufficient sea room, we should certainly have lain-to; but in the present exigency we were necessitated to carry both our courses and top-sails in order to keep clear of this lee shore. In one of these squalls, which was attended by several violent claps of thunder, a sudden flash of fire darted along our decks, which, dividing, exploded with a report like that of several pistols, and wounded many of our men and officers as it passed, marking them in different parts of the body. This flame was attended with a strong sulphureous stench, and was doubtless of the same nature with the larger and more violent blasts of lightning which then filled the air.

It were endless to recite minutely the various disasters, fatigues, and terrors which we encountered on this coast; all these went on increasing till the 22d of May, at which time the fury of all the storms which we had hitherto encountered seemed to be combined, and to have conspired our destruction. In this hurricane almost all our sails were split, and great part of our standing rigging broken; and, about eight in the evening, a mountainous over-grown sea took us upon our starboard-quarter, and gave us so prodigious a shock that several of our shrouds broke with the jerk, by which our masts were greatly endangered; our ballast and stores too were so strangely shifted that the ship heeled afterwards two streaks to port. Indeed it was a most tremendous blow, and we were thrown into the utmost consternation from the apprehension of instantly foundering; and though the wind abated in a few hours, yet, as we had no more sails left in a condition to bend to our yards, the ship laboured very much in a hollow sea, rolling gunwale to, for want of sail to steady her: so that we expected our masts, which were now very slenderly supported, to come by the board every moment. However, we exerted ourselves the best we could to stirrup our shrouds, to reeve new lanyards, and to mend our sails; but while these necessary operations were carrying on, we ran great risque of being driven on shore on the island of Chiloe, which was not far distant from us; but in the midst of our peril the wind happily shifted to the southward, and we steered off the land with the main-sail only, the master and myself undertaking the management of the helm, while every one else on board was busied in securing the masts, and bending the sails as fast as they could be repaired. This was the last effort of that stormy climate; for in a day or two after we got clear of the land, and found the weather more moderate than we had yet experienced since our passing Streights Le Maire. And now having cruized in vain for more than a fortnight in quest of the other ships of the squadron, it was resolved to take the advantage of the present favourable season, and the offing we had made from this terrible coast, and to make the best of our way for the island of Juan Fernandes. For though our next rendezvous was appointed off the harbour of Baldivia, yet as we had hitherto seen none of our companions at this first rendezvous, it was not to be supposed that any of them would be found at the second: indeed we had the greatest reason to suspect that all but ourselves had perished. Besides, we were by this time reduced to so low a condition, that instead of attempting to attack the places of the enemy, our utmost hopes could only suggest to us the possibility of saving the ship, and some part of the remaining enfeebled crew, by our speedy arrival at Juan Fernandes; for this was the only road in that part of the world where there was any probability of our recovering our sick, or refitting our vessel, and consequently our getting thither was the only chance we had left to avoid perishing at sea.

Our deplorable situation then allowing no room for deliberation, we stood for the island of Juan Fernandes; and to save time, which was now extremely precious (our men dying four, five, and six in a day), and likewise to avoid being engaged again with a lee shore, we resolved, if possible, to hit the island upon a meridian. And, on the 28th of May, being nearly in the parallel upon which it is laid down, we had great expectations of seeing it: but not finding it in the position in which the charts had taught us to expect it, we began to fear that we had gone too far to the westward; and therefore, though the commodore himself was strongly persuaded that he saw it on the morning of the 28th, yet his officers believing it to be only a cloud, to which opinion the haziness of the weather gave some kind of countenance, it was, on a consultation, resolved to stand to the eastward, in the parallel of the island, as it was certain that by this course we should either fall in with the island, if we were already to the westward of it, or should at least make the mainland of Chili, from whence we might take a new departure, and assure ourselves, by running to the westward afterwards, of not missing the island a second time.

On the 30th of May we had a view of the continent of Chili, distant about twelve or thirteen leagues; the land made exceeding high and uneven, and appeared quite white; what we saw being doubtless a part of the Cordilleras, which are always covered with snow. Though by this view of the land we ascertained our position, yet it gave us great uneasiness to find that we had so needlessly altered our course, when we were, in all probability, just upon the point of making the island; for the mortality amongst us was now increased to a most dreadful degree, and those who remained alive were utterly dispirited by this new disappointment, and the prospect of their longer continuance at sea. Our water too began to grow scarce; so that a general dejection prevailed amongst us, which added much to the virulence of the disease, and destroyed numbers of our best men; and to all these calamities there was added this vexatious circumstance, that when, after having got a sight of the main, we tacked and stood to the westward in quest of the island, we were so much delayed by calms and contrary winds, that it cost us nine days to regain the westing which, when we stood to the eastward, we ran down in two. In this desponding condition, with a crazy ship, a great scarcity of fresh water, and a crew so universally diseased that there were not above ten fore-mast men in a watch capable of doing duty, and even some of these lame, and unable to go aloft: under these disheartening circumstances, we stood to the westward; and, on the 9th of June, at daybreak, we at last discovered the long-wished-for island of Juan Fernandes. With this discovery I shall close this chapter and the first book, after observing (which will furnish a very strong image of our unparalleled distresses) that by our suspecting ourselves to be to the westward of the island on the 28th of May, and in consequence of this standing in for the main, we lost between seventy and eighty of our men, whom we should doubtless have saved had we made the island that day, which, had we kept on our course for a few hours longer, we could not have failed to have done.

BOOK II

CHAPTER I

THE ARRIVAL OF THE "CENTURION" AT THE ISLAND OF JUAN FERNANDES, WITH A DESCRIPTION OF THAT ISLAND

On the 9th of June, at daybreak, as is mentioned in the preceding chapter, we first descried the island of Juan Fernandes, bearing N. by E.½E., at eleven or twelve leagues distance. And though, on this first view, it appeared to be a very mountainous place, extremely ragged and irregular, yet as it was land, and the land we sought for, it was to us a most agreeable sight: because at this place only we could hope to put a period to those terrible calamities we had so long struggled with, which had already swept away above half our crew, and which, had we continued a few days longer at sea, would inevitably have compleated our destruction. For we were by this time reduced to so helpless a condition, that out of two hundred and odd men which remained alive, we could not, taking all our watches together, muster hands enough to work the ship on an emergency, though we included the officers, their servants, and the boys.

The wind being northerly when we first made the island, we kept plying all that day, and the next night, in order to get in with the land; and wearing the ship in the middle watch, we had a melancholy instance of the almost incredible debility of our people; for the lieutenant could muster no more than two quarter-masters and six fore-mast men capable of working; so that without the assistance of the officers, servants, and the boys, it might have proved impossible for us to have reached the island, after we had got sight of it; and even with this assistance they were two hours in trimming the sails: to so wretched a condition was a sixty-gun ship reduced, which had passed Streights Le Maire but three months before, with between four and five hundred men, almost all of them in health and vigour.

However, on the 10th in the afternoon we got under the lee of the island, and kept ranging along it, at about two miles distance, in order to look out for the proper anchorage, which was described to be in a bay on the north side. Being now nearer in with the shore, we could discover that the broken craggy precipices, which had appeared so unpromising at a distance, were far from barren, being in most places covered with woods, and that between them there were everywhere interspersed the finest vallies, clothed with a most beautiful verdure, and watered with numerous streams and cascades, no valley of any extent being unprovided of its proper rill. The water too, as we afterwards found, was not inferior to any we had ever tasted, and was constantly clear. The aspect of this country, thus diversified, would, at all times, have been extremely delightful; but in our distressed situation, languishing as we were for the land and its vegetable productions (an inclination constantly attending every stage of the sea-scurvy), it is scarcely credible with what eagerness and transport we viewed the shore, and with how much impatience we longed for the greens and other refreshments which were then in sight, and particularly the water, for of this we had been confined to a very sparing allowance a considerable time, and had then but five ton remaining on board. Those only who have endured a long series of thirst, and who can readily recal the desire and agitation which the ideas alone of springs and brooks have at that time raised in them, can judge of the emotion with which we eyed a large cascade of the most transparent water, which poured itself from a rock near a hundred feet high into the sea at a small distance from the ship. Even those amongst the diseased who were not in the very last stages of the distemper, though they had been long confined to their hammocks, exerted the small remains of strength that were left them, and crawled up to the deck to feast themselves with this reviving prospect. Thus we coasted the shore, fully employed in the contemplation of this enchanting landskip, which still improved upon us the farther we advanced. But at last the night closed upon us, before we had satisfied ourselves which was the proper bay to anchor in; and therefore we resolved to keep in soundings all night (we having then from sixty-four to seventy fathom), and to send our boat next morning to discover the road. However, the current shifted in the night, and set us so near the land that we were obliged to let go the best bower in fifty-six fathom, not half a mile from the shore. At four in the morning, the cutter was dispatched with our third lieutenant to find out the bay we were in search of, who returned again at noon with the boat laden with seals and grass; for though the island abounded with better vegetables, yet the boat's crew, in their short stay, had not met with them; and they well knew that even grass would prove a dainty, as indeed it was all soon and eagerly devoured. The seals too were considered as fresh provision, but as yet were not much admired, though they grew afterwards into more repute: for what rendered them less valuable at this juncture was the prodigious quantity of excellent fish which the people on board had taken during the absence of the boat.

The cutter, in this expedition, had discovered the bay where we intended to anchor, which we found was to the westward of our present station; and, the next morning, the weather proving favourable, we endeavoured to weigh, in order to proceed thither; but though, on this occasion, we mustered all the strength we could, obliging even the sick, who were scarce able to keep on their legs, to assist us, yet the capstan was so weakly manned that it was near four hours before we hove the cable right up and down: after which, with our utmost efforts, and with many surges and some purchases we made use of to increase our power, we found ourselves incapable of starting the anchor from the ground. However, at noon, as a fresh gale blew towards the bay, we were induced to set the sails, which fortunately tripped the anchor; and then we steered along shore till we came abreast of the point that forms the eastern part of the bay. On the opening of the bay, the wind, that had befriended us thus far, shifted and blew from thence in squalls; but by means of the headway we had got we loofed close in, till the anchor brought us up in fifty-six fathom. Soon after we had thus got to our new birth we discovered a sail, which we made no doubt was one of our squadron; and on its nearer approach we found it to be the *Tryal* sloop. We immediately sent some of our hands on board her, by whose assistance she was brought to an anchor between us and the land. We soon found that the sloop had not been exempted from the same calamities which we had so severely felt; for her commander, Captain Saunders, waiting on the commodore, informed him that out of his small complement he had buried thirty-four of his men; and those that remained were so universally afflicted with the scurvy, that only himself, his lieutenant, and three of his men, were able to stand by the sails. The *Tryal* came to an anchor within us on the 12th, about noon, and we carried our hawsers on board her, in order to moor ourselves nearer in shore; but the wind coming off the land in violent gusts, prevented our mooring in the birth we intended. Indeed our principal attention was employed on business rather of more importance: for we were now extremely occupied in sending on shore materials to raise tents for the reception of the sick, who died apace on board, and doubtless the distemper was considerably augmented by the stench and filthiness in which they lay; for the number of the deceased was so great, and so few could be spared from the necessary duty of the sails to look after them, that it was impossible to avoid a great relaxation in the article of cleanliness, which had rendered the ship extremely loathsome between decks. Notwithstanding our desire of freeing the sick from their hateful situation, and their own extreme impatience to get on shore, we had not hands enough to prepare the tents for their reception before the 16th; but on that and the two following days we sent them all on shore, amounting to a hundred and sixty-seven persons, besides twelve or fourteen who died in the boats, on their being exposed to the fresh air. The greatest part of our sick were so infirm that we were obliged to carry them out of the ship in their hammocks, and to convey them afterwards in the same manner from the water-side to their tents, over a stony beach. This was a work of considerable fatigue to the few who were healthy, and therefore the commodore, according to his accustomed humanity, not only assisted herein with his own labour, but obliged his officers, without distinction, to give their helping hand. The extreme weakness of our sick may in some measure be collected from the numbers who died after they had got on shore; for it had generally been found that the land, and the refreshments it produces, very soon recover most stages of the sea-scurvy; and we flattered ourselves, that those who had not perished on this first exposure to the open air, but had lived to be placed in their tents, would have been speedily restored to their health and vigour. Yet, to our great mortification, it was near twenty days after their landing before the mortality was tolerably ceased; and for the first ten or twelve days we buried rarely less than six each day, and many of those who survived recovered by very slow and insensible degrees. Indeed, those who were well enough at their first getting on shore to creep out of their tents, and crawl about, were soon relieved, and recovered their health and strength in a very short time; but in the rest, the disease seemed to have acquired a degree of inveteracy which was altogether without example.

Having proceeded thus far, and got our sick on shore, I think it necessary, before I enter into any longer detail of our transactions, to give a distinct account of this island of Juan Fernandes, its situation, productions, and all its conveniencies. These particulars we were well enabled to be minutely instructed in during our three months' stay there; and as it is the only commodious place in those seas where British cruisers can refresh and recover their men after their passage round Cape Horn, and where they may remain for some time without alarming the Spanish coast, these its advantages will merit a circumstantial description. Indeed Mr. Anson was particularly industrious in directing the roads and coasts to be surveyed, and other observations to be made, knowing from his own experience of how great consequence these materials might prove to any British vessels hereafter employed in those seas. For the uncertainty we were in of its position, and our standing in for the main on the 28th of May, in order to secure a sufficient easting, when we were indeed extremely near it, cost us the lives of between seventy and eighty of our men, by our longer continuance at sea: from which fatal accident we might have been exempted had we been furnished with such an account of its situation as we could fully have depended on.

The island of Juan Fernandes lies in the latitude of 33° 40' south, and is a hundred and ten leagues distant from the continent of Chili. It is said to have received its name from a Spaniard, who formerly procured a grant of it, and resided there some time with a view of settling on it, but afterwards abandoned it. The island itself is of an irregular figure. Its greatest extent is between four and five leagues, and its greatest breadth somewhat short of two leagues. The only safe anchoring at this island is on the north side, where are the three bays mentioned above, but the middlemost, known by the name of Cumberland Bay, is the widest and deepest, and in all respects much the best; for the other two, denominated the East and West bays, are scarcely more than good landing-places, where boats may conveniently put their casks on shore. Cumberland Bay is well secured to the southward, and is only exposed from the N. by W. to the E. by S.; and as the northerly winds seldom blow in that climate, and never with any violence, the danger from that quarter is not worth attending to.

As the bay last described, or Cumberland Bay, is by far the most commodious road in the island, so it is adviseable for all ships to anchor on the western side of this bay, within little more than two cables' lengths of the beach. Here they may ride in forty fathom of water, and be, in a great measure, sheltered from a large heavy sea which comes rolling in whenever an eastern or a western wind blows. It is however expedient, in this case, to cackle or arm the cables with an iron chain, or good rounding, for five or six fathom from the anchor, to secure them from being rubbed by the foulness of the ground.

I have before observed that a northerly wind, to which alone this bay is exposed, very rarely blew during our stay here; and as it was then winter, it may be supposed, in other seasons, to be less frequent. Indeed, in those few instances when it was in that quarter, it did not blow with any great force; but this perhaps might be owing to the highlands on the southward of the bay, which checked its current, and thereby abated its violence, for we had reason to suppose that a few leagues off it blew with considerable strength, since it sometimes drove before it a prodigious sea, in which we rode forecastle in. But though the northern winds are never to be apprehended, yet the southern winds, which generally prevail here, frequently blow off the land in violent gusts and squalls, which, however, rarely last longer than two or three minutes. This seems to be owing to the obstruction of the southern gale by the hills in the neighbourhood of the bay, for the wind being collected by this means, at last forces its passage through the narrow vallies, which, like so many funnels, both facilitate its escape and increase its violence. These frequent and sudden gusts make it difficult for ships to work in with the wind off shore, or to keep a clear hawse when anchored.

The northern part of this island is composed of high craggy hills, many of them inaccessible, though generally covered with trees. The soil of this part is loose and shallow, so that very large trees on the hills soon perish for want of root, and are then easily overturned; which occasioned the unfortunate death of one of our sailors, who being upon the hills in search of goats, caught hold of a tree upon a declivity to assist him in his ascent, and this giving way, he immediately rolled down the hill, and though in his fall he fastened on another tree of considerable bulk, yet that too gave way, and he fell amongst the rocks, and was dashed to pieces. Mr. Brett likewise met with an accident only by resting his back against a tree, near as large about as himself, which stood on a slope; for the tree giving way, he fell to a considerable distance, though without receiving any injury. Our prisoners (whom, as will be related in the sequel, we afterwards brought in here) remarked that the appearance of the hills in some part of the island resembled that of the mountains in Chili where the gold is found, so that it is not impossible but mines might be discovered here. We observed, in some places, several hills of a peculiar sort of red earth, exceeding vermilion in colour, which perhaps, on examination, might prove useful for many purposes. The southern, or rather the S.W., part of the island is widely different from the rest, being dry, stony, and destitute of trees, and very flat and low compared with the hills on the northern part. This part of the island is never frequented by ships, being surrounded by a steep shore, and having little or no fresh water; and, besides, it is exposed to the southerly wind, which generally blows here the whole year round, and in the winter solstice very hard.

The trees of which the woods on the northern side of the island are composed are most of them aromaticks, and of many different sorts. There are none of them of a size to yield any considerable timber, except the myrtle-trees, which are the largest on the island, and supplied us with all the timber we made use of, but even these would not work to a greater length than forty feet. The top of the myrtle-tree is circular, and appears as uniform and regular as if it had been clipped by art. It bears on its bark an excrescence like moss which in taste and smell resembles garlick, and was used by our people instead of it. We found here too the piemento-tree, and likewise the cabbage-tree, though in no great plenty. And besides a great number of plants of various kinds, which we were not botanists enough either to describe or attend to, we found here almost all the vegetables which are usually esteemed to be particularly adapted to the cure of those scorbutick disorders which are contracted by salt diet and long voyages. For here we had great quantities of water-cresses and purslain, with excellent wild sorrel, and a vast profusion of turnips and Sicilian radishes: these two last, having some resemblance to each other, were confounded by our people under the general name of turnips. We usually preferred the tops of the turnips to the roots, which were often stringy, though some of them were free from that exception, and remarkably good. These vegetables, with the fish and flesh we got here, and which I shall more particularly describe hereafter, were not only extremely grateful to our palates, after the long course of salt diet which we had been confined to, but were likewise of the most salutary consequence to our sick, in recovering and invigorating them, and of no mean service to us who were well, in destroying the lurking seeds of the scurvy, from which perhaps none of us were totally exempt, and in refreshing and restoring us to our wonted strength and activity.

To the vegetables I have already mentioned, of which we made perpetual use, I must add that we found many acres of ground covered with oats and clover. There were also some few cabbage-trees upon the island, as was observed before, but as they generally grew on the precipices, and in dangerous situations, and as it was necessary to cut down a large tree for every single cabbage, this was a dainty that we were able but rarely to indulge in.

The excellence of the climate and the looseness of the soil render this place extremely proper for all kinds of vegetation; for if the ground be anywhere accidentally turned up, it is immediately overgrown with turnips and Sicilian radishes. Mr. Anson therefore having with him garden-seeds of all kinds, and stones of different sorts of fruits, he, for the better accommodation of his countrymen who should hereafter touch here, sowed both lettuces, carrots, and other garden plants, and sett in the woods a great variety of plum, apricot, and peach stones: and these last, he has been informed, have since thriven to a very remarkable degree, for some gentlemen, who in their passage from Lima to Old Spain were taken and brought to England, having procured leave to wait upon Mr. Anson, to thank him for his generosity and humanity to his prisoners, some of whom were their relations, they, in casual discourse with him about his transactions in the South Seas, particularly asked him if he had not planted a great number of fruit trees on the island of Juan Fernandes, for they told him their late navigators had discovered there numbers of peach-trees and apricot-trees, which being fruits before unobserved in that place, they concluded them to have been produced from kernels sett by him.

This may in general suffice as to the soil and vegetable productions of this place; but the face of the country, at least of the north part of the island, is so extremely singular that I cannot avoid giving it a particular consideration. I have already taken notice of the wild, inhospitable air with which it first appeared to us, and the gradual improvement of this uncouth landskip as we drew nearer, till we were at last captivated by the numerous beauties we discovered on the shore. And I must now add that we found, during the time of our residence there, that the inland parts of the island did no ways fall short of the sanguine prepossessions which we first entertained in their favour: for the woods, which covered most of the steepest hills, were free from all bushes and underwood, and afforded an easy passage through every part of them, and the irregularities of the hills and precipicies, in the northern part of the island, necessarily traced out by their various combinations a great number of romantic vallies, most of which had a stream of the clearest water running through them, that tumbled in cascades from rock to rock, as the bottom of the valley, by the course of the neighbouring hills, was at any time broken into a sudden sharp descent. Some particular spots occurred in these vallies where the shade and fragrance of the contiguous woods, the loftiness of the overhanging rocks, and the transparency and frequent falls of the neighbouring streams, presented scenes of such elegance and dignity as would with difficulty be rivalled in any other part of the globe. It is in this place, perhaps, that the simple productions of unassisted nature may be said to excel all the fictitious descriptions of the most animated imagination. I shall finish this article with a short account of that spot where the commodore pitched his tent, and which he made choice of for his own residence, though I despair of conveying an adequate idea of its beauty. The piece of ground which he chose was a small lawn that lay on a little ascent, at the distance of about half a mile from the sea. In the front of his tent there was a large avenue cut through the woods to the seaside, which sloping to the water with a gentle descent, opened a prospect of the bay and the ships at anchor. This lawn was screened behind by a tall wood of myrtle sweeping round it, in the form of a theatre, the slope on which the wood stood, rising with a much sharper ascent than the lawn itself, though not so much but that the hills and precipices within land towered up considerably above the tops of the trees, and added to the grandeur of the view. There were, besides, two streams of crystal water, which ran on the right and left of the tent, within an hundred yards distance, and were shaded by the trees which skirted the lawn on either side, and compleated the symmetry of the whole.

It remains now only that we speak of the animals and provisions which we met with at this place. Former writers have related that this island abounded with vast numbers of goats, and their accounts are not to be questioned, this place being the usual haunt of the buccaneers and privateers who formerly frequented those seas. And there are two instances—one of a Musquito Indian, and the other of Alexander Selkirk, a Scotchman, who were left here by their respective ships, and lived alone upon this island for some years, and consequently were no strangers to its produce. Selkirk, who was the last, after a stay of between four and five years, was taken off the place by the *Duke* and *Duchess* privateers of Bristol, as may be seen at large in the journal of their voyage. His manner of life during his solitude was in most particulars very remarkable; but there is one circumstance he relates, which was so strangely verified by our own observation, that I cannot help reciting it. He tells us, amongst other things, that as he often caught more goats than he wanted, he sometimes marked their ears and let them go. This was about thirty-two years before our arrival at the island. Now it happened that the first goat that was killed by our people at their landing had his ears slit, whence we concluded that he had doubtless been formerly under the power of Selkirk. This was indeed an animal of a most venerable aspect, dignified with an exceeding majestic beard, and with many other symptoms of antiquity. During our stay on the island we met with others marked in the same manner, all the males being distinguished by an exuberance of beard and every other characteristick of extreme age.

But the great numbers of goats, which former writers describe to have been found upon this island, are at present very much diminished, as the Spaniards, being informed of the advantages which the buccaneers and privateers drew from the provisions which goat's flesh here furnished them with, have endeavoured to extirpate the breed, thereby to deprive their enemies of this relief. For this purpose they have put on shore great numbers of large dogs, who have increased apace and have destroyed all the goats in the accessible part of the country, so that there now remain only a few amongst the craggs and precipices, where the dogs cannot follow them. These are divided into separate herds of twenty or thirty each, which inhabit distinct fastnesses, and never mingle with each other. By this means we found it extremely difficult to kill them, and yet we were so desirous of their flesh, which we all agreed much resembled venison, that we got knowledge, I believe, of all their herds, and it was conceived, by comparing their numbers together, that they scarcely exceeded two hundred upon the whole island. I remember we had once an opportunity of observing a remarkable dispute betwixt a herd of these animals and a number of dogs; for going in our boat into the eastern bay, we perceived some dogs running very eagerly upon the foot, and being willing to discover what game they were after, we lay upon our oars some time to view them, and at last saw them take to a hill, where looking a little further, we observed upon the ridge of it an herd of goats, which seemed drawn up for their reception. There was a very narrow path skirted on each side by precipices, on which the master of the herd posted himself fronting the enemy, the rest of the goats being all behind him, where the ground was more open. As this spot was inaccessible by any other path, excepting where this champion had placed himself, the dogs, though they ran up-hill with great alacrity, yet when they came within about twenty yards of him, they found they durst not encounter him (for he would infallibly have driven them down the precipice), but gave over the chace, and quietly laid themselves down panting at a great rate. These dogs, who are masters of all the accessible parts of the island, are of various kinds, some of them very large, and are multiplied to a prodigious degree. They sometimes came down to our habitations at night, and stole our provision; and once or twice they set upon single persons, but assistance being at hand they were driven off without doing any mischief. As at present it is rare for goats to fall in their way, we conceived that they lived principally upon young seals, and indeed some of our people had the curiosity to kill dogs sometimes and dress them, and it seemed to be agreed that they had a fishy taste.

Goat's flesh, as I have mentioned, being scarce, we rarely being able to kill above one a day, and our people growing tired of fish (which, as I shall hereafter observe, abound at this place), they at last condescended to eat seals, which by degrees they came to relish, and called it lamb. The seal, numbers of which haunt this island, hath been so often mentioned by former writers that it is unnecessary to say anything particular about them in this place. But there is another amphibious creature to be met with here, called a sea-lion, that bears some resemblance to a seal, though it is much larger. This too we eat under the denomination of beef; and as it is so extraordinary an animal, I conceive it well merits a particular description. They are in size, when arrived at their full growth, from twelve to twenty feet in length, and from eight to fifteen in circumference. They are extremely fat, so that after having cut through the skin, which is about an inch in thickness, there is at least a foot of fat before you can come at either lean or bones; and we experienced more than once that the fat of some of the largest afforded us a butt of oil. They are likewise very full of blood, for if they are deeply wounded in a dozen places, there will instantly gush out as many fountains of blood, spouting to a considerable distance; and to try what quantity of blood they contained, we shot one first, and then cut its throat, and measuring the blood that came from him, we found that besides what remained in the vessels, which to be sure was considerable, we got at least two hogsheads. Their skins are covered with short hair, of a light dun colour, but their tails and their fins, which serve them for feet on shore, are almost black; their fins or feet are divided at the ends like fingers, the web which joins them not reaching to the extremities, and each of these fingers is furnished with a nail. They have a distant resemblance to an overgrown seal, though in some particulars there is a manifest difference between them, especially in the males. These have a large snout or trunk hanging down five or six inches below the end of the upper jaw, which the females have not, and this renders the countenance of the male and female easy to be distinguished from each other, and besides, the males are of a much larger size. The largest of these animals, which was found upon the island, was the master of the flock, and from his driving off the other males, and keeping a great number of females to himself, he was by the seamen ludicrously styled the Bashaw. These animals divide their time equally between the land and sea, continuing at sea all the summer, and coming on shore at the setting in of the winter, where they reside during that whole season. In this interval they engender and bring forth their young, and have generally two at a birth, which they suckle with their milk, they being at first about the size of a full-grown seal. During the time these sea-lions continue on shore, they feed on the grass and verdure which grows near the banks of the fresh-water streams; and, when not employed on feeding, sleep in herds in the most miry places they can find out. As they seem to be of a very lethargic disposition, and are not easily awakened, each herd was observed to place some of their males at a distance, in the nature of sentinels, who never failed to alarm them whenever any one attempted to molest or even to approach them; and they were very capable of alarming, even at a considerable distance, for the noise they make is very loud, and of different kinds, sometimes grunting like hogs, and at other times snorting like horses in full vigour. They often, especially the males, have furious battles with each other, principally about their females; and we were one day extremely surprized by the sight of two animals, which at first appeared different from all we had ever observed, but, on a nearer approach, they proved to be two sea-lions, who had been goring each other with their teeth, and were covered over with blood. And the Bashaw before mentioned, who generally lay surrounded with a seraglio of females, which no other male dared to approach, had not acquired that envied pre-eminence without many bloody contests, of which the marks still remained in the numerous scars which were visible in every part of his body. We killed many of them for food, particularly for their hearts and tongues, which we esteemed exceeding good eating, and preferable even to those of bullocks. In general there was no difficulty in killing them, for they were incapable either of escaping or resisting, as their motion is the most unwieldy that can be conceived, their blubber, all the time they are moving, being agitated in large waves under their skins. However, a sailor one day being carelessly employed in skinning a young sea-lion, the female from whence he had taken it came upon him unperceived, and getting his head in her mouth, she with her teeth scored his skull in notches in many places, and thereby wounded him so desperately, that, though all possible care was taken of him, he died in a few days.

These are the principal animals which we found upon the island: for we saw but few birds, and those chiefly hawks, blackbirds, owls, and humming birds. We saw not the pardela, which burrows in the ground, and which former writers have mentioned to be found here; but as we often met with their holes, we supposed that the dogs had destroyed them, as they have almost done the cats: for these were very numerous in Selkirk's time, but we saw not above one or two during our whole stay. However, the rats still keep their ground, and continue here in great numbers, and were very troublesome to us by infesting our tents nightly.

But that which furnished us with the most delicious repasts at this island remains still to be described. This was the fish with which the whole bay was most plentifully stored, and with the greatest variety: for we found here cod of a prodigious size, and by the report of some of our crew, who had been formerly employed in the Newfoundland fishery, not in less plenty than is to be met with on the banks of that island. We caught also cavallies, gropers, large breams, maids, silver fish, congers of a peculiar kind, and above all, a black fish which we most esteemed, called by some a chimney-sweeper, in shape resembling a carp. The beach indeed is everywhere so full of rocks and loose stones that there is no possibility of haling the Seyne; but with hooks and lines we caught what numbers we pleased, so that a boat with two or three lines would return loaded with fish in about two or three hours' time. The only interruption we ever met with arose from great quantities of dogfish and large sharks, which sometimes attended our boats and prevented our sport. Besides the fish we have already mentioned, we found here one delicacy in greater perfection, both as to size, flavour, and quantity, than is perhaps to be met with in any other part of the world. This was sea crayfish; they generally weighed eight or nine pounds apiece, were of a most excellent taste, and lay in such abundance near the water's edge, that the boat-hooks often struck into them in putting the boat to and from the shore.

These are the most material articles relating to the accommodations, soil, vegetables, animals, and other productions of the island of Juan Fernandes: by which it must appear how properly that place was adapted for recovering us from the deplorable situation to which our tedious and unfortunate navigation round Cape Horn had reduced us. And having thus given the reader some idea of the site and circumstances of this place, which was to be our residence for three months, I shall now proceed in the next chapter to relate all that occurred to us in that interval, resuming my narration from the 18th day of June, being the day in which the *Tryal* sloop, having by a squall been driven out to sea three days before, came again to her moorings, the day in which we finished the sending our sick on shore, and about eight days after our first anchoring at this island.

CHAPTER II

THE ARRIVAL OF THE "GLOUCESTER" AND THE "ANNA" PINK AT THE ISLAND OF JUAN FERNANDES, AND THE TRANSACTIONS AT THAT PLACE DURING THIS INTERVAL

The arrival of the *Tryal* sloop at this island so soon after we came there ourselves gave us great hopes of being speedily joined by the rest of the squadron, and we were for some days continually looking out, in expectation of their coming in sight. But near a fortnight being elapsed without any of them having appeared, we began to despair of ever meeting them again, as we knew that if our ship continued so much longer at sea we should every man of us have perished, and the vessel, occupied by dead bodies only, would have been left to the caprice of the winds and waves: and this we had great reason to fear was the fate of our consorts, as each hour added to the probability of these desponding suggestions.

But on the 21st of June, some of our people, from an eminence on shore, discerned a ship to leeward, with her courses even with the horizon; and they, at the same time, particularly observed that she had no sail abroad except her courses and her main top-sail. This circumstance made them conclude that it was one of our squadron, which had probably suffered in her sails and rigging as severely as we had done: but they were prevented from forming more definitive conjectures about her, for, after viewing her for a short time, the weather grew thick and hazy, and they lost sight of her. On this report, and no ship appearing for some days, we were all under the greatest concern, suspecting that her people were in the utmost distress for want of water, and so diminished and weakened by sickness as not to be able to ply up to windward; so that we feared that, after having been in sight of the island, her whole crew would notwithstanding perish at sea. However, on the 26th, towards noon, we discerned a sail in the north-east quarter, which we conceived to be the very same ship that had been seen before, and our conjectures proved true: and about one o'clock she approached so near that we could distinguish her to be the *Gloucester*. As we had no doubt of her being in great distress, the commodore immediately ordered his boat to her assistance, laden with fresh water, fish, and vegetables, which was a very seasonable relief to them; for our apprehensions of their calamities appeared to be but too well grounded, as perhaps there never was a crew in a more distressed situation. They had already thrown overboard two-thirds of their complement, and of those which remained alive, scarcely any were capable of doing duty, except the officers and their servants. They had been a considerable time at the small allowance of a pint of fresh water to each man for twenty-four hours, and yet they had so little left, that, had it not been for the supply we sent them, they must soon have died of thirst. The ship plied in within three miles of the bay; but, the winds and currents being contrary, she could not reach the road. However, she continued in the offing the next day; but as she had no chance of coming to an anchor, unless the winds and currents shifted, the commodore repeated his assistance, sending to her the *Tryal's* boat manned with the *Centurion's* people, and a farther supply of water and other refreshments. Captain Mitchel, the captain of the *Gloucester*, was under a necessity of detaining both this boat and that sent the preceding day; for without the help of their crews he had no longer strength enough to navigate the ship. In this tantalizing situation the *Gloucester* continued for near a fortnight, without being able to fetch the road, though frequently attempting it, and at some times bidding very fair for it. On the 9th of July, we observed her stretching away to the eastward at a considerable distance, which we supposed was with a design to get to the southward of the island; but as we soon lost sight of her, and she did not appear for near a week, we were prodigiously concerned, knowing that she must be again in extreme distress for want of water. After great impatience about her, we discovered her again on the 16th endeavouring to come round the eastern point of the island; but the wind, still blowing directly from the bay, prevented her getting nearer than within four leagues of the land. On this, Captain Mitchel made signals of distress, and our long-boat was sent to him with a store of water, and plenty of fish, and other refreshments. And the long-boat being not to be spared, the cockswain had positive orders from the commodore to return again immediately; but the weather proving stormy the next day, and the boat not appearing, we much feared she was lost, which would have proved an irretrievable misfortune to us all: however, the third day after, we were relieved from this anxiety by the joyful sight of the long-boat's sails upon the water; on which we sent the cutter immediately to her assistance, who towed her alongside in a few hours, when we found that the crew of our long-boat had taken in six of the *Gloucester's* sick men to bring them on shore, two of which had died in the boat. We now learnt that the *Gloucester* was in a most dreadful condition, having scarcely a man in health on board, except those they received from us: and, numbers of their sick dying daily, it appeared that, had it not been for the last supply sent by our long-boat, both the healthy and diseased must have all perished together for want of water. These calamities were the more terrifying as they appeared to be without remedy: for the *Gloucester* had already spent a month in her endeavours to fetch the bay, and she was now no farther advanced than at the first moment she made the island; on the contrary, the people on board her had worn out all their hopes of ever succeeding in it, by the many experiments they had made of its difficulty. Indeed, the same day her situation grew more desperate than ever, for after she had received our last supply of refreshments, we again lost sight of her; so that we in general despaired of her ever coming to an anchor.

Thus was this unhappy vessel bandied about within a few leagues of her intended harbour, whilst the neighbourhood of that place and of those circumstances which could alone put an end to the calamities they laboured under, served only to aggravate their distress by torturing them with a view of the relief it was not in their power to reach. But she was at last delivered from this dreadful situation at a time when we least expected it; for after having lost sight of her for several days, we were pleasingly surprized, on the morning of the 23d of July, to see her open the N.W. point of the bay with a flowing sail, when we immediately dispatched what boats we had to her assistance, and in an hour's time from our first perceiving her, she anchored safe within us in the bay. And now we were more particularly convinced of the importance of the assistance and refreshments we so often sent them, and how impossible it would have been for a man of them to have survived had we given less attention to their wants; for notwithstanding the water, the greens, and fresh provisions which we supplied them with, and the hands we sent them to navigate the ship, by which the fatigue of their own people was diminished, their sick relieved, and the mortality abated; notwithstanding this indulgent care of the commodore, they yet buried above three-fourths of their crew, and a very small proportion of the remainder were capable of assisting in the duty of the ship. On their coming to an anchor, our first endeavours were to assist them in mooring, and our next to send their sick on shore. These were now reduced by deaths to less than fourscore, of which we expected to lose the greatest part; but whether it was that those farthest advanced in the distemper were all dead, or that the greens and fresh provisions we had sent on board had prepared those which remained for a more speedy recovery, it happened, contrary to our expectations, that their sick were in general relieved and restored to their strength in a much shorter time than our own had been when we first came to the island, and very few of them died on shore.

I have thus given an account of the principal events relating to the arrival of the *Gloucester* in one continued narration. I shall only add, that we never were joined by any other of our ships, except our victualler, the *Anna* pink, who came in about the middle of August, and whose history I shall defer for the present, as it is now high time to return to the account of our own transactions on board and on shore, during the interval of the *Gloucester's* frequent and ineffectual attempts to reach the island.

Our next employment, after sending our sick on shore from the *Centurion*, was cleansing our ship and filling our water. The first of these measures was indispensably necessary to our future health, as the numbers of sick, and the unavoidable negligence arising from our deplorable situation at sea, had rendered the decks most intolerably loathsome. And the filling our water was a caution that appeared not less essential to our security, as we had reason to apprehend that accidents might intervene which would oblige us to quit the island at a very short warning; for some appearances we had discovered on shore upon our first landing, gave us grounds to believe that there were Spanish cruisers in these seas, which had left the island but a short time before our arrival, and might possibly return thither again, either for a recruit of water, or in search of us, since we could not doubt but that the sole business they had at sea was to intercept us, and we knew that this island was the likeliest place, in their own opinion, to meet with us. The circumstances which gave rise to these reflections (in part of which we were not mistaken, as shall be observed more at large hereafter) were our finding on shore several pieces of earthen jars, made use of in those seas for water and other liquids, which appeared to be fresh broken: we saw, too, many heaps of ashes, and near them fish-bones and pieces of fish, besides whole fish scattered here and there, which plainly appeared to have been but a short time out of the water, as they were but just beginning to decay. These were certain indications that there had been ships at this place but a short time before we came there; and as all Spanish merchantmen are instructed to avoid the island, on account of its being the common rendezvous of their enemies, we concluded those who had touched here to be ships of force; and not knowing that Pizarro was returned to Buenos Ayres, and ignorant what strength might have been fitted out at Callao, we were under some concern for our safety, being in so wretched and enfeebled a condition, that notwithstanding the rank of our ship, and the sixty guns she carried on board, which would only have aggravated our dishonour, there was scarcely a privateer sent to sea that was not an over-match for us. However, our fears on this head proved imaginary, and we were not exposed to the disgrace which might have been expected to have befallen us, had we been necessitated (as we must have been, had the enemy appeared) to fight our sixty-gun ship with no more than thirty hands.

Whilst the cleaning our ship and the filling our water went on, we set up a large copper oven on shore near the sick tents, in which we baked bread every day for our ship's company; for being extremely desirous of recovering our sick as soon as possible, we conceived that new bread, added to their greens and fresh fish, might prove a powerful article in their relief. Indeed we had all imaginable reason to endeavour at the augmenting our present strength, as every little accident, which to a full crew would be insignificant, was extremely alarming in our present helpless situation. Of this we had a troublesome instance on the 30th of June; for at five in the morning, we were astonished by a violent gust of wind directly off shore, which instantly parted our small bower cable about ten fathom from the ring of the anchor: the ship at once swung off to the best bower, which happily stood the violence of the jerk, and brought us up with two cables in eighty fathom. At this time we had not above a dozen seamen in the ship, and we were apprehensive, if the squall continued, that we should be driven to sea in this wretched condition. However, we sent the boat on shore to bring off all who were capable of acting; and the wind soon abating of its fury, gave us an opportunity of receiving the boat back again with a reinforcement. With this additional strength we immediately went to work to heave in what remained of the cable, which we suspected had received some damage from the foulness of the ground before it parted; and agreeable to our conjecture, we found that seven fathom and a half of the outer end had been rubbed, and rendered unserviceable. In the afternoon we bent the cable to the spare anchor, and got it over the ship's side; and the next morning, July 1, being favoured with the wind in gentle breezes, we warped the ship in again, and let go the anchor in forty-one fathom; the easternmost point now bearing from us E.½S.; the westermost N.W. by W.; and the bay as before, S.S.W.; a situation in which we remained secure for the future. However, we were much concerned for the loss of our anchor, and swept frequently for it, in hopes to have recovered it; but the buoy having sunk at the very instant that the cable parted, we were never able to find it.

And now as we advanced in July, some of our men being tolerably recovered, the strongest of them were put upon cutting down trees, and splitting them into billets; while others, who were too weak for this employ, undertook to carry the billets by one at a time to the water-side: this they performed, some of them with the help of crutches, and others supported by a single stick. We next sent the forge on shore, and employed our smiths, who were but just capable of working, in mending our chain-plates, and our other broken and decayed iron-work. We began too the repairs of our rigging; but as we had not junk enough to make spun-yarn, we deferred the general overhale, in hopes of the daily arrival of the *Gloucester*, who we knew had a great quantity of junk on board. However, that we might dispatch as fast as possible in our refitting, we set up a large tent on the beach for the sail-makers; and they were immediately employed in repairing our old sails, and making us new ones. These occupations, with our cleansing and watering the ship (which was by this time pretty well compleated), the attendance on our sick, and the frequent relief sent to the *Gloucester*, were the principal transactions of our infirm crew till the arrival of the *Gloucester* at an anchor in the bay. And then Captain Mitchel waiting on the commodore, informed him that he had been forced by the winds, in his last absence, as far as the small island called Masa Fuero, lying about twenty-two leagues to the westward of Juan Fernandes; and that he endeavoured to send his boat on shore there for water, of which he could observe several streams, but the wind blew so strong upon the shore, and occasioned such a surf, that it was impossible for the boat to land, though the attempt was not altogether useless, for his people returned with a boatload of fish. This island had been represented by former navigators as a barren rock; but Captain Mitchel assured the commodore that it was almost everywhere covered with trees and verdure, and was near four miles in length; and added, that it appeared to him far from impossible but some small bay might be found on it which might afford sufficient shelter for any ship desirous of refreshing there.

As four ships of our squadron were missing, this description of the island of Masa Fuero gave rise to a conjecture that some of them might possibly have fallen in with that island, and might have mistaken it for the true place of our rendezvous. This suspicion was the more plausible as we had no draught of either island that could be relied on: and therefore Mr. Anson determined to send the *Tryal* sloop thither, as soon as she could be fitted for the sea, in order to examine all its bays and creeks, that we might be satisfied whether any of our missing ships were there or not. For this purpose, some of our best hands were sent on board the *Tryal* the next morning, to overhale and fix her rigging; and our long-boat was employed in compleating her water, and whatever stores and necessaries she wanted were immediately supplied either from the *Centurion* or the *Gloucester*. But it was the 4th of August before the *Tryal* was in readiness to sail, when, having weighed, it soon after fell calm, and the tide set her very near the eastern shore. Captain Saunders hung out lights and fired several guns to acquaint us with his danger; upon which all the boats were sent to his relief, who towed the sloop into the bay, where she anchored until the next morning, and then weighing again, proceeded on her cruize with a fair breeze.

And now, after the *Gloucester's* arrival, we were employed in earnest in examining and repairing our rigging; but in the stripping our foremast, we were alarmed by discovering it was sprung just above the partners of the upper deck. The spring was two inches in depth, and twelve in circumference; however, the carpenters, on inspecting it, gave it as their opinion that fishing it with two leaves of an anchor-stock would render it as secure as ever. But, besides this defect in our mast, we had other difficulties in refitting, from the want of cordage and canvas; for though we had taken to sea much greater quantities of both than had ever been done before, yet the continued bad weather we met with had occasioned such a consumption of these stores that we were driven to great straits, as after working up all our junk and old shrouds to make twice-laid cordage, we were at last obliged to unlay a cable to work into running rigging. And with all the canvas, and remnants of old sails that could be mustered, we could only make up one compleat suit.

Towards the middle of August, our men being indifferently recovered, they were permitted to quit their sick tents, and to build separate huts for themselves, as it was imagined that by living apart they would be much cleanlier, and consequently likely to recover their strength the sooner; but at the same time particular orders were given, that on the firing of a gun from the ship they should instantly repair to the water-side. Their employment on shore was now either the procuring of refreshments, the cutting of wood, or the making of oil from the blubber of the sea-lions. This oil served us for several purposes, as burning in lamps, or mixing with pitch to pay the ship's sides, or, when worked up with wood-ashes, to supply the use of tallow (of which we had none left) to give the ship boot-hose tops. Some of the men too were occupied in salting of cod; for there being two Newfoundland fishermen in the *Centurion*, the commodore set them about laying in a considerable quantity of salted cod for a sea-store, though very little of it was used, as it was afterwards thought to be as productive of the scurvy as any other kind of salt provisions.

I have before mentioned that we had a copper oven on shore to bake bread for the sick; but it happened that the greatest part of the flour for the use of the squadron was embarked on board our victualler the *Anna* pink: and I should have mentioned that the *Tryal* sloop, at her arrival, had informed us, that on the 9th of May she had fallen in with our victualler not far distant from the continent of Chili, and had kept company with her for four days, when they were parted in a hard gale of wind. This afforded us some room to hope that she was safe, and that she might join us; but all June and July being past without any news of her, we then gave her over for lost, and at the end of July, the commodore ordered all the ships to a short allowance of bread. Nor was it in our bread only that we feared a deficiency; for since our arrival at this island we discovered that our former purser had neglected to take on board large quantities of several kinds of provisions which the commodore had expressly ordered him to receive; so that the supposed loss of our victualler was on all accounts a mortifying consideration. However, on Sunday, the 16th of August, about noon, we espied a sail in the northern quarter, and a gun was immediately fired from the *Centurion* to call off the people from shore, who readily obeyed the summons, repairing to the beach, where the boats waited to carry them on board. And being now prepared for the reception of this ship in view, whether friend or enemy, we had various speculations about her; at first, many imagined it to be the *Tryal* sloop returned from her cruize, though as she drew nearer this opinion was confuted, by observing she was a vessel with three masts. Then other conjectures were eagerly canvassed, some judging it to be the *Severn*, others the *Pearl*, and several affirming that it did not belong to our squadron: but about three in the afternoon our disputes were ended by an unanimous persuasion that it was our victualler the *Anna* pink. This ship, though, like the *Gloucester*, she had fallen in to the northward of the island, had yet the good fortune to come to an anchor in the bay, at five in the afternoon. Her arrival gave us all the sincerest joy; for each ship's company was immediately restored to their full allowance of bread, and we were now freed from the apprehensions of our provisions falling short before we could reach some amicable port; a calamity which in these seas is of all others the most irretrievable. This was the last ship that joined us; and the dangers she encountered, and the good fortune which she afterwards met with, being matters worthy of a separate narration, I shall refer them, together with a short account of the other missing ships of the squadron, to the ensuing chapter.

CHAPTER III

A SHORT NARRATIVE OF WHAT BEFEL THE "ANNA" PINK BEFORE SHE JOINED US, WITH AN ACCOUNT OF THE LOSS OF THE "WAGER," AND OF THE PUTTING BACK OF THE "SEVERN" AND "PEARL," THE TWO REMAINING SHIPS OF THE SQUADRON

On the first appearance of the *Anna* pink, it seemed wonderful to us how the crew of a vessel, which came to this rendezvous two months after us, should be capable of working their ship in the manner they did, with so little appearance of debility and distress. But this difficulty was soon solved when she came to an anchor; for we then found that they had been in harbour since the middle of May, which was near a month before we arrived at Juan Fernandes: so that their sufferings (the risk they had run of shipwreck only excepted) were greatly short of what had been undergone by the rest of the squadron. It seems, on the 16th of May, they fell in with the land, which was then but four leagues distant, in the latitude of 45° 15' south. On the first sight of it they wore ship and stood to the southward, but their fore top-sail splitting, and the wind being W.S.W., they drove towards the shore; and the captain at last, either unable to clear the land, or, as others say, resolved to keep the sea no longer, steered for the coast, with a view of discovering some shelter amongst the many islands which then appeared in sight: and about four hours after the first view of the land, the pink had the good fortune to come to an anchor to the eastward of the island of Inchin; but as they did not run sufficiently near to the east shore of that island, and had not hands enough to veer away the cable briskly, they were soon driven to the eastward, deepning their water from twenty-five fathom to thirty-five, and still continuing to drive, they, the next day, the 17th of May, let go their sheet-anchor. This, though it brought them up for a short time, yet, on the 18th, they drove again, till they came into sixty-five fathom water, and were now within a mile of the land, and expected to be forced on shore every moment, in a place where the coast was so very high and steep, too, that there was not the least prospect of saving the ship or cargo. As their boats were very leaky, and there was no appearance of a landing-place, the whole crew, consisting of sixteen men and boys, gave themselves over for lost, apprehending that if any of them by some extraordinary chance should get on shore, they would, in all probability, be massacred by the savages on the coast: for these, knowing no other Europeans but Spaniards, it might be expected they would treat all strangers with the same cruelty which they had so often and so signally exerted against their Spanish neighbours. Under these terrifying circumstances, the pink drove nearer and nearer to the rocks which formed the shore; but at last, when the crew expected each instant to strike, they perceived a small opening in the land, which raised their hopes; and immediately cutting away their two anchors, they steered for it, and found it to be a small channel betwixt an island and the main, that led them into a most excellent harbour, which, for its security against all winds and swells, and the smoothness of its water, may perhaps compare with any in the known world. And this place being scarcely two miles distant from the spot where they deemed their destruction inevitable, the horrors of shipwreck and of immediate death, which had so long and so strongly possessed them, vanished almost instantaneously, and gave place to the more joyous ideas of security, refreshment, and repose.

In this harbour, discovered in this almost miraculous manner, the pink came to an anchor in twenty-five fathom water, with only a hawser and a small anchor of about three hundredweight. Here she continued for near two months, and here her people, who were many of them ill of the scurvy, were soon restored to perfect health by the fresh provisions, of which they procured good store, and the excellent water with which the adjacent shore abounded. As this place may prove of the greatest importance to future navigators, who may be forced upon this coast by the westerly winds, which are almost perpetual in that part of the world, I shall, before I enter into any farther particulars of the adventures of the pink, give the best account I could collect of this port, its situation, conveniencies, and productions.

The latitude of this harbour, which is indeed a material point, is not well ascertained, the pink having no observation either the day before she came here, or within a day of her leaving it. But it is supposed that it is not very distant from 45° 30' south, and the large extent of the bay before the harbour renders this uncertainty of less moment. The island of Inchin lying before the bay is thought to be one of the islands of Chonos which are mentioned in the Spanish accounts as spreading all along that coast, and are said by them to be inhabited by a barbarous people, famous for their hatred of the Spaniards, and for their cruelties to such of that nation as have fallen into their hands; and it is possible too that the land, on which the harbour itself lies, may be another of those islands, and that the continent may be considerably farther to the eastward. The depths of water in the different parts of the port, and the channels by which it communicates with the bay, are sufficiently marked. But it must be remembered that there are two coves in it, where ships may conveniently heave down, the water being constantly smooth; and there are several fine runs of excellent fresh water which fall into the harbour, some of them so luckily situated that the casks may be filled in the long-boat with an hose. The most remarkable of these is the stream in the N.E. part of the port. This is a fresh-water river, where the pink's people got some few mullets of an excellent flavour, and they were persuaded that in a proper season (it being winter when they were there) it abounded with fish. The principal refreshments they met with in this port were greens, as wild celery, nettle-tops, etc. (which after so long a continuance at sea they devoured with great eagerness); shell-fish, as cockles and muscles of an extraordinary size, and extremely delicious; and good store of geese, shags, and penguins. The climate, though it was the depth of winter, was not remarkably rigorous, nor the trees and the face of the country destitute of verdure, whence in the summer many other species of fresh provisions, besides these here enumerated, might doubtless be found there. Notwithstanding the tales of the Spanish historians in relation to the violence and barbarity of the inhabitants, it doth not appear that their numbers are sufficient to give the least jealousy to any ship of ordinary force, or that their disposition is by any means so mischievous or merciless as hath hitherto been represented. With all these advantages, this place is so far removed from the Spanish frontier, and so little known to the Spaniards themselves, that there is reason to suppose that by proper precautions a ship might continue here undiscovered a long time. It is moreover a post of great defence, for by possessing the island that closes up the harbour, and which is accessible in very few places, a small force might secure this port against all the strength the Spaniards could muster in that part of the world, since this island towards the harbour is steep too, and has six fathom water close to the shore, so that the pink anchored within forty yards of it. Whence it is obvious how impossible it would prove, either to board or to cut out any vessel protected by a force posted on shore within pistol-shot, and where those who were thus posted could not themselves be attacked. All these circumstances seem to render this port worthy of a more accurate examination; and it is to be hoped that the important uses which this rude account of it seems to suggest may hereafter recommend it to the consideration of the public, and to the attention of those who are more immediately entrusted with the conduct of our naval affairs.

After this description of the place where the pink lay for two months, it may be expected that I should relate the discoveries made by the crew on the adjacent coast, and the principal incidents during their stay there; but here I must observe, that, being only a few in number, they did not dare to detach any of their people on distant searches, for they were perpetually terrified with the apprehension that they should be attacked either by the Spaniards or the Indians; so that their excursions were generally confined to that tract of land which surrounded the port, and where they were never out of view of the ship. Though had they at first known how little foundation there was for these fears, yet the country in the neighbourhood was so grown up with wood, and traversed with mountains, that it appeared impracticable to penetrate it: whence no account of the inland parts could be expected from them. Indeed they were able to disprove the relations given by Spanish writers, who have represented this coast as inhabited by a fierce and powerful people, for they were certain that no such inhabitants were there to be found, at least during the winter season, since all the time they continued there, they saw no more than one Indian family, which came into the harbour in a periagua, about a month after the arrival of the pink, and consisted of an Indian near forty years old, his wife, and two children, one three years of age and the other still at the breast. They seemed to have with them all their property, which was a dog and a cat, a fishing-net, a hatchet, a knife, a cradle, some bark of trees intended for the covering of a hut, a reel, some worsted, a flint and steel, and a few roots of a yellow hue and a very disagreeable taste which served them for bread. The master of the pink, as soon as he perceived them, sent his yawl, who brought them on board; and fearing, lest they might discover him, if they were permitted to go away, he took, as he conceived, proper precautions for securing them, but without any mixture of ill usage or violence, for in the daytime they were permitted to go where they pleased about the ship, but at night were locked up in the forecastle. As they were fed in the same manner with the rest of the crew, and were often indulged with brandy, which they seemed greatly to relish, it did not at first appear that they were much dissatisfied with their situation, especially as the master took the Indian on shore when he went a shooting (who always seemed extremely delighted when the master killed his game), and as all the crew treated them with great humanity; but it was soon perceived, that though the woman continued easy and chearful, yet the man grew pensive and restless at his confinement. He seemed to be a person of good natural parts, and though not capable of conversing with the pink's people, otherwise than by signs, was yet very curious and inquisitive, and shewed great dexterity in the manner of making himself understood. In particular, seeing so few people on board such a large ship, he let them know that he supposed they were once more numerous; and to represent to them what he imagined was become of their companions, he laid himself down on the deck closing his eyes, and stretching himself out motionless, to imitate the appearance of a dead body. But the strongest proof of his sagacity was the manner of his getting away, for after being in custody on board the pink eight days, the scuttle of the forecastle, where he and his family were locked up every night, happened to be unnailed, and the following night being extremely dark and stormy, he contrived to convey his wife and children through the unnailed scuttle, and then over the ship's side into the yawl; and to prevent being pursued, he cut away the long-boat and his own periagua, which were towing astern, and immediately rowed ashore. All this he conducted with so much diligence and secrecy, that though there was a watch on the quarter-deck with loaded arms, yet he was not discovered by them till the noise of his oars in the water, after he had put off from the ship, gave them notice of his escape; and then it was too late either to prevent him, or to pursue him, for, their boats being all adrift, it was a considerable time before they could contrive the means of getting on shore themselves to search for their boats. The Indian, too, by this effort, besides the recovery of his liberty, was in some sort revenged on those who had confined him, both by the perplexity they were involved in from the loss of their boats, and by the terror he threw them in at his departure, for on the first alarm of the watch, who cried out, "the Indians!" the whole ship was in the utmost confusion, believing themselves to be boarded by a fleet of armed periaguas.

The resolution and sagacity with which the Indian behaved upon this occasion, had it been exerted on a more extensive object than the retrieving the freedom of a single family, might perhaps have immortalized the exploit, and have given him a rank amongst the illustrious names of antiquity. Indeed his late masters did so much justice to his merit as to own that it was a most gallant enterprize, and that they were grieved they had ever been necessitated, by their attention to their own safety, to abridge the liberty of a person of whose prudence and courage they had now such a distinguished proof. As it was supposed by some of them that he still continued in the woods in the neighbourhood of the port, where it was feared he might suffer for want of provisions, they easily prevailed upon the master to leave a quantity of such food as they thought would be most agreeable to him in a particular part where they imagined he would be likely to find it, and there was reason to conjecture that this piece of humanity was not altogether useless to him, for, on visiting the place some time after, it was found that the provision was gone, and in a manner that made them conclude it had fallen into his hands.

But, however, though many of them were satisfied that this Indian still continued near them, yet others would needs conclude that he was gone to the island of Chiloe, where they feared he would alarm the Spaniards, and would soon return with a force sufficient to surprize the pink. On this occasion the master of the pink was prevailed on to omit firing the evening gun; for it must be remembered (and there is a particular reason hereafter for attending to this circumstance) that the master, from an ostentatious imitation of the practice of men-of-war, had hitherto fired a gun every evening at the setting of the watch. This he pretended was to awe the enemy, if there was any within hearing, and to convince them that the pink was always on her guard, but it being now represented to him that his great security was his concealment, and that the evening gun might possibly discover him and serve to guide the enemy to him, he was prevailed on to omit it for the future; and his crew being now well refreshed, and their wood and water sufficiently replenished, he, in a few days after the escape of the Indian, put to sea, and had a fortunate passage to the rendezvous at the island of Juan Fernandes, where he arrived on the 16th of August, as hath been already mentioned in the preceding chapter.

This vessel, the *Anna* pink, was, as I have observed, the last that joined the commodore at Juan Fernandes. The remaining ships of the squadron were the *Severn*, the *Pearl*, and the *Wager* store-ship. The *Severn* and *Pearl* parted company with the squadron off Cape Noir, and, as we afterwards learnt, put back to the Brazils, so that of all the ships which came into the South Seas, the *Wager*, Captain Cheap, was the only one that was missing. This ship had on board a few field-pieces mounted for land-service, together with some cohorn mortars, and several kinds of artillery stores and pioneers' tools intended for the operations on shore. Therefore, as the enterprise on Baldivia had been resolved on for the first undertaking of the squadron, Captain Cheap was extremely solicitous that these materials, which were in his custody, might be ready before Baldivia; that if the squadron should possibly rendezvous there (as he knew not the condition they were then reduced to), no delay nor disappointment might be imputed to him.

But whilst the *Wager*, with these views, was making the best of her way to her first rendezvous off the island of Socoro, whence (as there was little probability of meeting any of the squadron there) she proposed to steer directly for Baldivia, she made the land on the 14th of May, about the latitude of 47° south; and the captain exerting himself on this occasion, in order to get clear of it, he had the misfortune to fall down the after-ladder, and dislocated his shoulder, which rendered him incapable of acting. This accident, together with the crazy condition of the ship, which was little better than a wreck, prevented her from getting off to sea, and entangled her more and more with the land, insomuch that the next morning, at daybreak, she struck on a sunken rock, and soon after bilged and grounded between two small islands at about a musket-shot from the shore.

In this situation the ship continued entire a long time, so that all the crew had it in their power to get safe on shore; but a general confusion taking place, numbers of them, instead of consulting their safety, or reflecting on their calamitous condition, fell to pillaging the ship, arming themselves with the first weapons that came to hand, and threatening to murder all who should oppose them. This frenzy was greatly heightened by the liquors they found on board, with which they got so extremely drunk that some of them, falling down between decks, were drowned, as the water flowed into the wreck, being incapable of raising themselves up and retreating from it. The captain, therefore, having done his utmost to get the whole crew on shore, was at last obliged to leave the mutineers behind him, and to follow his officers, and such as he had been able to prevail on, but he did not fail to send back the boats to persuade those who remained to have some regard to their preservation, though all his efforts were for some time without success.

However, the weather next day proving stormy, and there being great danger of the ship's parting, they began to be alarmed with the fears of perishing, and were desirous of getting to land; but it seems their madness had not yet left them, for the boat not appearing to fetch them off so soon as they expected, they at last pointed a four-pounder, which was on the quarter-deck, against the hut, where they knew the captain resided on shore, and fired two shot, which passed but just over it.

From this specimen of the behaviour of part of the crew, it will not be difficult to frame some conjecture of the disorder and anarchy which took place when they at last got all on shore. For the men conceived that, by the loss of the ship, the authority of the officers was at an end, and they being now on a desolate coast, where scarcely any other provisions could be got except what should be saved out of the wreck, this was another unsurmountable source of discord, since the working upon the wreck, and the securing the provisions, so that they might be preserved for future exigences as much as possible, and the taking care that what was necessary for their present subsistance might be sparingly and equally distributed, were matters not to be brought about but by discipline and subordination; and the mutinous disposition of the people, stimulated by the impulses of immediate hunger, rendered every regulation made for this purpose ineffectual, so that there were continual concealments, frauds, and thefts, which animated each man against his fellow, and produced infinite feuds and contests. And hence there was a perverse and malevolent disposition constantly kept up amongst them, which rendered them utterly ungovernable.

Besides these heart-burnings occasioned by petulence and hunger, there was another important point which set the greatest part of the people at variance with the captain. This was their differing with him in opinion on the measures to be pursued in the present exigency: for the captain was determined, if possible, to fit up the boats in the best manner he could, and to proceed with them to the northward, since having with him above an hundred men in health, and having gotten some fire-arms and ammunition from the wreck, he did not doubt but they could master any Spanish vessel they should encounter with in those seas, and he thought he could not fail of meeting with one in the neighbourhood of Chiloe or Baldivia, in which, when he had taken her, he intended to proceed to the rendezvous at Juan Fernandes; and he farther insisted that should they light on no prize by the way, yet the boats alone would easily carry them thither. But this was a scheme that, however prudent, was no ways relished by the generality of his people; for, being quite jaded with the distresses and dangers they had already run through, they could not think of prosecuting an enterprise farther which had hitherto proved so disastrous. The common resolution therefore was to lengthen the long-boat, and with that and the rest of the boats to steer to the southward, to pass through the Streights of Magellan, and to range along the east side of South America till they should arrive at Brazil, where they doubted not to be well received, and to procure a passage to Great Britain. This project was at first sight infinitely more hazardous and tedious than what was proposed by the captain; but as it had the air of returning home, and flattered them with the hopes of bringing them once more to their native country, that circumstance alone rendered them inattentive to all its inconveniences, and made them adhere to it with insurmountable obstinacy; so that the captain himself, though he never changed his opinion, was yet obliged to give way to the torrent, and in appearance to acquiesce in this resolution, whilst he endeavoured underhand to give it all the obstruction he could, particularly in the lengthening the long-boat, which he contrived should be of such a size, that though it might serve to carry them to Juan Fernandes, would yet, he hoped, appear incapable of so long a navigation as that to the coast of Brazil.

But the captain, by his steady opposition at first to this favourite project, had much embittered the people against him, to which likewise the following unhappy accident greatly contributed. There was a midshipman whose name was Cozens, who had appeared the foremost in all the refractory proceedings of the crew. He had involved himself in brawls with most of the officers who had adhered to the captain's authority, and had even treated the captain himself with great abuse and insolence. As his turbulence and brutality grew every day more and more intolerable, it was not in the least doubted but there were some violent measures in agitation, in which Cozens was engaged as the ringleader: for which reason the captain, and those about him, constantly kept themselves on their guard. One day the purser, having, by the captain's order, stopped the allowance of a fellow who would not work, Cozens, though the man did not complain to him, intermeddled in the affair with great bitterness, and grossly insulted the purser, who was then delivering out provisions just by the captain's tent, and was himself sufficiently violent. The purser, enraged by his scurrility, and perhaps piqued by former quarrels, cried out, "A mutiny," adding, "The dog has pistols," and then himself fired a pistol at Cozens, which however mist him: but the captain, on this outcry and the report of the pistol, rushed out of his tent, and, not doubting but it had been fired by Cozens as the commencement of a mutiny, he immediately shot him in the head without farther deliberation, and though he did not kill him on the spot, yet the wound proved mortal, and he died about fourteen days after.

However, this incident, though sufficiently displeasing to the people, did yet, for a considerable time, awe them to their duty, and rendered them more submissive to the captain's authority; but at last, when towards the middle of October the long-boat was nearly compleated, and they were preparing to put to sea, the additional provocation he gave them by covertly traversing their project of proceeding through the Streights of Magellan, and their fears that he might at length engage a party sufficient to overturn this favourite measure, made them resolve to make use of the death of Cozens as a reason for depriving him of his command, under pretence of carrying him a prisoner to England, to be tried for murder; and he was accordingly confined under a guard. But they never intended to carry him with them, as they too well knew what they had to apprehend on their return to England, if their commander should be present to confront them: and therefore, when they were just ready to put to sea, they set him at liberty, leaving him and the few who chose to take their fortunes with him no other embarkation but the yawl, to which the barge was afterwards added, by the people on board her being prevailed on to return back.

When the ship was wreckt, there were alive on board the *Wager* near an hundred and thirty persons; of these above thirty died during their stay upon the place, and near eighty went off in the long-boat and the cutter to the southward: so that there remained with the captain, after their departure, no more than nineteen persons, which, however, were as many as the barge and the yawl, the only embarkations left them, could well carry off. It was the 13th of October, five months after the shipwreck, that the long-boat, converted into a schooner, weighed, and stood to the southward, giving the captain, who, with Lieutenant Hamilton of the land forces, and the surgeon, were then on the beach, three cheers at their departure: and on the 29th of January following they arrived at Rio Grande, on the coast of Brazil; but having, by various accidents, left about twenty of their people on shore at the different places they touched at, and a greater number having perished by hunger during the course of their navigation, there were no more than thirty of them remaining when they arrived in that port. Indeed, the undertaking of itself was a most extraordinary one; for (not to mention the length of the run) the vessel was scarcely able to contain the number that first put to sea in her, and their stock of provisions (being only what they had saved out of the ship) was extremely slender. They had this additional misfortune besides, that the cutter, the only boat they had with them, soon broke away from the stern, and was staved to pieces; so that when their provision and their water failed them, they had frequently no means of getting on shore to search for a fresh supply.

After the long-boat and cutter were gone, the captain, and those who were left with him, proposed to pass to the northward in the barge and yawl: but the weather was so bad, and the difficulty of subsisting so great, that it was two months from the departure of the long-boat before he was able to put to sea. It seems the place where the *Wager* was cast away was not a part of the continent, as was first imagined, but an island at some distance from the main, which afforded no other sorts of provision but shell-fish and a few herbs; and as the greatest part of what they had gotten from the ship was carried off in the long-boat, the captain and his people were often in extreme want of food, especially as they chose to preserve what little sea provisions remained, for their store when they should go to the northward. During their residence at this island, which was by the seamen denominated Wager's Island, they had now and then a straggling canoe or two of Indians, which came and bartered their fish and other provisions with our people. This was some little relief to their necessities, and at another season might perhaps have been greater: for as there were several Indian huts on the shore, it was supposed that in some years, during the height of summer, many of these savages might resort thither to fish: indeed, from what has been related in the account of the *Anna* pink, it should seem to be the general practice of those Indians to frequent this coast in the summer time for the benefit of fishing, and to retire in the winter into a better climate, more to the northward.

On this mention of the *Anna* pink, I cannot but observe how much it is to be lamented that the *Wager's* people had no knowledge of her being so near them on the coast; for as she was not above thirty leagues distant from them, and came into their neighbourhood about the same time the *Wager* was lost, and was a fine roomy ship, she could easily have taken them all on board, and have carried them to Juan Fernandes. Indeed, I suspect she was still nearer to them than what is here estimated; for several of the *Wager's* people, at different times, heard the report of a cannon, which I conceive could be no other than the evening gun fired from the *Anna* pink, especially as what was heard at Wager's Island was about the same time of the day. But to return to Captain Cheap.

Upon the 14th of December, the captain and his people embarked in the barge and the yawl, in order to proceed to the northward, taking on board with them all the provisions they could amass from the wreck of the ship; but they had scarcely been an hour at sea, when the wind began to blow hard, and the sea ran so high that they were obliged to throw the greatest part of their provisions overboard, to avoid immediate destruction. This was a terrible misfortune, in a part of the world where food is so difficult to be got: however, they persisted in their design, putting on shore as often as they could to seek subsistance. But about a fortnight after, another dreadful accident befel them, for the yawl sunk at an anchor, and one of the men in her was drowned; and as the barge was incapable of carrying the whole company, they were now reduced to the hard necessity of leaving four marines behind them on that desolate shore. Notwithstanding these disasters, they still kept on their course to the northward, though greatly delayed by the perverseness of the winds, and the frequent interruptions which their search after food occasioned, and constantly struggling with a series of the most sinister events, till at last, about the end of January, having made three unsuccessful attempts to double a headland, which they supposed to be what the Spaniards called Cape Tres Montes, it was unanimously resolved, finding the difficulties insurmountable, to give over this expedition, and to return again to Wager Island, where they got back about the middle of February, quite disheartened and dejected with their reiterated disappointments, and almost perishing with hunger and fatigue.

However, on their return they had the good luck to meet with several pieces of beef, which had been washed out of the wreck and were swimming in the sea. This was a most seasonable relief to them after the hardships they had endured: and to compleat their good fortune, there came, in a short time, two canoes of Indians, amongst which was a native of Chiloe, who spoke a little Spanish; and the surgeon, who was with Captain Cheap, understanding that language, he made a bargain with the Indian, that if he would carry the captain and his people to Chiloe in the barge, he should have her and all that belonged to her for his pains. Accordingly, on the 6th of March, the eleven persons to which the company was now reduced embarked in the barge on this new expedition; but after having proceeded for a few days, the captain and four of his principal officers being on shore, the six, who together with an Indian remained in the barge, put off with her to sea, and did not return again.

By this means there were left on shore Captain Cheap, Mr. Hamilton, lieutenant of marines, the Honourable Mr. Byron and Mr. Campbell, midshipmen, and Mr. Elliot the surgeon. One would have thought that their distresses had long before this time been incapable of augmentation; but they found, on reflection, that their present situation was much more dismaying than anything they had yet gone through, being left on a desolate coast without any provision, or the means of procuring any; for their arms, ammunition, and every conveniency they were masters of, except the tattered habits they had on, were all carried away in the barge.

But when they had sufficiently revolved in their own minds the various circumstances of this unexpected calamity, and were persuaded that they had no relief to hope for, they perceived a canoe at a distance, which proved to be that of the Indian who had undertaken to carry them to Chiloe, he and all his family being then on board it. He made no difficulty of coming to them; for it seems he had left Captain Cheap and his people a little before to go a-fishing, and had in the meantime committed them to the care of the other Indian, whom the sailors had carried to sea in the barge. When he came on shore, and found the barge gone and his companion missing, he was extremely concerned, and could with difficulty be persuaded that the other Indian was not murdered; yet being at last satisfied with the account that was given him, he still undertook to carry them to the Spanish settlements, and (as the Indians are well skilled in fishing and fowling) to procure them provisions by the way.

About the middle of March, Captain Cheap and the four who were left with him set out for Chiloe, the Indian having provided a number of canoes, and gotten many of his neighbours together for that purpose. Soon after they embarked, Mr. Elliot, the surgeon, died, so that there now remained only four of the whole company. At last, after a very complicated passage by land and water, Captain Cheap, Mr. Byron, and Mr. Campbell arrived in the beginning of June at the island of Chiloe, where they were received by the Spaniards with great humanity; but, on account of some quarrel among the Indians, Mr. Hamilton did not get there till two months later. Thus was it above a twelvemonth from the loss of the *Wager* before this fatiguing peregrination ended: and not till by a variety of misfortunes the company was diminished from twenty to no more than four, and those too brought so low that, had their distresses continued but a few days longer, in all probability none of them would have survived. For the captain himself was with difficulty recovered, and the rest were so reduced by the severity of the weather, their labour, their want of food, and of all kinds of necessaries, that it was wonderful how they supported themselves so long. After some stay at Chiloe, the captain and the three who were with him were sent to Valparaiso, and thence to St. Jago, the capital of Chili, where they continued above a year: but on the advice of a cartel being settled betwixt Great Britain and Spain, Captain Cheap, Mr. Byron, and Mr. Hamilton were permitted to return to Europe on board a French ship. The other midshipman, Mr. Campbell, having changed his religion whilst at St. Jago, chose to go back to Buenos Ayres with Pizarro and his officers, with whom he went afterwards to Spain on board the *Asia*; but having there failed in his endeavours to procure a commission from the court of Spain, he returned to England, and attempted to get reinstated in the British navy. He has since published a narration of his adventures, in which he complains of the injustice that had been done him, and strongly disavows his ever being in the Spanish service: but as the change of his religion, and his offering himself to the court of Spain (though he was not accepted), are matters which, he is conscious, are capable of being incontestably proved, on these two heads he has been entirely silent. And now, after this account of the accidents which befel the *Anna* pink, and the catastrophe of the *Wager*, I shall again resume the thread of our own story.

CHAPTER IV

CONCLUSION OF OUR PROCEEDINGS AT JUAN FERNANDES, FROM THE ARRIVAL OF THE "ANNA" PINK TO OUR FINAL DEPARTURE FROM THENCE

About a week after the arrival of our victualler, the *Tryal* sloop, that had been sent to the island of Masa Fuero, returned to an anchor at Juan Fernandes, having been round that island without meeting any part of our squadron. As upon this occasion the island of Masa Fuero was more particularly examined than I dare say it had ever been before, or perhaps ever will be again, and as the knowledge of it may, in certain circumstances, be of great consequence hereafter, I think it incumbent on me to insert the accounts given of this place by the officers of the *Tryal* sloop.

The Spaniards have generally mentioned two islands under the name of Juan Fernandes, styling them the greater and the less: the greater being that island where we anchored, and the less being the island we are now describing, which, because it is more distant from the continent, they have distinguished by the name of Masa Fuero. The *Tryal* sloop found that it bore from the greater Juan Fernandes W. by S., and was about twenty-two leagues distant. It is a much larger and better spot than has been generally reported; for former writers have represented it as a small barren rock, destitute of wood and water, and altogether inaccessible; whereas our people found it was covered with trees, and that there were several fine falls of water pouring down its sides into the sea. They found, too, that there was a place where a ship might come to an anchor on the north side of it, though indeed the anchorage is inconvenient; for the bank extends but a little way, is steep too, and has very deep water upon it, so that you must come to an anchor very near the shore, and there lie exposed to all the winds but a southerly one. And besides the inconvenience of the anchorage, there is also a reef of rocks running off the eastern point of the island, about two miles in length, though there is little danger to be feared from them, because they are always to be seen by the seas breaking over them. This place has at present one advantage beyond the island of Juan Fernandes; for it abounds with goats, who, not being accustomed to be disturbed, were no ways shy or apprehensive of danger till they had been frequently fired at. These animals reside here in great tranquillity, the Spaniards having not thought the island considerable enough to be frequented by their enemies, and have not therefore been solicitous to destroy the provisions upon it, so that no dogs have been hitherto set on shore there. Besides the goats, our people found there vast numbers of seals and sea-lions: and upon the whole, they seemed to imagine that though it was not the most eligible place for a ship to refresh at, yet in case of necessity it might afford some sort of shelter, and prove of considerable use, especially to a single ship, who might apprehend meeting with a superior force at Fernandes.

The latter part of the month of August was spent in unlading the provisions from the *Anna* pink, when we had the mortification to find that great quantities of our provisions, as bread, rice, grots, etc., were decayed, and unfit for use. This was owing to the water the pink had made by her working and straining in bad weather; for hereby several of her casks had rotted, and her bags were soaked through. And now, as we had no farther occasion for her service, the commodore, pursuant to his orders from the Board of Admiralty, sent notice to Mr. Gerard, her master, that he discharged the *Anna* pink from attending the squadron, and gave him, at the same time, a certificate specifying how long she had been employed. In consequence of this dismission, her master was at liberty either to return directly to England, or to make the best of his way to any port where he thought he could take in such a cargo as would answer the interest of his owners. But the master being sensible of the bad condition of the ship, and of her unfitness for any such voyage, wrote the next day an answer to the commodore's message, acquainting Mr. Anson, that from the great quantity of water the pink had made in her passage round Cape Horn, and since, that in the tempestuous weather she had met with on the coast of Chili, he had reason to apprehend that her bottom was very much decayed. He added that her upper works were rotten abaft; that she was extremely leaky; that her fore beam was broke; and that, in his opinion, it was impossible to proceed to sea with her before she had been thoroughly refitted; and he therefore requested the commodore that the carpenters of the squadron might be directed to survey her, that their judgment of her condition might be known. In compliance with this desire, Mr. Anson immediately ordered the carpenters to take a careful and strict survey of the *Anna* pink, and to give him a faithful report, under their hands, of the condition in which they found her, directing them at the same time to proceed herein with such circumspection that, if they should be hereafter called upon, they might be able to make oath of the veracity of their proceedings. Pursuant to these orders, the carpenters immediately set about the examination, and the next day made their report; which was, that the pink had no less than fourteen knees and twelve beams broken and decayed; that one breast hook was broken, and another rotten; that her water-ways were open and decayed; that two standards and several clamps were broken, besides others which were rotten; that all her iron-work was greatly decayed; that her spirkiting and timbers were very rotten; and that, having ripped off part of her sheathing, they found her wales and outside planks extremely defective, and her bows and decks very leaky; and in consequence of these defects and decays, they certified that in their opinion she could not depart from the island without great hazard, unless she was first of all thoroughly refitted.

The thorough refitting of the *Anna* pink, proposed by the carpenters, was, in our present situation, impossible to be complied with, as all the plank and iron in the squadron was insufficient for that purpose. And now the master, finding his own sentiments confirmed by the opinion of all the carpenters, he offered a petition to the commodore in behalf of his owners, desiring that, since it appeared he was incapable of leaving the island, Mr. Anson would please to purchase the hull and furniture of the pink for the use of the squadron. Hereupon the commodore ordered an inventory to be taken of every particular belonging to the pink, with its just value; and as by this inventory it appeared that there were many stores which would be useful in refitting the other ships, and which were at present very scarce in the squadron, by reason of the great quantities that had been already expended, he agreed with Mr. Gerard to purchase the whole together for £300. The pink being thus broken up, Mr. Gerard, with the hands belonging to the pink, were sent on board the *Gloucester*, as that ship had buried the greatest number of men in proportion to her complement. But afterwards, one or two of them were received on board the *Centurion*, on their own petition, they being extremely averse to sailing in the same ship with their old master, on account of some particular ill-usage they conceived they had suffered from him.

This transaction brought us down to the beginning of September, and our people by this time were so far recovered of the scurvy, that there was little danger of burying any more at present; and therefore I shall now sum up the total of our loss since our departure from England, the better to convey some idea of our past sufferings, and of our present strength. We had buried on board the *Centurion*, since our leaving St. Helens, two hundred and ninety-two, and had now remaining on board two hundred and fourteen. This will doubtless appear a most extraordinary mortality: but yet on board the *Gloucester* it had been much greater, for out of a much smaller crew than ours they had lost the same number, and had only eighty-two remaining alive. It might be expected that on board the *Tryal* the slaughter would have been the most terrible, as her decks were almost constantly knee-deep in water; but it happened otherwise, for she escaped more favourably than the rest, since she only buried forty-two, and had now thirty-nine remaining alive. The havock of this disease had fallen still severer on the invalids and marines than on the sailors; for on board the *Centurion*, out of fifty invalids and seventy-nine marines, there remained only four invalids, including officers, and eleven marines: and on board the *Gloucester* every invalid perished, and out of forty-eight marines only two escaped. From this account it appears that the three ships together departed from England with nine hundred and sixty-one men on board, of whom six hundred and twenty-six were dead before this time;

so that the whole of our remaining crews, which were now to be distributed amongst three ships, amounted no more than three hundred and thirty-five men and boys: a number greatly insufficient for the manning the *Centurion* alone, and barely capable of navigating all the three, with the utmost exertion of their strength and vigour. This prodigious reduction of our men was still the more terrifying as we were hitherto uncertain of the fate of Pizarro's squadron, and had reason to suppose that some part of it at least had got round into these seas. Indeed, we were satisfied from our own experience that they must have suffered greatly in their passage; but then every port in the South Seas was open to them, and the whole power of Chili and Peru would doubtless be united in refreshing and refitting them, and recruiting the numbers they had lost. Besides, we had some obscure knowledge of a force to be sent out from Callao; and, however contemptible the ships and sailors of this part of the world may have been generally esteemed, it was scarcely possible for anything bearing the name of a ship of force to be feebler or less considerable than ourselves. And had there been nothing to be apprehended from the naval power of the Spaniards in this part of the world, yet our enfeebled condition would nevertheless give us the greatest uneasiness, as we were incapable of attempting any of their considerable places; for the risquing of twenty men, weak as we then were, was risquing the safety of the whole: so that we conceived we should be necessitated to content ourselves with what few prizes we could pick up at sea before we were discovered; after which we should in all probability be obliged to depart with precipitation, and esteem ourselves fortunate to regain our native country, leaving our enemies to triumph on the inconsiderable mischief they had received from a squadron whose equipment had filled them with such dreadful apprehensions. This was a subject on which we had reason to imagine the Spanish ostentation would remarkably exert itself, though the causes of our disappointment and their security were neither to be sought for in their valour nor our misconduct.

Such were the desponding reflections which at that time arose on the review and comparison of our remaining strength with our original numbers. Indeed, our fears were far from being groundless, or disproportioned to our feeble and almost desperate situation; for though the final event proved more honourable than we had foreboded, yet the intermediate calamities did likewise greatly surpass our most gloomy apprehensions, and could they have been predicted to us at this island of Juan Fernandes, they would doubtless have appeared insurmountable. But to return to our narration.

In the beginning of September, as has been already mentioned, our men were tolerably well recovered; and now, the season for navigation in this climate drawing near, we exerted ourselves in getting our ships in readiness for the sea. We converted the fore-mast of the victualler into a main-mast for the *Tryal* sloop; and still flattering ourselves with the possibility of the arrival of some other ships of our squadron, we intended to leave the main-mast of the victualler to make a mizen-mast for the *Wager*. Thus all hands being employed in forwarding our departure, we, on the 8th, about eleven in the morning, espied a sail to the N.E. which continued to approach us till her courses appeared even with the horizon. Whilst she advanced, we had great hopes she might prove one of our own squadron; but as at length she steered away to the eastward without haling in for the island, we thence concluded she must be a Spaniard. And now great disputes were set on foot about the possibility of her having discovered our tents on shore, some of us strongly insisting that she had doubtless been near enough to have perceived something that had given her a jealousy of an enemy, which had occasioned her standing to the eastward without haling in. However, leaving these contests to be settled afterwards, it was resolved to pursue her, and, the *Centurion* being in the greatest forwardness, we immediately got all our hands on board, set up our rigging, bent our sails, and by five in the afternoon got under sail. We had at this time very little wind, so that all the boats were employed to tow us out of the bay; and even what wind there was, lasted only long enough to give us an offing of two or three leagues, when it flatted to a calm. The night coming on, we lost sight of the chace, and were extremely impatient for the return of daylight, in hopes to find that she had been becalmed as well as we, though I must confess that her greater distance from the land was a reasonable ground for suspecting the contrary, as we indeed found in the morning to our great mortification, for though the weather continued perfectly clear, we had no sight of the ship from the mast-head. But as we were now satisfied that it was an enemy, and the first we had seen in these seas, we resolved not to give over the search lightly; and, a small breeze springing up from the W.N.W., we got up our top-gallant masts and yards, set all the sails, and steered to the S.E. in hopes of retrieving our chace, which we imagined to be bound to Valparaiso. We continued on this course all that day and the next, and then, not getting sight of our chace, we gave over the pursuit, conceiving that by that time she must, in all probability, have reached her port. Being therefore determined to return to Juan Fernandes, we haled up to the S.W. with that view, having but very little wind till the 12th, when, at three in the morning, there sprung up a fresh gale from the W.S.W. which obliged us to tack and stand to the N.W. At daybreak we were agreeably surprized with the sight of a sail on our weather-bow, between four and five leagues distant. We immediately crouded all the sail we could, and stood after her, and soon perceived it not to be the same ship we originally gave chace to. She at first bore down upon us, shewing Spanish colours, and making a signal as to her consort; but observing that we did not answer her signal, she instantly loofed close to the wind, and stood to the southward. Our people were now all in spirits, and put the ship about with great briskness; and as the chace appeared to be a large ship, and had mistaken us for her consort, we conceived that she was a man-of-war, and probably one of Pizarro's squadron. This induced the commodore to order all the officers' cabins to be knocked down and thrown overboard, with several casks of water and provisions which stood between the guns, so that we had soon a clear ship, ready for an engagement. About nine o'clock we had thick hazy weather and a shower of rain, during which we lost sight of the chace; and we were apprehensive, if this dark weather should continue, that by going upon the other tack, or by some other artifice, she might escape us; but it clearing up in less than an hour, we found that we had both weathered and fore-reached upon her considerably, and were then near enough to discover that she was only a merchantman, without so much as a single tier of guns. About half an hour after twelve, being got within a reasonable distance of her, we fired four shot amongst her rigging; on which they lowered their top-sails, and bore down to us, but in very great confusion, their top-gallant sails and stay-sails all fluttering in the winds: this was owing to their having let run their sheets and halyards just as we fired at them, after which not a man amongst them had courage enough to venture aloft (for there the shot had passed but just before) to take them in. As soon as the vessel came within hale of us, the commodore ordered them to bring-to under his lee quarter, and then hoisted out the boat, and sent Mr. Saumarez, his first lieutenant, to take possession of the prize, with directions to send all the prisoners on board the *Centurion*, but first the officers and passengers. When Mr. Saumarez came on board them, they received him at the side with the strongest tokens of the most abject submission, for they were all of them (especially the passengers, who were twenty-five in number) extremely terrified, and under the greatest apprehensions of meeting with very severe and cruel usage; but the lieutenant endeavoured, with great courtesy, to dissipate their fright, assuring them that their fears were altogether groundless, and that they would find a generous enemy in the commodore, who was not less remarkable for his lenity and humanity than for his resolution and courage. The prisoners, who were first sent on board the *Centurion*, informed us that our prize was called *Neustra Senora del Monte Carmelo*, and was commanded by Don Manuel Zamorra. Her cargo consisted chiefly of sugar, and great quantities of blue cloth made in the province of Quito, somewhat resembling our English coarse broadcloths, but inferior to them. They had besides several bales of a coarser sort of cloth, of different colours, somewhat like Colchester bays, called by them Pannia da Tierra, with a few bales of cotton, and some tobacco, which, though strong, was not ill flavoured. These were the principal goods on board her; but we found besides what was to us much more valuable than the rest of the cargoe: this was some trunks of wrought plate, and twenty-three serons of dollars, each weighing upwards of 200 lb. averdupois. The ship's burthen was about four hundred and fifty tons; she had fifty-three sailors on board, both whites and blacks; she came from Callao, and had been twenty-seven days at sea before she fell into our hands. She was bound to the port of Valparaiso, in the kingdom of Chili, and proposed to have returned from thence loaded with corn and Chili wine, some gold, dried beef, and small cordage, which at Callao they convert into larger rope. Our prize had been built upwards of thirty years; yet, as they lie in harbour all the winter months, and the climate is favourable, they esteemed it no very great age. Her rigging was very indifferent, as were likewise her sails, which were made of cotton. She had only three four-pounders, which were altogether unserviceable, their carriages being scarcely able to support them: and there were no small arms on board, except a few pistols belonging to the passengers. The prisoners informed us that they left Callao in company with two other ships, whom they had parted with some days before, and that at first they conceived us to be one of their company; and by the description we gave them of the ship we had chased from Juan Fernandes, they assured us she was of their number, but that the coming in sight of that island was directly repugnant to the merchants' instructions, who had expressly forbid it, as knowing that if any English squadron was in those seas, the island of Fernandes was most probably the place of their rendezvous.

After this short account of the ship and her cargoe, it is necessary that I should relate the important intelligence which we met with on board her, partly from the information of the prisoners, and partly from the letters and papers which fell into our hands. We here first learnt with certainty the force and destination of that squadron which cruized off the Maderas at our arrival there, and afterwards chased the *Pearl* in our passage to Port St. Julian. This we now knew was a squadron composed of five large Spanish ships, commanded by Admiral Pizarro, and purposely fitted out to traverse our designs, as hath been already more amply related in the third chapter of the first book. We had at the same time, too, the satisfaction to find that Pizarro, after his utmost endeavours to gain his passage into these seas, had been forced back again into the river of Plate, with the loss of two of his largest ships. And besides this disappointment of Pizarro, which, considering our great debility, was no unacceptable intelligence, we farther learnt, that though an embargo had been laid upon all shipping in these seas by the Viceroy of Peru, in the month of May preceding, on a supposition that about that time we might arrive upon the coast, yet it now no longer subsisted: for on the account sent overland by Pizarro of his own distresses, part of which they knew we must have encountered, as we were at sea during the same time, and on their having no news of us in eight months after we were known to set sail from St. Catherine's, they were fully satisfied that we were either shipwrecked, or had perished at sea, or, at least, had been obliged to put back again, as it was conceived impossible for any ships to continue at sea during so long an interval: and therefore, on the application of the merchants, and the firm persuasion of our having miscarried, the embargo had been lately taken off.

This last article made us flatter ourselves that, as the enemy was still a stranger to our having got round Cape Horn, and the navigation of these seas was restored, we might meet with some valuable captures, and might thereby indemnify ourselves for the incapacity we were under of attempting any of their considerable settlements on shore. And thus much we were certain of, from the information of our prisoners, that, whatever our success might be, as to the prizes we might light on, we had nothing to fear, weak as we were, from the Spanish force in this part of the world, though we discovered that we had been in most imminent peril from the enemy when we least apprehended it, and when our other distresses were at the greatest height; for we learnt, from the letters on board, that Pizarro, in the express he dispatched to the Viceroy of Peru, after his return to the river of Plate, had intimated to him that it was possible some part at least of the English squadron might get round; but that, as he was certain from his own experience that if they did arrive in those seas it must be in a very weak and defenceless condition, he advised the viceroy, in order to be secure at all events, to send what ships of war he had to the southward, where, in all probability, they would intercept us singly, before we had an opportunity of touching at any port for refreshment; in which case he doubted not but we should prove an easy conquest. The Viceroy of Peru approved of this advice, and as he had already fitted out four ships of force from Callao—one of fifty guns, two of forty guns, and one of twenty-four guns, which were intended to join Pizarro when he arrived on the coast of Chili—the viceroy now stationed three of these off the Port of Conception, and one of them at the island of Fernandes, where they continued cruizing for us till the 6th of June, and then not seeing anything of us, and conceiving it to be impossible that we could have kept the seas so long, they quitted their cruise and returned to Callao, fully persuaded that we had either perished, or at least had been driven back. Now, as the time of their quitting their stations was but a few days before our arrival at the island of Fernandes, it is evident that had we made that island on our first search for it, without haling in for the main to secure our easting (a circumstance which at that time we considered as very unfortunate to us, on account of the numbers which we lost by our longer continuance at sea)— had we, I say, made the island on the 28th of May, when we first expected to see it, and were in reality very near it, we had doubtless fallen in with some part of the Spanish squadron; and in the distressed condition we were then in, the meeting with a healthy, well-provided enemy was an incident that could not but have been perplexing, and might perhaps have proved fatal, not only to us, but to the *Tryal*, the *Gloucester*, and the *Anna* pink, who separately joined us, and who were each of them less capable than we were of making any considerable resistance. I shall only add, that these Spanish ships sent out to intercept us had been greatly shattered by a storm during their cruise, and that, after their arrival at Callao, they had been laid up. And our prisoners assured us that whenever intelligence was received at Lima of our being in these seas, it would be at least two months before this armament could be again fitted out.

The whole of this intelligence was as favourable as we, in our reduced circumstances, could wish for. And now we were no longer at a loss as to the broken jars, ashes, and fishbones which we had observed at our first landing at Juan Fernandes, these things being doubtless the relicts of the cruisers stationed off that port. Having thus satisfied ourselves in the material articles of our inquiry, and having gotten on board the *Centurion* most of the prisoners, and all the silver, we, at eight in the same evening, made sail to the northward, in company with our prize, and at six the next morning discovered the island of Fernandes, where, the following day, both we and our prize came to an anchor.

And here I cannot omit one remarkable incident which occurred when the prize and her crew came into the bay where the rest of the squadron lay. The Spaniards in the *Carmelo* had been sufficiently informed of the distresses we had gone through, and were greatly surprized that we had ever surmounted them; but when they saw the *Tryal* sloop at anchor, they were still more astonished that after all our fatigues we had the industry (besides refitting our other ships) to complete such a vessel in so short time, they taking it for granted that we had built her upon the spot: nor was it without great difficulty they were at last prevailed upon to believe that she came from England with the rest of the squadron, they long insisting that it was impossible such a bauble as that could pass round Cape Horn, when the best ships of Spain were obliged to put back.

By the time we arrived at Juan Fernandes, the letters found on board our prize were more minutely examined: and, it appearing from them, and from the accounts of our prisoners, that several other merchantmen were bound from Callao to Valparaiso, Mr. Anson dispatched the *Tryal* sloop the very next morning to cruise off the last-mentioned port, reinforcing her with ten hands from on board his own ship. Mr. Anson likewise resolved, on the intelligence recited above, to separate the ships under his command, and employ them in distinct cruises, as he thought that by this means we should not only increase our chance for prizes, but that we should likewise run a less risque of alarming the coast, and of being discovered. And now the spirits of our people being greatly raised, and their despondency dissipated by this earnest of success, they forgot all their past distresses, and resumed their wonted alacrity, and laboured indefatigably in completing our water, receiving our lumber, and in preparing to take our farewell of the island: but as these occupations took us up four or five days with all our industry, the commodore, in that interval, directed that the guns belonging to the *Anna* pink, being four six-pounders, four four-pounders, and two swivels, should be mounted on board the *Carmelo*, our prize: and having sent on board the *Gloucester* six passengers and twenty-three seamen to assist in navigating the ship, he directed Captain Mitchel to leave the island as soon as possible, the service demanding the utmost dispatch, ordering him to proceed to the latitude of five degrees south, and there to cruise off the highland of Paita, at such a distance from shore as should prevent his being discovered. On this station he was to continue till he should be joined by the commodore, which would be whenever it should be known that the viceroy had fitted out the ships at Callao, or on Mr. Anson's receiving any other intelligence that should make it necessary to unite our strength. These orders being delivered to the captain of the *Gloucester*, and all our business compleated, we, on the Saturday following, being the 19th of September, weighed our anchor, in company with our prize, and got out of the bay, taking our last leave of the island of Juan Fernandes, and steering to the eastward, with an intention of joining the *Tryal* sloop in her station off Valparaiso.

CHAPTER V

OUR CRUISE FROM THE TIME OF OUR LEAVING JUAN FERNANDES TO THE TAKING THE TOWN OF PAITA

Although the *Centurion*, with her prize, the *Carmelo*, weighed from the bay of Juan Fernandes on the 19th of September, leaving the *Gloucester* at anchor behind her, yet, by the irregularity and fluctuation of the winds in the offing, it was the 22d of the same month, in the evening, before we lost sight of the island: after which we continued our course to the eastward, in order to reach our station, and to join the *Tryal* off Valparaiso. The next night the weather proved squally, and we split our main top-sail, which we handed for the present, but got it repaired, and set it again the next morning. In the evening, a little before sunset, we saw two sail to the eastward; on which our prize stood directly from us, to avoid giving any suspicion of our being cruisers, whilst we, in the meantime, made ourselves ready for an engagement, and steered with all our canvas towards the two ships we had discovered. We soon perceived that one of these, which had the appearance of being a very stout ship, made directly for us, whilst the other kept at a great distance. By seven o'clock we were within pistol-shot of the nearest, and had a broadside ready to pour into her, the gunners having their matches in their hands, and only waiting for orders to fire; but, as we knew it was now impossible for her to escape us, Mr. Anson, before he permitted us to fire, ordered the master to hale the ship in Spanish; on which the commanding officer on board her, who proved to be Mr. Hughes, lieutenant of the *Tryal*, answered us in English, and informed us that she was a prize taken by the *Tryal* a few days before, and that the other sail at a distance was the *Tryal* herself disabled in her masts. We were soon after joined by the *Tryal*, and Captain Saunders, her commander, came on board the *Centurion*. He acquainted the commodore that he had taken this ship the 18th instant; that she was a prime sailor, and had cost him thirty-six hours' chace before he could come up with her; that for some time he gained so little upon her that he began to despair of taking her; and the Spaniards, though alarmed at first with seeing nothing but a cloud of sail in pursuit of them, the *Tryal's* hull being so low in the water that no part of it appeared, yet knowing the goodness of their ship, and finding how little the *Tryal* neared them, they at length laid aside their fears, and recommending themselves to the blessed Virgin for protection, began to think themselves secure. Indeed their success was very near doing honour to their Ave Marias, for, altering their course in the night, and shutting up their windows to prevent any of their lights from being seen, they had some chance of escaping; but a small crevice in one of the shutters rendered all their invocations ineffectual, for through this crevice the people on board the *Tryal* perceived a light, which they chased till they arrived within gunshot, and then Captain Saunders alarmed them unexpectedly with a broadside, when they flattered themselves they were got out of his reach. However, for some time after they still kept the same sail abroad, and it was not observed that this first salute had made any impression on them; but, just as the *Tryal* was preparing to repeat her broadside, the Spaniards crept from their holes, lowered their sails, and submitted without any opposition. She was one of the largest merchantmen employed in those seas, being about six hundred tuns burthen, and was called the *Arranzazu*. She was bound from Callao to Valparaiso, and had much the same cargo with the *Carmelo* we had taken before, except that her silver amounted only to about £5000 sterling.

But to balance this success, we had the misfortune to find that the *Tryal* had sprung her main-mast, and that her main top-mast had come by the board; and as we were all of us standing to the eastward the next morning, with a fresh gale at south, she had the additional ill-luck to spring her fore-mast: so that now she had not a mast left on which she could carry sail. These unhappy incidents were still aggravated by the impossibility we were just then under of assisting her; for the wind blew so hard, and raised such a hollow sea, that we could not venture to hoist out our boat, and consequently could have no communication with her; so that we were obliged to lie to for the greatest part of forty-eight hours to attend her, as we could have no thought of leaving her to herself in her present unhappy situation. It was no small accumulation to these misfortunes that we were all the while driving to the leeward of our station, at the very time too, when, by our intelligence, we had reason to expect several of the enemy's ships would appear upon the coast, who would now gain the port of Valparaiso without obstruction. And I am verily persuaded that the embarrassment we received from the dismasting of the *Tryal*, and our absence from our intended station, occasioned thereby, deprived us of some very considerable captures.

The weather proving somewhat more moderate on the 27th, we sent our boat for the captain of the *Tryal*, who, when he came on board us, produced an instrument, signed by himself and all his officers, representing that the sloop, besides being dismasted, was so very leaky in her hull that even in moderate weather it was necessary to ply the pumps constantly, and that they were then scarcely sufficient to keep her free; so that in the late gale, though they had all been engaged at the pumps by turns, yet the water had increased upon them; and, upon the whole, they apprehended her at present to be so very defective, that if they met with much bad weather they must all inevitably perish; and therefore they petitioned the commodore to take some measures for their future safety. But the refitting of the *Tryal*, and the repairing of her defects, was an undertaking that in the present conjuncture greatly exceeded our power; for we had no masts to spare her, we had no stores to complete her rigging, nor had we any port where she might be hove down and her bottom examined: besides, had a port and proper requisites for this purpose been in our possession, yet it would have been extreme imprudence, in so critical a conjuncture, to have loitered away so much time as would have been necessary for these operations. The commodore therefore had no choice left him, but was under a necessity of taking out her people and destroying her. However, as he conceived it expedient to keep up the appearance of our force, he appointed the *Tryal's* prize (which had been often employed by the Viceroy of Peru as a man-of-war) to be a frigate in his Majesty's service, manning her with the *Tryal's* crew, and giving commissions to the captain and all the inferior officers accordingly. This new frigate, when in the Spanish service, had mounted thirty-two guns; but she was now to have only twenty, which were the twelve that were on board the *Tryal*, and eight that had belonged to the *Anna* pink. When this affair was thus resolved on, Mr. Anson gave orders to Captain Saunders to put it in execution, directing him to take out of the sloop the arms, stores, ammunition, and everything that could be of any use to the other ships, and then to scuttle her and sink her. After Captain Saunders had seen her destroyed, he was to proceed with his new frigate (to be called the *Tryal's* prize) and to cruise off the highland of Valparaiso, keeping it from him N.N.W. at the distance of twelve or fourteen leagues: for as all ships bound from Valparaiso to the northward steer that course, Mr. Anson proposed by this means to stop any intelligence that might be dispatched to Callao of two of their ships being missing, which might give them apprehensions of the English squadron being in their neighbourhood. The *Tryal's* prize was to continue on this station twenty-four days, and, if not joined by the commodore at the expiration of that term, she was then to proceed down the coast to Pisco or Nasca, where she would be certain to meet with Mr. Anson. The commodore likewise ordered Lieutenant Saumarez, who commanded the *Centurion's* prize, to keep company with Captain Saunders, both to assist him in unloading the sloop, and also that by spreading in their cruise there might be less danger of any of the enemy's ships slipping by unobserved. These orders being dispatched, the *Centurion* parted from the other vessels at eleven in the evening, on the 27th of September, directing her course to the southward, with a view of cruising for some days to the windward of Valparaiso.

And now by this distribution of our ships we flattered ourselves that we had taken all the advantages of the enemy that we possibly could with our small force, since our disposition was doubtless the most prudent that could be projected. For, as we might suppose the *Gloucester* by this time to be drawing near the highland of Paita, we were enabled, by our separate stations, to intercept all vessels employed either betwixt Peru and Chili to the southward, or betwixt Panama and Peru to the northward: since the principal trade from Peru to Chili being carried on to the port of Valparaiso, the *Centurion* cruising to the windward of Valparaiso would, in all probability, meet with them, as it is the constant practice of those ships to fall in with the coast to the windward of that port. The *Gloucester* would, in like manner, be in the way of the trade bound from Panama or to the northward, to any part of Peru, since the highland off which she was stationed is constantly made by every ship in that voyage. And whilst the *Centurion* and *Gloucester* were thus situated for interrupting the enemy's trade, the *Tryal's* prize and *Centurion's* prize were as conveniently posted for preventing all intelligence, by intercepting all ships bound from Valparaiso to the northward; for it was on board these vessels that it was to be feared some account of us might possibly be sent to Peru.

But the most prudent dispositions carry with them only a probability of success, and can never ensure its certainty, since those chances which it was reasonable to overlook in deliberation are sometimes of most powerful influence in execution. Thus in the present case, the distress of the *Tryal*, and our quitting our station to assist her (events which no degree of prudence could either foresee or obviate), gave an opportunity to all the ships bound to Valparaiso to reach that port without molestation during this unlucky interval. So that though, after leaving Captain Saunders, we were very expeditious in regaining our station, where we got the 29th at noon, yet in plying on and off till the 6th of October we had not the good fortune to discover a sail of any sort: and then having lost all hopes of meeting with better fortune by a longer stay, we made sail to the leeward of the port, in order to join our prizes; but when we arrived off the highland where they were directed to cruise, we did not find them, though we continued there four or five days. We supposed that some chace had occasioned their leaving their station, and therefore we proceeded down the coast to the highland of Nasca, which was the second rendezvous, where Captain Saunders was directed to join us. Here we got on the 21st, and were in great expectation of falling in with some of the enemy's vessels, as both the accounts of former voyages and the information of our prisoners assured us that all ships bound to Callao constantly make this land, to prevent the danger of running to the leeward of the port. But notwithstanding the advantages of this station, we saw no sail till the 2d of November, when two ships appeared in sight together; we immediately gave them chace, and soon perceived that they were the *Tryal's* and *Centurion's* prizes. As they had the wind of us, we brought to and waited their coming up, when Captain Saunders came on board us, and acquainted the commodore that he had cleared the *Tryal* pursuant to his orders, and having scuttled her, he remained by her till she sunk, but that it was the 4th of October before this was effected; for there ran so large and hollow a sea, that the sloop, having neither masts nor sails to steady her, rolled and pitched so violently, that it was impossible for a boat to lay alongside of her for the greatest part of the time: and during this attendance on the sloop, they were all driven so far to the north-west that they were afterwards obliged to stretch a long way to the westward to regain the ground they had lost; which was the reason that we had not met with them on their station, as we expected. We found they had not been more fortunate in their cruise than we were, for they had seen no vessel since they separated from us. The little success we all had, and our certainty that had any ships been stirring in these seas for some time past we must have met with them, made us believe that the enemy at Valparaiso, on the missing of the two ships we had taken, had suspected us to be in the neighbourhood, and had consequently laid an embargo on all the trade in the southern parts. We likewise apprehended that they might by this time be fitting out the men-of-war at Callao, as we knew that it was no uncommon thing for an express from Valparaiso to reach Lima in twenty-nine or thirty days, and it was now more than fifty since we had taken our first prize. These apprehensions of an embargo along the coast, and of the equipment of the Spanish squadron at Callao, determined the commodore to hasten down to the leeward of Callao, and to join Captain Michel (who was stationed off Paita) as soon as possible, that our strength being united we might be prepared to give the ships from Callao a warm reception, if they dared to put to sea. With this view we bore away the same afternoon, taking particular care to keep at such a distance from the shore that there might be no danger of our being discovered from thence; for we knew that all the country ships were commanded, under the severest penalty, not to sail by the port of Callao without stopping; and as this order was constantly complied with, we should undoubtedly be known for enemies if we were seen to act contrary to it. In this new navigation, not being certain whether we might not meet the Spanish squadron in our route, the commodore took on board the *Centurion* part of his crew with which he had formerly manned the *Carmelo*. And now standing to the northward, we, before night came on, had a view of the small island called St. Gallen, which bore from us N.N.E.½E., about seven leagues distant. This island lies in the latitude of about fourteen degrees south, and about five miles to the northward of a highland called Morro Veijo, or the old man's head. I mention this island and the highland near it more particularly because between them is the most eligible station on that coast for cruising upon the enemy, as hereabouts all ships bound to Callao, whether from the northward or the southward, run well in with the land. By the 5th of November, at three in the afternoon, we were advanced within view of the highland of Barranca, lying in the latitude of 10° 36' south, bearing from us N.E. by E., distant eight or nine leagues; and an hour and an half afterwards we had the satisfaction so long wished for, of seeing a sail. She first appeared to leeward, and we all immediately gave her chace; but the *Centurion* so much outsailed the two prizes, that we soon ran them out of sight, and gained considerably on the chace. However, night coming on before we came up with her, we, about seven o'clock, lost sight of her, and were in some perplexity what course to steer; but at last Mr. Anson resolved, as we were then before the wind, to keep all his sails set, and not to change his course: for though we had no doubt but the chace would alter her course in the night, yet, as it was uncertain what tack she would go upon, it was thought prudent to keep on our course, as we must by this means unavoidably come near her, rather than to change it on conjecture, when, if we should mistake, we must infallibly lose her. Thus then we continued the chace about an hour and an half in the dark, some one or other on board us constantly imagining they discerned her sails right ahead of us; but at length Mr. Brett, our second lieutenant, did really discover her about four points on the larboard-bow, steering off to the seaward. We immediately clapped the helm a-weather, and stood for her, and in less than an hour came up with her, and having fired fourteen shot at her, she struck. Our third lieutenant, Mr. Dennis, was sent in the boat with sixteen men to take possession of the prize, and to return the prisoners to our ship. This vessel was named the *Santa Teresa de Jesus*, built at Guaiaquil, of about three hundred tuns burthen, and was commanded by Bartolome Urrunaga, a Biscayer. She was bound from Guaiaquil to Callao; her loading consisted of timber, cocao, coconuts, tobacco, hides, Pito thread (which is very strong, and is made of a species of grass), Quito cloth, wax, etc. The specie on board her was inconsiderable, being principally small silver money, and not amounting to more than £170 sterling. It is true her cargoe was of great value, could we have disposed of it: but the Spaniards having strict orders never to ransom their ships, all the goods that we took in these seas, except what little we had occasion for ourselves, were of no advantage to us. Indeed, though we could make no profit thereby ourselves, it was some satisfaction to us to consider that it was so much really lost to the enemy, and that the despoiling them was no contemptible branch of that service in which we were now employed by our country.

Besides our prize's crew, which amounted to forty-five hands, there were on board her ten passengers, consisting of four men and three women, who were natives of the country, born of Spanish parents, together with three black slaves that attended them. The women were a mother and her two daughters, the eldest about twenty-one, and the youngest about fourteen. It is not to be wondered at that women of these years should be excessively alarmed at the falling into the hands of an enemy, whom, from the former outrages of the buccaneers, and by the artful insinuations of their priests, they had been taught to consider as the most terrible and brutal of all mankind. These apprehensions too were in the present instance exaggerated by the singular beauty of the youngest of the women, and the riotous disposition which they might well expect to find in a set of sailors who had not seen a woman for near a twelvemonth. Full of these terrors, the women all hid themselves upon our officer's coming on board, and when they were found out, it was with great difficulty that he could persuade them to approach the light. However, he soon satisfied them, by the humanity of his conduct, and by his assurances of their future security and honourable treatment, that they had nothing to fear. Nor were these assurances of the officer invalidated in the sequel: for the commodore being informed of the matter, sent directions that they should be continued on board their own ship, with the use of the same apartments, and with all the other conveniencies they had enjoyed before, giving strict orders that they should receive no kind of inquietude or molestation whatever: and that they might be the more certain of having these orders complied with, or have the means of complaining, if they were not, the commodore permitted the pilot, who in Spanish ships is generally the second person on board, to stay with them, as their guardian and protector. The pilot was particularly chosen for this purpose by Mr. Anson, as he seemed to be extremely interested in all that concerned the women, and had at first declared that he was married to the youngest of them, though it afterwards appeared, both from the information of the rest of the prisoners, and other circumstances, that he asserted this with a view the better to secure them from the insults they expected on their first falling into our hands. By this compassionate and indulgent behaviour of the commodore, the consternation of our female prisoners entirely subsided, and they continued easy and cheerful during the whole time they were with us, as I shall have occasion to mention more particularly hereafter.

I have before observed, that at the beginning of this chace the *Centurion* ran her two consorts out of sight, on which account we lay by all the night, after we had taken the prize, for Captain Saunders and Lieutenant Saumarez to join us, firing guns and making false fires every half-hour, to prevent their passing by us unobserved; but they were so far astern that they neither heard nor saw any of our signals, and were not able to come up with us till broad daylight. When they had joined us, we proceeded together to the northward, being now four sail in company. We here found the sea, for many miles round us, of a beautiful red colour. This, upon examination, we imputed to an immense quantity of spawn spread upon its surface; for, taking up some of the water in a wine glass, it soon changed from a dirty aspect to a clear crystal, with only some red globules of a slimy nature floating on the top. At present having a supply of timber on board our new prize, the commodore ordered our boats to be repaired, and a swivel gun-stock to be fixed in the bow both of the barge and pinnace, in order to encrease their force, in case we should be obliged to have recourse to them for boarding ships, or for any attempts on shore.

As we stood from hence to the northward, nothing remarkable occurred for two or three days, though we spread our ships in such a manner that it was not probable any vessel of the enemy could escape us. In our run along this coast we generally observed that there was a current which set us to the northward at the rate of ten or twelve miles each day. And now, being in about eight degrees of south latitude, we began to be attended with vast numbers of flying fish and bonitos, which were the first we saw after our departure from the coast of Brazil. But it is remarkable that on the east side of South America they extended to a much higher latitude than they do on the west side, for we did not lose them on the coast of Brazil till we approached the southern tropic. The reason for this diversity is doubtless the different degrees of heat obtaining in the same latitude on different sides of that continent. And on this occasion, I must beg leave to make a short digression on the heat and cold of different climates, and on the varieties which occur in the same place in different parts of the year, and in different places in the same degree of latitude.

The ancients conceived that of the five zones into which they divided the surface of the globe, two only were habitable, supposing that the heat between the tropics, and the cold within the polar circles, were too intense to be supported by mankind. The falsehood of this reasoning has been long evinced; but the particular comparisons of the heat and cold of these various climates has as yet been very imperfectly considered. However, enough is known safely to determine this position, that all places between the tropics are far from being the hottest on the globe, as many of those within the polar circles are far from enduring that extreme degree of cold to which their situation should seem to subject them: that is to say, that the temperature of a place depends much more upon other circumstances than upon its distance from the pole, or its proximity to the equinoctial.

This proposition relates to the general temperature of places, taking the whole year round; and in this sense it cannot be denied that the city of London, for instance, enjoys much warmer seasons than the bottom of Hudson's Bay, which is nearly in the same latitude with it, but where the severity of the winter is so great that it will scarcely permit the hardiest of our garden plants to live. And if the comparison be made between the coast of Brazil and the western shore of South America, as, for example, betwixt Bahia and Lima, the difference will be still more considerable; for though the coast of Brazil is extremely sultry, yet the coast of the South Seas in the same latitude is perhaps as temperate and tolerable as any part of the globe, since in ranging along it we did not once meet with so warm weather as is frequent in a summer's day in England: which was still the more remarkable as there never fell any rains to refresh and cool the air.

The causes of this temperature in the South Seas are not difficult to be assigned, and shall be hereafter mentioned. I am now only solicitous to establish the truth of this assertion, that the latitude of a place alone is no rule whereby to judge of the degree of heat and cold which obtains there.

Perhaps this position might be more briefly confirmed by observing, that on the tops of the Andes, though under the equinoctial, the snow never melts the whole year round: a criterion of cold stronger than what is known to take place in many parts far removed within the polar circle.

I have hitherto considered the temperature of the air all the year through, and the gross estimations of heat and cold which every one makes from his own sensation. If this matter be examined by means of thermometers, which in respect to the absolute degree of heat and cold are doubtless the most unerring evidences—if this be done, the result will be indeed most wonderful, since it will hence appear that the heat in very high latitudes, as at Petersburgh, for instance, is at particular times much greater than any that has been hitherto observed between the tropics; and that even at London, in the year 1746, there was the part of one day considerably hotter than what was at any time felt by a ship of Mr. Anson's squadron in running from hence to Cape Horn and back again, and passing twice under the sun; for in the summer of that year, the thermometer in London (being one of those graduated according to the method of Farenheit) stood once at 78°; and the greatest height at which a thermometer of the same kind stood in the foregoing ship I find to be 76°: this was at St. Catherine's, in the latter end of December, when the sun was within about three degrees of the vertex. And as to Petersburgh, I find, by the acts of the academy established there, that in the year 1734, on the 20th and 25th of July, the thermometer rose to 98° in the shade, that is, it was twenty-two divisions higher than it was found to be at St. Catherine's; which is a degree of heat that, were it not authorised by the regularity and circumspection with which the observations seem to have been made, would appear altogether incredible.

If it should be asked how it comes to pass, then, that the heat in many places between the tropics is esteemed so violent and insufferable, when it appears, by these instances, that it is sometimes rivalled or exceeded in very high latitudes not far from the polar circle? I should answer that the estimation of heat in any particular place ought not to be founded upon that degree of heat which may now and then obtain there, but is rather to be deduced from the medium observed in a whole season, or perhaps in a whole year; and in this light it will easily appear how much more intense the same degree of heat may prove by being long continued without remarkable variation. For instance, in comparing together St. Catherine's and Petersburgh, we will suppose the summer heat at St. Catherine's to be 76°, and the winter heat to be twenty divisions short of it. I do not make use of this last conjecture upon sufficient observation, but I am apt to suspect that the allowance is full large. Upon this supposition, then, the medium heat all the year round will be 66°, and this perhaps by night as well as day, with no great variation. Now those who have attended to thermometers will readily own that a continuation of this degree of heat for a length of time would, by the generality of mankind, be stiled violent and suffocating; but at Petersburgh, though a few times in the year the heat by the thermometer may be considerably greater than at St. Catherine's, yet, as at other times the cold is immensely sharper, the medium for a year, or even for one season only, would be far short of 66°. For I find that the thermometer at Petersburgh is at least five times greater, from its highest to its lowest point, than what I have supposed to take place at St. Catherine's.

Besides this estimation of the heat of a place, by taking the medium for a considerable time together, there is another circumstance which will still augment the apparent heat of the warmer climates, and diminish that of the colder, though I do not remember to have seen it remarked in any author. To explain myself more distinctly upon this head, I must observe that the measure of absolute heat marked by the thermometer is not the certain criterion of the sensation of heat with which human bodies are affected: for as the presence and perpetual succession of fresh air is necessary to our respiration, so there is a species of tainted or stagnated air often produced by the continuance of great heats, which, being less proper for respiration, never fails to excite in us an idea of sultriness and suffocating warmth much beyond what the heat of the air alone, supposing it pure and agitated, would occasion. Hence it follows, that the mere inspection of the thermometer will never determine the heat which the human body feels from this cause; and hence it follows, too, that the heat in most places between the tropics must be much more troublesome and uneasy than the same degree of absolute heat in a high latitude: for the equability and duration of the tropical heat contribute to impregnate the air with a multitude of steams and vapours from the soil and water, and these being, many of them, of an impure and noxious kind, and being not easily removed, by reason of the regularity of the winds in those parts, which only shift the exhalations from place to place without dispersing them, the atmosphere is by this means rendered less capable of supporting the animal functions, and mankind are consequently affected with what they stile a most intense and stifling heat: whereas in the higher latitudes these vapours are probably raised in smaller quantities, and the irregularity and violence of the winds frequently disperse them, so that, the air being in general pure and less stagnant, the same degree of absolute heat is not attended with that uneasy and suffocating sensation. This may suffice in general with respect to the present speculation; but I cannot help wishing, as it is a subject in which mankind, especially travellers of all sorts, are very much interested, that it were more thoroughly and accurately examined, and that all ships bound to the warmer climates would furnish themselves with thermometers of a known fabric, and would observe them daily, and register their observations; for considering the turn to philosophical inquiries which has obtained in Europe for the last fourscore years, it is incredible how very rarely anything of this kind hath been attended to. As to my own part, I do not recollect that I have ever seen any observations of the heat and cold, either in the East or West Indies, which were made by mariners or officers of vessels, except those made by Mr. Anson's order on board the *Centurion*, and by Captain Legg on board the *Severn*, which was another ship of our squadron.

This digression I have been in some measure drawn into by the consideration of the fine weather we met with on the coast of Peru, even under the equinoctial itself, but the particularities of this weather I have not yet described: I shall now therefore add, that in this climate every circumstance concurred that could make the open air and the daylight desirable. For in other countries the scorching heat of the sun in summer renders the greater part of the day unapt either for labour or amusement; and the frequent rains are not less troublesome in the more temperate parts of the year. But in this happy climate the sun rarely appears. Not that the heavens have at any time a dark and gloomy look; for there is constantly a chearful grey sky, just sufficient to screen the sun, and to mitigate the violence of its perpendicular rays, without obscuring the air, or tinging the daylight with an unpleasant or melancholy hue. By this means all parts of the day are proper for labour or exercise abroad, nor is there wanting that refreshment and pleasing refrigeration of the air which is sometimes produced in other climates by rains; for here the same effect is brought about by the fresh breezes from the cooler regions to the southward. It is reasonable to suppose that this fortunate complexion of the heavens is principally owing to the neighbourhood of those vast hills called the Andes, which, running nearly parallel to the shore, and at a small distance from it, and extending themselves immensely higher than any other mountains upon the globe, form upon their sides and declivities a prodigious tract of country, where, according to the different approaches to the summit, all kinds of climates may at all seasons of the year be found. These mountains, by intercepting great part of the eastern winds which generally blow over the continent of South America, and by cooling that part of the air which forces its way over their tops, and by keeping besides a large portion of the atmosphere perpetually cool, from its contiguity to the snows with which they are covered—these hills, thus spreading the influence of their frozen crests to the neighbouring coasts and seas of Peru, are doubtless the cause of the temperature and equability which constantly prevail there. For when we were advanced beyond the equinoctial, where these mountains left us, and had nothing to screen us to the eastward but the highlands on the isthmus of Panama, which are but mole-hills to the Andes, we then soon found that in a short run we had totally changed our climate, passing in two or three days from the temperate air of Peru to the sultry burning atmosphere of the West Indies. But it is time to return to our narration.

On the 10th of November we were three leagues south of the southermost island of Lobos, lying in the latitude of 6° 27' south. There are two islands of this name: this called Lobos de la Mar, and another, which is situated to the northward of it, very much resembling it in shape and appearance, and often mistaken for it, called Lobos de Tierra. We were now drawing near to the station appointed to the *Gloucester*, for which reason, fearing to miss her, we made an easy sail all night. The next morning, at daybreak, we saw a ship in shore, and to windward, plying up the coast. She had passed by us with the favour of the night, and we soon perceiving her not to be the *Gloucester*, got our tacks on board and gave her chace; but it proving very little wind, so that neither of us could make much way, the commodore ordered the barge, his pinnace, and the *Tryal's* pinnace to be manned and armed, and to pursue the chace and board her. Lieutenant Brett, who commanded the barge, came up with her first, about nine o'clock, and running alongside of her, he fired a volley of small shot between the masts, just over the heads of the people on board, and then instantly entered with the greatest part of his men; but the enemy made no resistance, being sufficiently frighted by the dazzling of the cutlasses, and the volley they had just received. Lieutenant Brett ordered the sails to be trimmed, and bore down to the commodore, taking up in his way the two pinnaces. When he was got within about four miles of us, he put off in the barge, bringing with him a number of the prisoners, who had given him some material intelligence, which he was desirous the commodore should be acquainted with as soon as possible. On his arrival we learnt that the prize was called *Nuestra Senora del Carmin*, of about two hundred and seventy tuns burthen; she was commanded by Marcos Morena, a native of Venice, and had on board forty-three mariners. She was deep laden with steel, iron, wax, pepper, cedar, plank, snuff, rosarios, European bale goods, powder-blue, cinnamon, Romish indulgencies, and other species of merchandize; and though this cargo, in our present circumstances, was but of little value to us, yet with respect to the Spaniards it was the most considerable capture we made in this part of the world, for it amounted to upwards of 400,000 dollars prime cost at Panama. This ship was bound to Callao, and had stopped at Paita in her passage, to take in a recruit of water and provisions, having left that place not above twenty-four hours before she fell into our hands.

I have mentioned that Mr. Brett had received some important intelligence, which he endeavoured to let the commodore know immediately. The first person he learnt it from (though upon further examination it was confirmed by the other prisoners) was one John Williams, an Irishman, whom he found on board the Spanish vessel. Williams was a Papist, who worked his passage from Cadiz, and had travelled over all the kingdom of Mexico as a pedlar. He pretended that by this business he had once got 4000 or 5000 dollars, but that he was embarrassed by the priests, who knew he had money, and was at last stript of everything he had. He was indeed at present all in rags, being but just got out of Paita gaol, where he had been confined for some misdemeanor; he expressed great joy upon seeing his countrymen, and immediately told them that, a few days before, a vessel came into Paita, where the master of her informed the governor that he had been chased in the offing by a very large ship, which, from her size and the colour of her sails, he was persuaded must be one of the English squadron. This we then conjectured to have been the *Gloucester*, as we afterwards found it was. The governor, upon examining the master, was fully satisfied of his relation, and immediately sent away an express to Lima to acquaint the viceroy therewith; and the royal officer residing at Paita, apprehensive of a visit from the English, had, from his first hearing of this news, been busily employed in removing the king's treasure and his own to Piura, a town within land about fourteen leagues distant. We further learnt from our prisoners that there was a very considerable sum of money belonging to some merchants of Lima that was now lodged in the custom-house at Paita, and that this was intended to be shipped on board a vessel, which was then in the port of Paita, and was preparing to sail with the utmost expedition, being bound for the bay of Sonsonnate, on the coast of Mexico, in order to purchase a part of the cargo of the Manila ship. As the vessel on which the money was to be shipped was esteemed a prime sailor, and had just received a new coat of tallow on her bottom, and might, in the opinion of the prisoners, be able to sail the succeeding morning, the character they gave of her left us little reason to believe that our ship, which had been in the water near two years, could have any chance of coming up with her if we once suffered her to escape out of the port. Therefore, as we were now discovered, and the coast would be soon alarmed, and as our cruising in these parts any longer would answer no purpose, the commodore resolved to endeavour to surprize the place, having first minutely informed himself of its strength and condition, and being fully satisfied that there was little danger of losing many of our men in the attempt. This attack on Paita, besides the treasure it promised us, and its being the only enterprize it was in our power to undertake, had these other advantages attending it, that we should in all probability supply ourselves with great quantities of live provision, of which we were at this time in want: and that we should likewise have an opportunity of setting our prisoners on shore, who were now very numerous, and made a greater consumption of our food than our stock that remained was capable of furnishing long. In all these lights the attempt was a most eligible one, and what our necessities, our situation, and every prudential consideration prompted us to. How it succeeded, and how far it answered our expectations, shall be the subject of the following chapter.

CHAPTER VI

THE TAKING OF PAITA, AND OUR PROCEEDINGS THERE

The town of Paita is situated in the latitude of 5° 12' south, on a most barren soil, composed only of sand and slate. The extent of it is but small, containing in all less than two hundred families. The houses are only ground floors, the walls built of split cane and mud, and the roofs thatched with leaves. These edifices, though extremely slight, are abundantly sufficient for a climate where rain is considered as a prodigy, and is not seen in many years: so that it is said a small quantity of rain falling in this country in the year 1728 ruined a great number of buildings, which mouldered away, and as it were melted before it. The inhabitants of Paita are principally Indians and black slaves, or at least a mixed breed, the whites being very few. The port of Paita, though in reality little more than a bay, is esteemed the best on that part of the coast, and is indeed a very secure and commodious anchorage. It is greatly frequented by all vessels coming from the north, since here only the ships from Acapulco, Sonsonnate, Realeijo, and Panama can touch and refresh in their passage to Callao: and the length of these voyages (the wind for the greatest part of the year being full against them) renders it impossible to perform them without calling upon the coast for a recruit of fresh water. It is true Paita is situated on so parched a spot that it does not itself furnish a drop of fresh water, or any kind of greens or provisions, except fish and a few goats; but there is an Indian town called Colan, about two or three leagues distant to the northward, from whence water, maize, greens, fowls, etc., are conveyed to Paita on balsas or floats, for the conveniency of the ships that touch here; and cattle are sometimes brought from Piura, a town which lies about fourteen leagues up in the country. The water fetched from Colan is whitish, and of a disagreeable appearance, but is said to be very wholesome, for it is pretended by the inhabitants that it runs through large woods of sarsaparilla, and is sensibly impregnated therewith. This port of Paita, besides furnishing the northern trade bound to Callao with water and necessaries, is the usual place where passengers from Acapulco or Panama, bound to Lima, disembark; for, as it was two hundred leagues from hence to Callao, the port of Lima, and as the wind is generally contrary, the passage by sea is very tedious and fatiguing, but by land there is a tolerable good road parallel to the coast, with many stations and villages for the accommodation of travellers.

It appears that the town of Paita is itself an open place, so that its sole protection and defence is a fort. It was of consequence to us to be well informed of the fabrick and strength of this fort; and from the examination of our prisoners we found that there were eight pieces of cannon mounted in it, but that it had neither ditch nor outwork, being surrounded by a plain brick wall; and that the garrison consisted of only one weak company, though the town itself might possibly arm three hundred men more.

Mr. Anson having informed himself of the strength of the place, resolved (as hath been said in the preceding chapter) to attempt it that very night. We were then about twelve leagues distant from the shore, far enough to prevent our being discovered, yet not so far but that by making all the sail we could, we might arrive in the bay with our ships long before daybreak. However, the commodore prudently considered that this would be an improper method of proceeding, as our ships, being such large bodies, might be easily seen at a distance even in the night, and might thereby alarm the inhabitants, and give them an opportunity of removing their valuable effects. He therefore, as the strength of the place did not require our whole force, resolved to attempt it with our boats only, ordering the eighteen-oared barge and our own and the *Tryal's* pinnaces on that service; and having picked out fifty-eight men to mann them, well furnished with arms and ammunition, he entrusted the command of the expedition to Lieutenant Brett, and gave him his necessary orders. And the better to prevent the disappointment and confusion which might arise from the darkness of the night, and from the ignorance of the streets and passages of the place, two of the Spanish pilots were ordered to attend the lieutenant, who were to conduct him to the most convenient landing-place, and were afterwards to be his guides on shore; and that we might have the greater security for their behaviour on this occasion, the commodore took care to assure our prisoners that they should all of them be released and set on shore at this place, provided the pilots acted faithfully; but in case of any misconduct or treachery, he threatened that the pilots should be instantly shot, and that he would carry the rest of the Spaniards who were on board him prisoners to England. So that the prisoners themselves were interested in our success, and therefore we had no reason to suspect our conductors either of negligence or perfidy.

On this occasion I cannot but remark a singular circumstance of one of the pilots employed by us in this business. It seems (as we afterwards learnt) he had been taken by Captain Clipperton above twenty years before, and had been obliged to lead Clipperton and his people to the surprize of Truxillo, a town within land to the southward of Paita, where, however, he contrived to alarm his countrymen and to save them, though the place was carried and pillaged. Now that the only two attempts on shore, which were made at so long an interval from each other, should be guided by the same person, and he too a prisoner both times, and forced upon the employ contrary to his inclination, is an incident so very extraordinary that I could not help mentioning it. But to return to the matter in hand.

During our preparations, the ships themselves stood towards the port with all the sail they could make, being secure that we were yet at too great a distance to be seen. But about ten o'clock at night, the ships being then within five leagues of the place, Lieutenant Brett, with the boats under his command, put off, and arrived at the mouth of the bay without being discovered, though no sooner had he entered it than some of the people on board a vessel riding at anchor there perceived him, who instantly getting into their boat, rowed towards the fort, shouting and crying, "The English, the English dogs," etc., by which the whole town was suddenly alarmed, and our people soon observed several lights hurrying backwards and forwards in the fort, and other marks of the inhabitants being in great motion. Lieutenant Brett, on this, encouraged his men to pull briskly up, that they might give the enemy as little time as possible to prepare for their defence. However, before our boats could reach the shore, the people in the fort had got ready some of their cannon, and pointed them towards the landing-place; and though in the darkness of the night it might be well supposed that chance had a greater share than skill in their direction, yet the first shot passed extremely near one of the boats, whistling just over the heads of the crew. This made our people redouble their efforts, so that they had reached the shore and were in part disembarked by the time the second gun fired. As soon as our men landed, they were conducted by one of the Spanish pilots to the entrance of a narrow street, not above fifty yards distant from the beach, where they were covered from the fire of the fort; and being formed in the best manner the shortness of the time would allow, they immediately marched for the parade, which was a large square at the end of this street, the fort being one side of the square, and the governor's house another. In this march (though performed with tolerable regularity) the shouts and clamours of threescore sailors, who had been confined so long on shipboard, and were now for the first time on shore in an enemy's country, joyous as they always are when they land, and animated besides in the present case with the hopes of an immense pillage— the huzzas, I say, of this spirited detachment, joined with the noise of their drums, and favoured by the night, had augmented their numbers, in the opinion of the enemy, to at least three hundred, by which persuasion the inhabitants were so greatly intimidated that they were much more solicitous about the means of flight than of resistance: so that though upon entering the parade our people received a volley from the merchants who owned the treasure then in the town, and who, with a few others, had ranged themselves in a gallery that ran round the governor's house, yet that post was immediately abandoned upon the first fire made by our people, who were thereby left in quiet possession of the parade.

On this success Lieutenant Brett divided his men into two parties, ordering one of them to surround the governor's house, and, if possible, to secure the governor, whilst he himself at the head of the other marched to the fort, with an intent to force it. But, contrary to his expectation, he entered it without opposition; for the enemy, on his approach, abandoned it, and made their escape over the walls. By this means the whole place was mastered in less than a quarter of an hour's time from the first landing, and with no other loss than that of one man killed on the spot, and two wounded, one of which was the Spanish pilot of the *Teresa*, who received a slight bruise by a ball which grazed on his wrist. Indeed another of the company, the Honourable Mr. Kepple, son to the Earl of Albemarle, had a very narrow escape; for having on a jockey cap, one side of the peak was shaved off close to his temple by a ball, which, however, did him no other injury.

Lieutenant Brett, when he had thus far happily succeeded, placed a guard at the fort, and another at the governor's house, and appointed centinels at all the avenues of the town, both to prevent any surprize from the enemy, and to secure the effects in the place from being embezzled. This being done, his next care was to seize on the custom-house, where the treasure lay, and to examine if any of the inhabitants remained in the town, that he might know what further precautions it was necessary to take; but he soon found that the numbers left behind were no ways formidable, for the greatest part of them (being in bed when the place was surprized) had run away with so much precipitation that they had not given themselves time to put on their cloaths. In this general rout the governor was not the last to secure himself, for he fled betimes half naked, leaving his wife, a young lady of about seventeen years of age, to whom he had been married but three or four days, behind him, though she too was afterwards carried off in her shift by a couple of centinels, just as the detachment ordered to invest the house arrived before it. This escape of the governor was an unpleasing circumstance, as Mr. Anson had particularly recommended it to Lieutenant Brett to secure his person if possible, in hopes that by that means we might be able to treat for the ransom of the place: but it seems his alertness rendered the execution of these orders impracticable. The few inhabitants who remained were confined in one of the churches under a guard, except some stout negroes which were found in the town; these, instead of being shut up, were employed the remaining part of the night to assist in carrying the treasure from the custom-house and other places to the fort; however, there was care taken that they should be always attended by a file of musqueteers.

The transporting the treasure from the custom-house to the fort was the principal occupation of Mr. Brett's people after he had got possession of the place. But the sailors, while they were thus busied, could not be prevented from entering the houses which lay near them in search of private pillage: where the first things which occurred to them being the cloaths that the Spaniards in their flight had left behind them, and which, according to the custom of the country, were most of them either embroidered or laced, our people eagerly seized these glittering habits, and put them on over their own dirty trowsers and jackets, not forgetting, at the same time, the tye or bag-wig and laced hat which were generally found with the cloaths; and when this practice was once begun, there was no preventing the whole detachment from imitating it: but those who came latest into the fashion not finding men's cloaths sufficient to equip themselves, were obliged to take up with women's gowns and petticoats, which (provided there was finery enough) they made no scruple of putting on and blending with their own greasy dress. So that when a party of them thus ridiculously metamorphosed first appeared before Mr. Brett, he was extremely surprized at the grotesque sight, and could not immediately be satisfied they were his own people.

These were the transactions of our detachment on shore at Paita the first night: but to return to what was done on board the *Centurion* in that interval. I must observe that after the boats were gone off, we lay by till one o'clock in the morning, and then supposing our detachment to be near landing, we made an easy sail for the bay. About seven in the morning we began to open the bay, and soon after had a view of the town; and though we had no reason to doubt of the success of the enterprize, yet it was with great joy that we first discovered an infallible signal of the certainty of our hopes; this was by means of our perspectives, for through them we saw an English flag hoisted on the flagstaff of the fort, which to us was an incontestable proof that our people were in possession of the place. We plied into the bay with as much expedition as the wind, which then blew off shore, would permit us: and at eleven the *Tryal's* boat came on board us, loaden with dollars and church-plate, when the officer who commanded her informed us of the preceding night's transactions, such as we have already related them. About two in the afternoon we anchored in ten fathom and a half, at a mile and a half distance from the town, and were consequently near enough to have a more immediate intercourse with those on shore. And now we found that Mr. Brett had hitherto gone on in collecting and removing the treasure without interruption; but that the enemy had rendezvouzed from all parts of the country on a hill, at the back of the town, where they made no inconsiderable appearance: for amongst the rest of their force there were two hundred horse, seemingly very well armed and mounted, and, as we conceived, properly trained and regimented, being furnished with trumpets, drums, and standards. These troops paraded about the hill with great ostentation, sounding their military musick, and practising every art to intimidate us (as our numbers on shore were by this time not unknown to them), in hopes that we might be induced by our fears to abandon the place before the pillage was compleated. But we were not so ignorant as to believe that this body of horse, which seemed to be what the enemy principally depended on, would dare to venture in streets and amongst houses, even had their numbers been three times as large; and, therefore, notwithstanding their menaces, we went on calmly as long as the daylight lasted, in sending off the treasure, and in employing the boats to carry on board the refreshments, such as hogs, fowls, etc., which we found here in great abundance. However, at night, to prevent any surprize, the commodore sent on shore a reinforcement, who posted themselves in all the passages leading to the parade, and, for their further security, traversed the streets with barricadoes six feet high, but the enemy continuing quiet all night, we, at daybreak, returned again to our labour of loading the boats, and sending them off.

By this time we were convinced of what consequence it would have been to us had fortune seconded the prudent views of the commodore, by permitting us to have secured the governor. For as we found in the place many storehouses full of valuable effects, which were useless to us at present, and such as we could not find room for on board, had the governor been in our power, he would in all probability have treated for the ransom of this merchandise, which would have been extremely advantageous both to him and us; whereas, he being now at liberty, and having collected all the force of the country for many leagues round, and having even got a body of militia from Piura, which was fourteen leagues distant, he was so elated with his numbers, and so fond of his new military command, that he seemed not to trouble himself about the fate of his government. So that though Mr. Anson sent several messages to him by some of the inhabitants whom he had taken prisoners, offering to enter into a treaty for the ransom of the town and goods, giving him, at the same time, an intimation that we should be far from insisting on a rigorous equivalent, but perhaps might be satisfied with some live cattle, and a few necessaries for the use of the squadron, threatening, too, that if he would not condescend at least to treat, we would set fire to the town and all the warehouses. Yet the governor was so imprudent and arrogant that he despised all these reiterated overtures, and did not deign even to return the least answer to them.

On the second day of our being in possession of the place, several negro slaves deserted from the enemy on the hill, and coming into the town, voluntarily engaged in our service. One of these was well known to a gentleman on board who remembered him formerly at Panama. We now learnt that the Spaniards without the town were in extreme want of water, for many of their slaves crept into the place by stealth, and carried away several jars of water to their masters on the hill; and though some of them were seized by our men in the attempt, yet the thirst among the enemy was so pressing that they continued this practice till we left the place. On this second day we were assured, both by the deserters and by these prisoners we took, that the Spaniards on the hill, who were by this time encreased to a formidable number, had resolved to storm the town and fort the succeeding night; and that one Gordon, a Scots papist, and captain of a ship in those seas, was to have the command of this enterprize. However, we, notwithstanding, continued sending off our boats, and prosecuted our work without the least hurry or precipitation till the evening; when a reinforcement was again sent on shore by the commodore, and Lieutenant Brett doubled his guards at each of the barricadoes; and our posts being connected by the means of centinels placed within call of each other, and the whole being visited by frequent rounds, attended with a drum, these marks of our vigilance, which the enemy could not be ignorant of, as they could doubtless hear the drum, if not the calls of the centinels; these marks, I say, of our vigilance, and of our readiness to receive them, cooled their resolution, and made them forget the vaunts of the preceding day, so that we passed this second night with as little molestation as we had done the first.

We had finished sending the treasure on board the *Centurion* the evening before, so that the third morning, being the 15th of November, the boats were employed in carrying off the most valuable part of the effects that remained in the town. And the commodore intending to sail in the afternoon, he, about ten o'clock, pursuant to his promise, sent all his prisoners, amounting to eighty-eight, on shore, giving orders to Lieutenant Brett to secure them in one of the churches under a strict guard till the men were ready to be embarked. Mr. Brett was at the same time ordered to burn the whole town, except the two churches (which by good fortune stood at some distance from the houses), and then he was to abandon the place, and to return on board. These orders were punctually complied with, for Mr. Brett immediately set his men to work to distribute pitch, tar, and other combustibles (of which great quantities were found here) into houses situated in different streets of the town, so that the place being fired in many quarters at the same time, the destruction might be more violent and sudden, and the enemy, after our departure, might not be able to extinguish it. When these preparations were made, he, in the next place, commanded the cannon, which he found in the fort, to be nailed up; and then setting fire to those houses which were most to the windward, he collected his men and marched towards the beach, where the boats waited to carry them off. As that part of the beach whence he intended to embark was an open place without the town, the Spaniards on the hill perceiving he was retreating, resolved to try if they could not precipitate his departure, and thereby lay some foundation for their future boasting. To this end a small squadron of their horse, consisting of about sixty, picked out, as I suppose, for this service, marched down the hill with much seeming resolution, so that had we not entertained an adequate opinion of their prowess, we might have imagined that now we were on the open beach with no advantage of situation, they would certainly have charged us, but we presumed (and we were not mistaken) that this was mere ostentation. For, notwithstanding the pomp and parade they at first came on with, Mr. Brett had no sooner ordered his men to halt and face about, than the enemy stopped their career, and never dared to advance a step further.

When our people were arrived at their boats, and were ready to go on board, they were for some time retarded by missing one of their number; and being unable, on their mutual enquiries amongst each other, to inform themselves where he was left, or by what accident he was detained, they, after a considerable delay, resolved to get into their boats, and to depart without him. But when the last man was actually embarked, and the boats were just putting off, they heard him calling to them to take him in. The place was by this time so thoroughly on fire, and the smoke covered the beach so effectually, that they could scarcely discern him, though they heard his voice. However, the lieutenant instantly ordered one of the boats to his relief, who found him up to the chin in water, for he had waded as far as he durst, being extremely frightened with the apprehensions of falling into the hands of an enemy, enraged, as they doubtless were, at the pillage and destruction of their town. On enquiring into the cause of his staying behind, it was found that he had taken that morning too large a dose of brandy, which had thrown him into so sound a sleep that he did not awake till the fire came near enough to scorch him. He was strangely amazed at first opening his eyes to see the houses all in a blaze on one side, and several Spaniards and Indians not far from him on the other. The greatness and suddenness of his fright instantly reduced him to a state of sobriety, and gave him sufficient presence of mind to push through the thickest of the smoke, as the likeliest means to escape the enemy, and making the best of his way to the beach, he ran as far into the water as he durst (for he could not swim) before he ventured to look back.

I cannot but observe here, to the honour of our people, that though there were great quantities of wine and spirituous liquors found in the place, yet this man was the only one who was known to have so far neglected his duty as to get drunk. Indeed, their whole behaviour while they were ashore was much more regular than could well have been expected from sailors who had been so long confined to a ship, and though part of this prudent demeanour must doubtless be imputed to the diligence of their officers, and to the excellent discipline to which they had been constantly inured on board by the commodore, yet it was doubtless no small reputation to the men, that they should generally refrain from indulging themselves in those intoxicating liquors which they found ready to their hands at almost every warehouse.

Having mentioned this single instance of drunkenness, I cannot pass by another oversight, which was likewise the only one of its kind, and which was attended with very particular circumstances. There was an Englishman, who had formerly wrought as a ship-carpenter in the yard at Portsmouth, but leaving his country, had afterwards entered into the Spanish service, and was employed by them at the port of Guaiaquil; and it being well known to his friends in England that he was then in that part of the world, they put letters on board the *Centurion*, directed to him. This man being then by accident amongst the Spaniards, who were retired to the hill at Paita, he was ambitious (as it should seem) of acquiring some reputation amongst his new masters. With this view he came down unarmed to a centinel of ours, placed at some distance from the fort towards the enemy, to whom he pretended that he was desirous of surrendering himself, and of entering into our service. Our centinel had a cocked pistol in his hand, but being deceived by the other's fair speeches, he was so imprudent as to let him approach much nearer than he ought, so that the shipwright, watching his opportunity, rushed on the centinel, and seizing his pistol, wrenched it out of his hand, and instantly ran away with it up the hill. By this time two of our people, who, seeing the fellow advance, had suspected his intention, were making towards him, and were thereby prepared to pursue him, but he got to the top of the hill before they could reach him, and then turning about, fired the pistol, whereupon his pursuers immediately returned the fire, and though he was at a great distance, and the crest of the hill hid him as soon as they had fired, so that they took it for granted they had missed him, yet we afterwards learnt that he was shot through the body, and had fallen down dead the very next step he took after he was out of sight. The centinel, too, who had been thus grossly imposed upon, did not escape unpunished, since he was ordered to be severely whipt for being thus shamefully surprized upon his post, and having thereby given an example of carelessness which, if followed in other instances, might prove fatal to us all. But to return.

By the time our people had helped their comrade out of the water, and were making the best of their way to the squadron, the flames had taken possession of every part of the town, and had got such hold, both by means of the combustibles that had been distributed for that purpose, and by the slightness of the materials of which the houses were composed, and their aptitude to take fire, that it was sufficiently apparent no efforts of the enemy (though they flocked down in great numbers) could possibly put a stop to it, or prevent the entire destruction of the place and all the merchandize contained therein. A whole town on fire at once, especially where the buildings burnt with such facility and violence, being a very singular spectacle, Mr. Brett had the curiosity to delineate its appearance, together with that of the ships in the harbour.

Our detachment under Lieutenant Brett having safely joined the squadron, the commodore prepared to leave the place the same evening. He found when he first came into the bay, six vessels of the enemy at anchor; one whereof was the ship, which, according to our intelligence, was to have sailed with the treasure to the coast of Mexico, and which, as we were persuaded she was a good sailor, we resolved to take with us. The others were two snows, a bark, and two row gallies of thirty-six oars a-piece. These last, as we were afterwards informed, with many others of the same kind built at divers ports, were intended to prevent our landing in the neighbourhood of Callao, for the Spaniards, on the first intelligence of our squadron and its force, expected that we would attempt the city of Lima. The commodore, having no occasion for these other vessels, had ordered the masts of all five of them to be cut away at his first arrival, and on his leaving the place they were towed out of the harbour and scuttled and sunk; and the command of the remaining ship, called the *Solidad*, being given to Mr. Hughes, the lieutenant of the *Tryal*, who had with him a crew of ten men to navigate her, the squadron, towards midnight, weighed anchor and sailed out of the bay, being at present augmented to six sail, that is, the *Centurion*, and the *Tryal's* prize, together with the *Carmelo*, the *Teresa*, the *Carmin*, and our last acquired vessel the *Solidad*.

And now, before I entirely quit the account of our transactions at this place, it may not perhaps be improper to give a succinct relation of the booty we got here, and of the loss the Spaniards sustained. I have before observed that there were great quantities of valuable effects in the town, but as most of them were what we could neither dispose of nor carry away, the total amount of this merchandize can only be rudely guessed at. The Spaniards, in their representations sent to the Court of Madrid (as we were afterwards assured) estimated their whole loss at a million and a half of dollars, and when it is considered that no small part of the goods we burnt there were of the richest and most expensive species, as broad cloths, silks, cambricks, velvets, etc., I cannot but think their valuation sufficiently moderate. As to ourselves, the acquisition we made, though inconsiderable in comparison of what we destroyed, was yet far from despicable, for the wrought plate, dollars, and other coin which fell into our hands, amounted to upwards of £30,000 sterling, besides several rings, bracelets, and jewels, whose intrinsic value we could not then determine; and over and above all this, the plunder which became the property of the immediate captors was very great, so that upon the whole it was by much the most important booty we met with upon that coast.

There remains still another matter to be related, which on account of the signal honour which our national character in those parts has thence received, and the reputation which our commodore in particular has thereby acquired, merits a distinct and circumstantial discussion. It has been already observed that all the prisoners taken by us in our preceding prizes were here put on shore and discharged, amongst whom there were some persons of considerable distinction, especially a youth of about seventeen years of age, son of the vice-president of the Council of Chili. As the barbarity of the buccaneers, and the artful use the ecclesiasticks had made of it, had filled the natives of those countries with the most terrible ideas of the English cruelty, we always found our prisoners, at their first coming on board us, to be extremely dejected, and under great horror and anxiety. Particularly this youth whom I last mentioned, having never been from home before, lamented his captivity in the most moving manner, regretting, in very plaintive terms, his parents, his brothers, his sisters, and his native country, of all which he was fully persuaded he had taken his last farewel, believing that he was now devoted for the remaining part of his life to an abject and cruel servitude. Indeed his companions on board, and all the Spaniards that came into our power, had the same desponding opinion of their situation. Mr. Anson constantly exerted his utmost endeavours to efface these terrifying impressions they had received of us, always taking care that as many of the principal people among them as there was room for should dine at his table by turns, and giving the strictest orders, too, that they should at all times, and in every circumstance, be treated with the utmost decency and humanity. But notwithstanding this precaution, it was generally observed that the first day or two they did not quit their fear, suspecting the gentleness of their usage to be only preparatory to some unthought-of calamity. However, being at length convinced of our sincerity, they grew perfectly easy in their situation, and remarkably chearful, so that it was often disputable whether or no they considered their being detained by us as a misfortune. For the youth I have above mentioned, who was near two months on board us, had at last so far conquered his melancholy surmises, and had taken such an affection to Mr. Anson, and seemed so much pleased with the manner of life, totally different from all he had ever seen before, that it is doubtful to me whether, if his own opinion had been asked, he would not have preferred a voyage to England in the *Centurion* to the being set on shore at Paita, where he was at liberty to return to his country and friends.

This conduct of the commodore to his prisoners, which was continued without interruption or deviation, gave them all the highest idea of his humanity and benevolence, and induced them likewise (as mankind are fond of forming general opinions) to entertain very favourable thoughts of the whole English nation. But whatever they might be disposed to think of Mr. Anson before the capture of the *Teresa*, their veneration for him was prodigiously increased by his conduct towards those women whom (as I have already mentioned) he took in that vessel: for the leaving them in the possession of their apartments, the strict orders given to prevent all his people on board from approaching them, and the permitting the pilot to stay with them as their guardian, were measures that seemed so different from what might be expected from an enemy and an heretick, that the Spaniards on board, though they had themselves experienced his beneficence, were surprized at this new instance of it, and the more so as all this was done without his ever seeing the women, though the two daughters were both esteemed handsome, and the youngest was celebrated for her uncommon beauty. The women themselves, too, were so sensible of the obligations they owed him for the care and attention with which he had protected them, that they absolutely refused to go on shore at Paita till they had been permitted to wait on him on board the *Centurion*, to return him thanks in person. Indeed, all the prisoners left us with the strongest assurances of their grateful remembrance of his uncommon treatment. A Jesuit in particular, whom the commodore had taken, and who was an ecclesiastick of some distinction, could not help expressing himself with great thankfulness for the civilities he and his countrymen had found on board, declaring that he should consider it as his duty to do Mr. Anson justice at all times; adding that his usage of the men prisoners was such as could never be forgot, and such as he could never fail to acknowledge and recite upon all occasions: but that his behaviour to the women was so extraordinary, and so extremely honourable, that he doubted all the regard due to his own ecclesiastical character would be scarcely sufficient to render it credible. Indeed we were afterwards informed that he and the rest of our prisoners had not been silent on this head, but had, both at Lima and at other places, given the greatest encomiums to our commodore; the Jesuit in particular, as we were told, having on his account interpreted in a lax and hypothetical sense that article of his church which asserts the impossibility of hereticks being saved.

Nor let it be imagined that the impressions which the Spaniards hence received to our advantage is a matter of small import; for, not to mention several of our countrymen who have already felt the good effects of these prepossessions, the Spaniards are a nation whose good opinion of us is doubtless of more consequence than that of all the world besides, not only as the commerce we had formerly carried on with them, and perhaps may again hereafter, is so extremely valuable, but also as the transacting it does so immediately depend on the honour and good faith of those who are intrusted with its management. However, had no national conveniences attended it, the commodore's equity and good temper would not less have deterred him from all tyranny and cruelty to those whom the fortune of war had put into his hands. I shall only add, that by his constant attachment to these humane and prudent maxims he has acquired a distinguished reputation amongst the Creolian Spaniards, which is not confined merely to the coast of the South Seas, but is extended through all the Spanish settlements in America; so that his name is frequently to be met with in the mouths of most of the Spanish inhabitants of that prodigious empire.

CHAPTER VII

FROM OUR DEPARTURE FROM PAITA TO OUR ARRIVAL AT QUIBO

When we got under sail from the coast of Paita (which, as I have already observed, was about midnight on the 16th of November) we stood to the westward, and in the morning the commodore gave orders that the whole squadron should spread themselves to look out for the *Gloucester*. For we then drew near the station where Captain Mitchel had been directed to cruise, and we hourly expected to get sight of him; but the whole day passed without seeing him.

And now a jealousy, which had taken its rise at Paita, between those who had been commanded on shore for the attack, and those who had continued on board, grew to such a height that the commodore, being made acquainted with it, thought it necessary to interpose his authority to appease it. The ground of this animosity was the plunder gotten at Paita, which those who had acted on shore had appropriated to themselves, considering it as a reward for the risques they had run, and the resolution they had shown in that service. But those who had remained on board looked on this as a very partial and unjust procedure, urging that had it been left to their choice, they should have preferred the action on shore to the continuing on board; that their duty, while their comrades were on shore, was extremely fatiguing; for besides the labour of the day, they were constantly under arms all night to secure the prisoners, whose numbers exceeded their own, and of whom it was then necessary to be extremely watchful, to prevent any attempts they might have formed in that critical conjuncture: that upon the whole it could not be denied but that the presence of a sufficient force on board was as necessary to the success of the enterprize as the action of the others on shore, and therefore those who had continued on board maintained that they could not be deprived of their share of the plunder without manifest injustice. These were the contests amongst our men, which were carried on with great heat on both sides: and though the plunder in question was a very trifle in comparison of the treasure taken in the place (in which there was no doubt but those on board had an equal right), yet as the obstinacy of sailors is not always regulated by the importance of the matter in dispute, the commodore thought it necessary to put a stop to this ferment betimes. Accordingly, the morning after our leaving Paita, he ordered all hands upon the quarter-deck, where, addressing himself to those who had been detached on shore, he commended their behaviour, and thanked them for their services on that occasion: but then representing to them the reasons urged by those who had continued on board, for an equal distribution of the plunder, he told them that he thought these reasons very conclusive, and that the expectations of their comrades were justly founded; and therefore he insisted, that not only the men, but all the officers likewise, who had been employed in taking the place, should produce the whole of their plunder immediately upon the quarter-deck, and that it should be impartially divided among the whole crew, in proportion to each man's rank and commission: and to prevent those who had been in possession of the plunder from murmuring at this diminution of their share, the commodore added, that as an encouragement to others who might be hereafter employed on like services, he would give his entire share to be distributed amongst those who had been detached for the attack of the place. Thus this troublesome affair, which, if permitted to have gone on, might perhaps have been attended with mischievous consequences, was by the commodore's prudence soon appeased, to the general satisfaction of the ship's company: not but there were some few whose selfish dispositions were uninfluenced by the justice of this procedure, and who were incapable of discerning the force of equity, however glaring, when it tended to deprive them of any part of what they had once got into their hands.

This important business employed the best part of the day after we came from Paita. And now, at night, having no sight of the *Gloucester*, the commodore ordered the squadron to bring to, that we might not pass her in the dark. The next morning we again looked out for her, and at ten we saw a sail, to which we gave chace; and at two in the afternoon we came near enough to discover her to be the *Gloucester*, with a small vessel in tow. About an hour after we were joined by them; and then we learnt that Captain Mitchel, in the whole time of his cruise, had only taken two prizes; one of them being a small snow, whose cargoe consisted chiefly of wine, brandy, and olives in jars, with about £7000 in specie; and the other a large boat or launch, which the *Gloucester's* barge came up with near the shore. The prisoners on board this last vessel alledged that they were very poor, and that their loading consisted only of cotton, though the circumstances in which the barge surprized them seemed to insinuate that they were more opulent than they pretended to be, for the *Gloucester's* people found them at dinner upon pigeon-pye, served up in silver dishes. However, the officer who commanded the barge having opened several of the jars on board, to satisfy his curiosity, and finding nothing in them but cotton, he was inclined to believe the account the prisoners gave him: but the cargoe being taken into the *Gloucester*, and there examined more strictly, they were agreeably surprized to find that the whole was a very extraordinary piece of false package, and that there was concealed among the cotton, in every jar, a considerable quantity of double doubloons and dollars, to the amount on the whole of near £12,000. This treasure was going to Paita, and belonged to the same merchants who were the proprietors of the greatest part of the money we had taken there; so that had this boat escaped the *Gloucester*, it is probable her cargoe would have fallen into our hands. Besides these two prizes which we have mentioned, the *Gloucester's* people told us that they had been in sight of two or three other ships of the enemy, which had escaped them; and one of them we had reason to believe, from some of our intelligence, was of immense value.

Being now joined by the *Gloucester* and her prize, it was resolved that we should stand to the northward, and make the best of our way either to Cape St. Lucas on California, or to Cape Corientes on the coast of Mexico. Indeed the commodore, when at Juan Fernandes, had determined with himself to touch in the neighbourhood of Panama, and to endeavour to get some correspondence overland with the fleet under the command of Admiral Vernon. For when we departed from England, we left a large force at Portsmouth which was intended to be sent to the West Indies, there to be employed in an expedition against some of the Spanish settlements. And Mr. Anson taking it for granted that this enterprize had succeeded, and that Porto Bello perhaps might be then garrisoned by British troops, he hoped that on his arrival at the isthmus he should easily procure an intercourse with our countrymen on the other side, either by the Indians, who were greatly disposed in our favour, or even by the Spaniards themselves, some of whom, for proper rewards, might be induced to carry on this intelligence, which, after it was once begun, might be continued with very little difficulty; so that Mr. Anson flattered himself that he might by this means have received a reinforcement of men from the other side, and that by settling a prudent plan of operations with our commanders in the West Indies, he might have taken even Panama itself, which would have given to the British nation the possession of that isthmus, whereby we should have been in effect masters of all the treasures of Peru, and should have had in our hands an equivalent for any demands, however extraordinary, which we might have been induced to have made on either of the branches of the House of Bourbon.

Such were the projects which the commodore revolved in his thoughts at the island of Juan Fernandes, notwithstanding the feeble condition to which he was then reduced. And indeed, had the success of our force in the West Indies been answerable to the general expectation, it cannot be denied but these views would have been the most prudent that could have been thought of. But in examining the papers which were found on board the *Carmelo*, the first prize we took, we learnt (though I then omitted to mention it) that our attempt against Carthagena had failed, and that there was no probability that our fleet in that part of the world would engage in any new enterprize that would at all facilitate this plan. Mr. Anson therefore gave over all hopes of being reinforced across the isthmus, and consequently had no inducement at present to proceed to Panama, as he was incapable of attacking the place;

and there was great reason to believe that by this time there was a general embargo on all the coast.

The only feasible measure then which was left us was to steer as soon as possible to the southern parts of California, or to the adjacent coast of Mexico, there to cruise for the Manila galeon, which we knew was now at sea, bound to the port of Acapulco. And we doubted not to get on that station time enough to intercept her; for this ship does not usually arrive at Acapulco till towards the middle of January, and we were now but in the middle of November, and did not conceive that our passage thither would cost us above a month or five weeks; so that we imagined we had near twice as much time as was necessary for our purpose. Indeed there was a business which we foresaw would occasion some delay, but we flattered ourselves that it would be dispatched in four or five days, and therefore could not interrupt our project. This was the recruiting of our water; for the number of prisoners we had entertained on board since our leaving the island of Fernandes had so far exhausted our stock, that it was impossible to think of venturing upon this passage to the coast of Mexico till we had procured a fresh supply, especially as at Paita, where we had some hopes of getting a quantity, we did not find enough for our consumption during our stay there. It was for some time a matter of deliberation where we should take in this necessary article; but by consulting the accounts of former navigators, and examining our prisoners, we at last resolved for the island of Quibo, situated at the mouth of the bay of Panama: nor was it but on good grounds that the commodore conceived this to be the properest place for watering the squadron. Indeed, there was a small island called Cocos, which was less out of our way than Quibo, where some of the buccaneers have pretended they found water: but none of our prisoners knew anything of it, and it was thought too dangerous to risque the safety of the squadron, by exposing ourselves to the hazard of not meeting with water when we came there, on the mere authority of these legendary writers, of whose misrepresentations and falsities we had almost daily experience. Besides, by going to Quibo we were not without hopes that some of the enemies ships bound to or from Panama might fall into our hands, particularly such of them as were put to sea before they had any intelligence of our squadron.

Determined therefore by these reasons for Quibo, we directed our course northward, being eight sail in company, and consequently having the appearance of a very formidable fleet; and on the 19th, at daybreak, we discovered Cape Blanco, bearing S.S.E.½E. seven miles distant. This cape lies in the latitude of 40° 15' south, and is always made by ships bound either to windward or to leeward; so that off this cape is a most excellent station to cruise upon the enemy. By this time we found that our last prize, the *Solidad*, was far from answering the character given her of a good sailor; and she and the *Santa Teresa* delaying us considerably, the commodore commanded them both to be cleared of everything that might prove useful to the rest of the ships, and then to be burnt; and having given proper instructions, and a rendezvous to the *Gloucester* and the other prizes, we proceeded in our course for Quibo, and on the 22d, in the morning, saw the island of Plata, bearing east, distant four leagues. Here one of our prizes was ordered to stand close in with it, both to discover if there were any ships between that island and the continent, and likewise to look out for a stream of fresh water which was reported to be there, and which would have saved us the trouble of going to Quibo; but she returned without having seen any ship, or finding any water. At three in the afternoon Point Manta bore S.E. by E. seven miles distant; and there being a town of the same name in the neighbourhood, Captain Mitchel took this opportunity of sending away several of his prisoners from the *Gloucester* in the Spanish launch. The boats were now daily employed in distributing provisions on board our prizes to complete their stock for six months: and that the *Centurion* might be the better prepared to give the Manila ships (one of which we were told was of an immense size) a warm reception, the carpenters were ordered to fix eight stocks in the main and fore-tops, which were properly fitted for the mounting of swivel guns.

On the 25th we had a sight of the island of Gallo, bearing E.S.E.½E. four leagues distant; and from hence we crossed the bay of Panama with a N.W. course, hoping that this would have carried us in a direct line to the island of Quibo. But we afterwards found that we ought to have stood more to the westward, for the winds in a short time began to incline to that quarter, and made it difficult to gain the island. After passing the equinoctial (which we did on the 22d) and leaving the neighbourhood of the Cordilleras, and standing more and more towards the isthmus, where the communication of the atmosphere to the eastward and the westward was no longer interrupted, we found in very few days an extraordinary alteration in the climate. For instead of that uniform temperature where neither the excess of heat or cold was to be complained of, we had now, for several days together, close and sultry weather, resembling what we had before met with on the coast of Brazil, and in other parts between the tropics on the eastern side of America. We had besides frequent calms and heavy rains, which we at first ascribed to the neighbourhood of the line, where this kind of weather is generally found to prevail at all seasons of the year; but observing that it attended us to the latitude of seven degrees north, we were at length induced to believe that the stormy season, or, as the Spaniards call it, the Vandevals, was not yet over; though many writers, particularly Captain Shelvocke, positively assert that this season begins in June, and is ended in November, and our prisoners all affirmed the same thing. But perhaps its end may not be always constant, and it might last this year longer than usual.

On the 27th, Captain Mitchel having finished the clearing of his largest prize, she was scuttled and set on fire; but we still consisted of five ships, and were fortunate enough to find them all good sailors, so that we never occasioned any delay to each other. Being now in a rainy climate, which we had been long disused to, we found it necessary to caulk the decks and sides of the *Centurion*, to prevent the rain water from running into her.

On the 3d of December we had a view of the island of Quibo, the east end of which then bore from us N.N.W. four leagues distant, and the island of Quicara W.N.W. about the same distance. Here we struck ground with sixty-five fathom of line, the bottom consisting of grey sand with black specks. When we had thus got sight of the land, we found the wind to hang westerly; and therefore, night coming on, we thought it adviseable to stand off till morning, as there are said to be some shoals in the entrance of the channel. At six the next morning Point Mariato bore N.E.½N. three or four leagues distant. In weathering this point all the squadron except the *Centurion* were very near it; and the *Gloucester* being the leewardmost ship, was forced to tack and stand to the southward, so that we lost sight of her. At nine, the island of Sebaco bore N.W. by N. four leagues distant; but the wind still proving unfavourable, we were obliged to ply on and off for the succeeding twenty-four hours, and were frequently taken aback. However, at eleven the next morning, the wind happily settled in the S.S.W., and we bore away for the S.S.E. end of the island, and about three in the afternoon entered the Canal Bueno, passing round a shoal which stretches off about two miles from the south point of the island. This Canal Bueno, or Good Channel, is at least six miles in breadth; and as we had the wind large, we kept in a good depth of water, generally from twenty-eight or thirty-three fathom, and came not within a mile and a half distance of the breakers, though, in all probability, if it had been necessary, we might have ventured much nearer without incurring the least danger. At seven in the evening we anchored in thirty-three fathom muddy ground; the south point of the island bearing S.E. by S., a remarkable high part of the island W. by N., and the island Sebaco E. by N. Being thus arrived at this island of Quibo, the account of the place, and of our transactions there, shall be referred to the ensuing chapter.

CHAPTER VIII

OUR PROCEEDINGS AT QUIBO, WITH AN ACCOUNT OF THE PLACE

The next morning after our anchoring, an officer was dispatched on shore to discover the watering-place, who, having found it, returned before noon; and then we sent the long-boat for a load of water, and at the same time we weighed and stood farther in with our ships. At two we came again to an anchor in twenty-two fathom, with a rough bottom of gravel intermixed with broken shells, the watering-place now bearing from us N.W.½N. only three-quarters of a mile distant.

This island of Quibo is extremely convenient for wooding and watering, since the trees grow close to the high-water mark, and a large rapid stream of fresh water runs over the sandy beach into the sea: so that we were little more than two days in laying in all the wood and water we wanted. The whole island is of a very moderate height, excepting one part. It consists of a continued wood spread all over the whole surface of the country, which preserves its verdure the year round. Amongst the other wood, we found there abundance of cassia, and a few lime-trees. It appeared singular to us, that considering the climate and the shelter, we should see no other birds than parrots, parroquets, and mackaws; indeed, of these last there were prodigious flights. Next to these birds, the animals we found in most plenty were monkeys and guanos, and these we frequently killed for food; for notwithstanding there were many herds of deer upon the place, yet the difficulty of penetrating the woods prevented our coming near them, so that though we saw them often, we killed only two during our stay. Our prisoners assured us that this island abounded with tygers; and we did once discover the print of a tyger's paw upon the beach, but the tygers themselves we never saw. The Spaniards too informed us that there was frequently found in the woods a most mischievous serpent, called the flying snake, which, they said, darted itself from the boughs of trees on either man or beast that came within its reach, and whose sting they believed to be inevitable death. Besides these dangerous land animals, the sea hereabouts is infested with great numbers of alligators of an extraordinary size; and we often observed a large kind of flat fish jumping a considerable height out of the water, which we supposed to be the fish that is said frequently to destroy the pearl divers by clasping them in its fins as they rise from the bottom; and we were told that the divers, for their security, are now always armed with a sharp knife, which, when they are entangled, they stick into the belly of the fish, and thereby disengage themselves from its embraces.

Whilst the ship continued here at anchor, the commodore, attended by some of his officers, went in a boat to examine a bay which lay to the northward, and they afterwards ranged all along the eastern side of the island. And in the places where they put on shore in the course of this expedition, they generally found the soil to be extremely rich, and met with great plenty of excellent water. In particular, near the N.E. point of the island they discovered a natural cascade, which surpassed, as they conceived, everything of this kind which human art or industry hath hitherto produced. It was a river of transparent water, about forty yards wide, which rolled down a declivity of near a hundred and fifty in length. The channel it fell in was very irregular, for it was entirely composed of rock, both its sides and bottom being made up of large detached blocks; and by these the course of the water was frequently interrupted, for in some parts it ran sloping with a rapid but uniform motion, while in others it tumbled over the ledges of rocks with a perpendicular descent. All the neighbourhood of this stream was a fine wood; and even the huge masses of rock which overhung the water, and which, by their various projections, formed the inequalities of the channel, were covered with lofty forest trees. Whilst the commodore with those accompanying him were attentively viewing this place, and were remarking the different blendings of the water, the rocks, and the wood, there came in sight (as it were still to heighten and animate the prospect) a prodigious flight of mackaws, which, hovering over this spot, and often wheeling and playing on the wing about it, afforded a most brilliant appearance by the glittering of the sun on their variegated plumage; so that some of the spectators cannot refrain from a kind of transport when they recount the complicated beauties which occurred in this extraordinary waterfall.

In this expedition which the boat made along the eastern side of the island, though they discovered no inhabitants, yet they saw many huts upon the shore, and great heaps of shells of fine mother-of-pearl scattered up and down in different places. These were the remains left by the pearl-fishers from Panama, who often frequent this place in the summer season; for the pearl oysters, which are to be met with everywhere in the bay of Panama, do so abound at Quibo, that by advancing a very little way into the sea you might stoop down and reach them from the bottom. They are usually very large, and out of curiosity we opened some of them with a view of tasting them, but we found them extremely tough and unpalatable. And having mentioned these oysters and the pearl-fishery, I must beg leave to recite a few particulars relating to that subject.

The oysters most productive of pearls are those found in considerable depths; for though what are taken up by wading near shore are of the same species, yet the pearls they contain are few in number, and very small. It is said, too, that the pearl partakes, in some degree, of the quality of the bottom on which the oyster is lodged; so that if the bottom be muddy, the pearl is dark and ill coloured.

The taking up oysters from great depths for the sake of their pearls is a work performed by negro slaves, of which the inhabitants of Panama and the neighbouring coast formerly kept vast numbers, which were carefully trained to this business. These are said not to be esteemed compleat divers till they have by degrees been able to protract their stay under water so long that the blood gushes out from their nose, mouth, and ears. And it is the tradition of the country, that when this accident has once befallen them, they dive for the future with much greater facility than before; and they have no apprehension either that any inconvenience can attend it, the bleeding generally stopping of itself, or that there is any probability of their being ever subject to it a second time. But to return from this digression.

Though the pearl oyster, as hath been said, was incapable of being eaten, yet that defect was more than repaid by the turtle, a dainty which the sea at this place furnished us with in the greatest plenty and perfection. There are generally reckoned four species of turtle; that is, the trunk turtle, the loggerhead, the hawksbill, and the green turtle. The two first are rank and unwholesome; the hawksbill (which affords the tortoise-shell) is but indifferent food, though better than the other two; but the green turtle is generally esteemed, by the greatest part of those who are acquainted with its taste, to be the most delicious of all eatables; and that it is a most wholesome food we are amply convinced by our own experience, for we fed on this last species, or the green turtle, near four months, and consequently, had it been in any degree noxious, its ill effects could not possibly have escaped us. At this island we caught what quantity we pleased with great facility; for as they are an amphibious animal, and get on shore to lay their eggs, which they generally deposit in a large hole in the sand just above the high-water mark, covering them up, and leaving them to be hatched by the heat of the sun, we usually dispersed several of our men along the beach, whose business it was to turn them on their backs when they came to land, and the turtle being thereby prevented from getting away, we brought them off at our leisure. By this means we not only secured a sufficient stock for the time we stayed on the island, but we carried a number of them with us to sea, which proved of great service both in lengthening out our store of provision, and in heartening the whole crew with an almost constant supply of fresh and palatable food. For the turtle being large, they generally weighing about 200 lb. weight each, those we took with us lasted near a month: so that before our store was spent, we met with a fresh recruit on the coast of Mexico, where in the heat of the day we often saw great numbers of them fast asleep, floating on the surface of the water. Upon discovering them, we usually sent out our boat with a man in the bow who was a dextrous diver; and as the boat came within a few yards of the turtle, the diver plunged into the water, taking care to rise close upon it, when seizing the shell near the tail, and pressing down the hinder parts, the turtle was thereby awakened, and began to strike with its claws, which motion supported both it and the diver till the boat came up and took them in. By this management we never wanted turtle for the succeeding four months in which we continued at sea; and though, when at the island of Quibo, we had already been three months on board, without otherwise putting our feet on shore than in the few days we stayed there (except those employed in the attack at Paita), yet in the whole seven months from our leaving Juan Fernandes to our anchoring in the harbour of Chequetan, we buried no more in the whole squadron than two men; a most incontestable proof that the turtle, on which we fed for the last four months of this term, was at least innocent, if not something more.

Considering the scarcity of other provisions on some part of the coast of the South Seas, it appears wonderful that a species of food so very palatable and salubrious as turtle, and there so much abounding, should be proscribed by the Spaniards as unwholesome, and little less than poisonous. Perhaps the strange appearance of this animal may have been the foundation of this ridiculous and superstitious aversion, which is strongly rooted in the inhabitants of those countries, and of which we had many instances during the course of this navigation. I have already observed that we put our Spanish prisoners on shore at Paita, and that the *Gloucester* sent theirs to Manta; but as we had taken in our prizes some Indian and negro slaves, we did not dismiss them with their masters, but continued them on board, as our crews were thin, to assist in navigating our ships. These poor people being possessed with the prejudices of the country they came from, were astonished at our feeding on turtle, and seemed fully persuaded that it would soon destroy us; but finding that none of us died, nor even suffered in our health by a continuation of this diet, they at last got so far the better of their aversion as to be persuaded to taste it, to which the absence of all other kinds of fresh provisions might not a little contribute. However, it was with great reluctance, and very sparingly, that they first began to eat of it: but the relish improving upon them by degrees, they at last grew extremely fond of it, and preferred it to every other kind of food, and often felicitated each other on the happy experience they had acquired, and the luxurious and plentiful repasts it would always be in their power to procure when they should again return back to their country. Those who are acquainted with the manner of life of these unhappy wretches need not be told that, next to large draughts of spirituous liquors, plenty of tolerable food is the greatest joy they know, and consequently the discovering the means of being always supplied with what quantity they pleased of a food more delicious to the palate than any their haughty lords and masters could indulge in, was doubtless a circumstance which they considered as the most fortunate that could befall them. After this digression, which the prodigious quantity of turtle on this island of Quibo, and the store of it we thence took to sea, in some measure led me into, I shall now return to our own proceedings.

In three days' time we had compleated our business at this place, and were extremely impatient to depart, that we might arrive time enough on the coast of Mexico to intercept the Manila galeon. But the wind being contrary, detained us a night; and the next day, when we got into the offing, which we did through the same channel by which we entered, we were obliged to keep hovering about the island, in hopes of getting sight of the *Gloucester*, who, as I have in the last chapter mentioned, was separated from us on our first arrival. It was the 9th of December, in the morning, when we put to sea; and continuing to the southward of the island, looking out for the *Gloucester*, we, on the 10th, at five in the afternoon, discerned a small sail to the northward of us, to which we gave chace, and coming up with her took her. She proved to be a bark from Panama called the *Jesu Nazareno*. She had nothing on board but some oakum, about a ton of rock salt, and between £30 and £40 in specie, most of it consisting of small silver money intended for purchasing a cargoe of provisions at Cheripe, an inconsiderable village on the continent.

And on occasion of this prize I cannot but observe for the use of future cruisers that, had we been in want of provisions, we had by this capture an obvious method of supplying ourselves. For at Cheripe there is a constant store of provisions prepared for the vessels who go thither every week from Panama, the market of Panama being chiefly supplied from thence: so that by putting a few of our hands on board our prize, we might easily have seized a large quantity without any hazard, since Cheripe is a place of no strength. As provisions are the staple commodity of that place and of its neighbourhood, the knowledge of this circumstance may be of great use to such cruisers as find their provisions grow scant and yet are desirous of continuing on that coast as long as possible. But to return.

On the 12th of December we were at last relieved from the perplexity we had suffered occasioned by the separation of the *Gloucester*; for on that day she joined us, and informed us that in tacking to the southward on our first arrival she had sprung her fore top-mast, which had disabled her from working to windward, and prevented her from joining us sooner. And now we scuttled and sunk the *Jesu Nazareno*, the prize we took last; and having the greatest impatience to get into a proper station for intercepting the Manila galeon, we stood all together to the westward, leaving the island of Quibo, notwithstanding all the impediments we met with, about nine days after our first coming in sight of it.

CHAPTER IX

FROM QUIBO TO THE COAST OF MEXICO

On the 12th of December we stood from Quibo to the westward, and the same day the commodore delivered fresh instructions to the captains of the men-of-war, and the commanders of our prizes, appointing them the rendezvouses they were to make and the courses they were to steer in case of a separation. And first, they were directed to use all possible dispatch in getting to the northward of the harbour of Acapulco, where they were to endeavour to fall in with the land between the latitudes of 18 and 19 degrees; from thence they were to beat up the coast at eight or ten leagues distance from the shore, till they came abreast of Cape Corientes, in the latitude of 20° 20'. After they arrived there, they were to continue cruising on that station till the 14th of February, when they were to depart for the middle island of the Tres Marias, in the latitude of 21° 25', bearing from Cape Corientes N.W. by N., twenty-five leagues distant. And if at this island they did not meet the commodore, they were there to recruit their wood and water, and then immediately to proceed for the island of Macao, on the coast of China. These orders being distributed to all the ships, we had little doubt of arriving soon upon our intended station, as we expected upon the increasing our offing from Quibo to fall in with the regular trade-wind. But, to our extreme vexation, we were baffled for near a month, either by tempestuous weather from the western quarter, or by dead calms and heavy rains, attended with a sultry air; so that it was the 25th of December before we saw the island of Cocos, which according to our reckoning was only a hundred leagues from the continent; and even then we had the mortification to make so little way that we did not lose sight of it again in five days.

This island we found to be in the latitude of 5° 20' N. It has a high hummock towards the western part, which descends gradually, and at last terminates in a low point to the eastward. From the island of Cocos we stood W. by N., and were till the 9th of January in running an hundred leagues more. We had at first flattered ourselves that the uncertain weather and western gales we met with were owing to the neighbourhood of the continent, from which, as we got more distant, we expected every day to be relieved, by falling in with the eastern trade-wind. But as our hopes were so long baffled, and our patience quite exhausted, we began at length to despair of succeeding in the great purpose we had in view, that of intercepting the Manila galeon. This produced a general dejection amongst us, as we had at first considered the project as almost infallible, and had indulged ourselves in the most boundless hopes of the advantages we should thence receive. However, our despondency was at last somewhat alleviated by a favourable change of the wind; for on the 9th of January a gale sprung up the first time from the N.E., and on this we took the *Carmelo* in tow, as the *Gloucester* did the *Carmin*, making all the sail we could to improve the advantage, because we still suspected that it was only a temporary gale which would not last long, though the next day we had the satisfaction to find that the wind did not only continue in the same quarter, but blew with so much briskness and steadiness that we no longer doubted of its being the true trade-wind. As we now advanced apace towards our station, our hopes began again to revive, and our former despair by degrees gave place to more sanguine prejudices; insomuch that though the customary season of the arrival of the galeon at Acapulco was already elapsed, yet we were by this time unreasonable enough to flatter ourselves that some accidental delay might, for our advantage, lengthen out her passage beyond its usual limits.

When we got into the trade-wind, we found no alteration in it till the 17th of January, when we were advanced to the latitude of 12° 50', but on that day it shifted to the westward of the north. This change we imputed to our having haled up too soon, though we then esteemed ourselves full seventy leagues from the coast; whence, and by our former experience, we were fully satisfied that the trade-wind doth not take place, but at a considerable distance from the continent. After this the wind was not so favourable to us as it had been. However, we still continued to advance, and, on the 26th of January, being then to the northward of Acapulco, we tacked and stood to the eastward, with a view of making the land.

In the preceding fortnight we caught some turtle on the surface of the water, and several dolphins, bonitoes, and albicores. One day, as one of the sailmaker's mates was fishing from the end of the gib-boom, he lost his hold and dropped into the sea, and the ship, which was then going at the rate of six or seven knots, went directly over him; but as we had the *Carmelo* in tow, we instantly called out to the people on board her, who threw him over several ends of ropes, one of which he fortunately caught hold of, and twisting it round his arm, he was thereby haled into the ship without having received any other injury than a wrench in the arm, of which he soon recovered.

When, on the 26th of January, we stood to the eastward, we expected, by our reckonings, to have fallen in with the land on the 28th, yet though the weather was perfectly clear, we had no sight of it at sunset, and therefore we continued our course, not doubting but we should see it by the next morning. About ten at night we discovered a light on the larboard bow, bearing from us N.N.E. The *Tryal's* prize, too, who was about a mile ahead of us, made a signal at the same time for seeing a sail. As we had none of us any doubt but what we saw was a ship's light, we were all extremely animated with a firm persuasion that it was the Manila galeon, which had been so long the object of our wishes. And what added to our alacrity was our expectation of meeting with two of them instead of one, for we took it for granted that the light in view was carried in the top of one ship for a direction to her consort. We immediately cast off the *Carmelo*, and pressed forward with all our canvas, making a signal for the *Gloucester* to do the same. Thus we chased the light, keeping all our hands at their respective quarters, under an expectation of engaging within half an hour, as we sometimes conceived the chace to be about a mile distant, and at other times to be within reach of our guns; for some on board us positively averred that besides the light they could plainly discern her sails. The commodore himself was so fully persuaded that we should be soon alongside of her that he sent for his first lieutenant, who commanded between decks, and directed him to see all the great guns loaded with two round shot for the first broadside, and after that with one round shot and one grape, strictly charging him, at the same time, not to suffer a gun to be fired till he, the commodore, should give orders, which, he informed the lieutenant, would not be till we arrived within pistol-shot of the enemy. In this constant and eager attention we continued all night, always presuming that another quarter of an hour would bring us up with this Manila ship, whose wealth, and that of her supposed consort, we now estimated by round millions. But when the morning broke, and daylight came on, we were most strangely and vexatiously disappointed, by finding that the light which had occasioned all this bustle and expectancy, was only a fire on the shore. It must be owned, the circumstances of this deception were so extraordinary as to be scarcely credible, for, by our run during the night, and the distance of the land in the morning, there was no doubt to be made but this fire, when we first discovered it, was above twenty-five leagues from us; and yet, I believe, there was no person on board who doubted of its being a ship's light, or of its being near at hand. It was indeed upon a very high mountain, and continued burning for several days afterwards; however, it was not a vulcano, but rather, as I suppose, a tract of stubble or heath, set on fire for some purpose of agriculture.

At sun-rising, after this mortifying delusion, we found ourselves about nine leagues off the land, which extended from the N.W. to E.½N. On this land we observed two remarkable hummocks, such as are usually called paps, which bore north from us: these a Spanish pilot and two Indians, who were the only persons amongst us that pretended to have traded in this part of the world, affirmed to be over the harbour of Acapulco. Indeed, we very much doubted their knowledge of the coast, for we found these paps to be in the latitude of 17° 56', whereas those over Acapulco are said to be 17 degrees only; and we afterwards found our suspicions of their skill to be well grounded. However, they were very confident, and assured us that the height of the mountains was itself an infallible mark of the harbour, the coast, as they pretended, though falsly, being generally low to the eastward and westward of it.

Being now in the track of the Manila galeon, it was a great doubt with us, as it was near the end of January, whether she was or was not arrived; but examining our prisoners about it, they assured us that she was sometimes known to come in after the middle of February, and they endeavoured to persuade us that the fire we had seen on shore was a proof that she was yet at sea, it being customary, as they said, to make use of these fires as signals for her direction when she continued longer out than ordinary. On this reasoning of our prisoners, strengthened by our propensity to believe them in a matter which so pleasingly flattered our wishes, we resolved to cruise for her some days, and we accordingly spread our ships at the distance of twelve leagues from the coast in such a manner that it was impossible she should pass us unobserved. However, not seeing her soon, we were at intervals inclined to suspect that she had gained her port already, and as we now began to want a harbour to refresh our people, the uncertainty of our present situation gave us great uneasiness, and we were very solicitous to get some positive intelligence, which might either set us at liberty to consult our necessities, if the galeon was arrived, or might animate us to continue our present cruise with chearfulness, if she was not. With this view, the commodore, after examining our prisoners very particularly, resolved to send a boat, under colour of the night, into the harbour of Acapulco, to see if the Manila ship was there or not, one of the Indians being very positive that this might be done without the boat itself being discovered. To execute this enterprize, the barge was dispatched the 6th of February, carrying a sufficient crew and two officers, as also a Spanish pilot, with the Indian who had insisted on the facility of this project, and had undertaken to conduct it. Our barge did not return to us again till the 11th, when the officers acquainted Mr. Anson that, agreeable to our suspicions, there was nothing like a harbour in the place where the Spanish pilots had at first asserted Acapulco to lie; that after they had satisfied themselves in this particular, they steered to the eastward, in hopes of discovering it, and had coasted along shore thirty-two leagues; that in this whole range they met chiefly with sandy beaches of a great length, over which the sea broke with so much violence that it was impossible for a boat to land; that at the end of their run they could just discover two paps at a very great distance to the eastward, which from their appearance and their latitude they concluded to be those in the neighbourhood of Acapulco; but that not having a sufficient quantity of fresh water and provision for their passage thither and back again, they were obliged to return to the commodore, to acquaint him with their disappointment. On this intelligence we all made sail to the eastward, in order to get into the neighbourhood of that port, the commodore being determined to send the barge a second time upon the same enterprize, when we were arrived within a moderate distance. Accordingly, the next day, which was the 12th of February, we being by that time considerably advanced, the barge was again dispatched, and particular instructions given to the officers to preserve themselves from being seen from the shore. On the 13th we espied a high land to the eastward, which was first imagined to be that over the harbour of Acapulco; but we afterwards found that it was the high land of Seguateneio, where there is a small harbour, of which we shall have occasion to make more ample mention hereafter. We waited six days, from the departure of our barge, without any news of her, so that we began to be uneasy for her safety; but on the 7th day, that is, on the 19th of February, she returned: when the officers informed the commodore that they had discovered the harbour of Acapulco, which they esteemed to bear from us E.S.E. at least fifty leagues distant; that on the 17th, about two in the morning, they were got within the island that lies at the mouth of the harbour, and yet neither the Spanish pilot, nor the Indian, could give them any information where they then were; but that while they were lying upon their oars in suspence what to do, being ignorant that they were then at the very place they sought for, they discerned a small light near the surface of the water, on which they instantly plied their paddles, and moving as silently as possible towards it, they found it to be in a fishing canoe, which they surprized, with three negroes that belonged to it. It seems the negroes at first attempted to jump overboard, and being so near the shore they would easily have swam to land, but they were prevented by presenting a piece at them, on which they readily submitted, and were taken into the barge. The officers further added that they had immediately turned the canoe adrift against the face of a rock, where it would inevitably be dashed to pieces by the fury of the sea. This they did to deceive those who perhaps might be sent from the town to search after the canoe, for upon seeing several remains of a wreck, they would immediately conclude that the people on board her had been drowned, and would have no suspicion of their having fallen into our hands. When the crew of the barge had taken this precaution, they exerted their utmost strength in pulling out to sea, and by dawn of the day had gained such an offing as rendered it impossible for them to be seen from the coast.

Having now gotten the three negroes in our possession, who were not ignorant of the transactions at Acapulco, we were soon satisfied about the most material points which had long kept us in suspence. On examining them we found that we were indeed disappointed in our expectation of intercepting the galeon before her arrival at Acapulco; but we learnt other circumstances which still revived our hopes, and which, we then conceived, would more than balance the opportunity we had already lost, for though our negroe prisoners informed us that the galeon arrived at Acapulco on our 9th of January, which was about twenty days before we fell in with this coast, yet they at the same time told us that the galeon had delivered her cargo, and was taking in water and provisions in order to return, and that the Viceroy of Mexico had by proclamation fixed her departure from Acapulco to the 14th of March, N.S. This last news was most joyfully received by us, since we had no doubt but she must certainly fall into our hands, and it was much more eligible to seize her on her return than it would have been to have taken her before her arrival, as the species for which she had sold her cargoe, and which she would now have on board, would be prodigiously more to be esteemed by us than the cargoe itself; great part of which would have perished on our hands, and none of it could have been disposed of by us at so advantageous a mart as Acapulco.

Thus we were a second time engaged in an eager expectation of meeting with this Manila ship, which, by the fame of its wealth, we had been taught to consider as the most desirable capture that was to be made on any part of the ocean. But since all our future projects will be in some sort regulated with a view to the possession of this celebrated galeon, and since the commerce which is carried on by means of these vessels between the city of Manila and the port of Acapulco is perhaps the most valuable, in proportion to its quantity, of any in the known world, I shall endeavour, in the ensuing chapter, to give as circumstantial an account as I can of all the particulars relating thereto, both as it is a matter in which I conceive the public to be in some degree interested, and as I flatter myself, that from the materials which have fallen into my hands, I am enabled to describe it with more distinctness than has hitherto been done, at least in our language.

CHAPTER X

AN ACCOUNT OF THE COMMERCE CARRIED ON BETWEEN THE CITY OF MANILA ON THE ISLAND OF LUCONIA, AND THE PORT OF ACAPULCO ON THE COAST OF MEXICO

About the end of the fifteenth century and the beginning of the sixteenth, the searching after new countries, and new branches of commerce, was the reigning passion among several of the European princes. But those who engaged most deeply and fortunately in these pursuits were the kings of Spain and Portugal, the first of them having discovered the immense and opulent continent of America and its adjacent islands, whilst the other, by doubling the Cape of Good Hope, had opened to his fleets a passage to the southern coast of Asia, usually called the East Indies, and by his settlements in that part of the globe, became possessed of many of the manufactures and natural productions with which it abounded, and which, for some ages, had been the wonder and delight of the more polished and luxurious part of mankind.

In the meantime, these two nations of Spain and Portugal, who were thus prosecuting the same views, though in different quarters of the world, grew extremely jealous of each other, and became apprehensive of mutual encroachments. And, therefore, to quiet their jealousies, and to enable them with more tranquillity to pursue the propagation of the Catholick faith in these distant countries (they having both of them given distinguished marks of their zeal for their mother church, by their butchery of innocent pagans), Pope Alexander VI. granted to the Spanish crown the property and dominion of all places, either already discovered, or that should be discovered, an hundred leagues to the westward of the islands of Azores, leaving all the unknown countries to the eastward of this limit to the industry and disquisition of the Portuguese; and this boundary being afterwards removed two hundred and fifty leagues more to the westward, by the agreement of both nations, it was imagined that this regulation would have suppressed all the seeds of future contests. For the Spaniards presumed that the Portuguese would be thereby prevented from meddling with their colonies in America, and the Portuguese supposed that their East Indian settlements, and particularly the Spice Islands, which they had then newly found out, were for ever secured from any attempts of the Spanish nation.

But it seems the infallibility of the Holy Father had, on this occasion, deserted him, and for want of being more conversant in geography, he had not foreseen that the Spaniards, by pursuing their discoveries to the west, and the Portuguese to the east, might at last meet with each other, and be again embroiled, as it actually happened within a few years afterwards. For

Ferdinand Magellan, an officer in the King of Portugal's service, having received some disgust from the court, either by the defalcation of his pay, or by having his parts, as he conceived, too cheaply considered, he entered into the service of the King of Spain. As he appears to have been a man of ability, he was desirous of signalizing his talents in some enterprize which might prove extremely vexatious to his former masters, and might teach them to estimate his worth from the greatness of the mischief he brought upon them, this being the most obvious and natural turn of all fugitives, more especially of those who, being really men of capacity, have quitted their country by reason of the small account that has been made of them. Magellan, in pursuance of these vindictive views, knowing that the Portuguese considered their traffic to the Spice Islands as their most important acquisition in the east, resolved with himself to instigate the court of Spain to an attempt, which, by still pushing their discoveries to the westward, would give them a right to interfere both in the property and commerce of those renowned countries; and the King of Spain approving of this project, Magellan, in the year 1519, set sail from the port of Sevil in order to carry this enterprize into execution. He had with him a considerable force, consisting of five ships and two hundred and thirty-four men, with which he stood for the coast of South America, and ranging along shore, he at length, towards the end of October 1520, had the good fortune to discover those streights which have since been denominated from him, and which opened him a passage into the South Seas. This, which was the first part of his scheme, being thus happily accomplished, he, after some stay on the coast of Peru, set sail again to the westward, with a view of falling in with the Spice Islands. In this extensive run across the Pacific Ocean, he first discovered the Ladrones or Marian Islands, and continuing on his course, he at length reached the Philippine Islands, which are the most eastern part of Asia, where, venturing on shore in an hostile manner, and skirmishing with the Indians, he was slain.

By the death of Magellan, his original project of securing some of the Spice Islands was defeated; for those who were left in command contented themselves with ranging through them, and purchasing some spices from the natives, after which they returned home round the Cape of Good Hope, being the first ships which had ever surrounded this terraqueous globe, and thereby demonstrated, by a palpable experiment obvious to the grossest and most vulgar capacity, the reality of its long-disputed spherical figure.

But though Spain did not hereby acquire the property of any of the Spice Islands, yet the discovery of the Philippines, made in this expedition, was thought too considerable to be neglected, since these were not far distant from those places which produced spices, and were very well situated for the Chinese trade, and for the commerce of other parts of India. A communication, therefore, was soon established and carefully supported between these islands and the Spanish colonies on the coast of Peru: whence the city of Manila (which was built on the island of Luconia, the chief of the Philippines) became in a short time the mart for all Indian commodities, which were brought up by the inhabitants, and were annually sent to the South Seas, to be there vended on their account; and the returns of this commerce to Manila being principally made in silver, the place by degrees grew extremely opulent, and its trade so far increased as to engage the attention of the court of Spain, and to be frequently controlled and regulated by royal edicts.

In the infancy of this trade it was carried on from the port of Callao to the city of Manila, in which navigation the trade-wind continually favoured them; so that notwithstanding these places were distant between three and four thousand leagues, yet the voyage was often made in little more than two months. But then the return from Manila was extremely troublesome and tedious, and is said to have sometimes lasted above a twelvemonth; which, if they pretend to ply up within the limits of the trade-wind, is not at all to be wondered at. Indeed, though it is asserted that in their first voyages they were so imprudent and unskilful as to attempt this course, yet that route was soon laid aside, by the advice, as it is said, of a Jesuit, who persuaded them to steer to the northward till they got clear of the trade-winds, and then by the favour of the westerly winds, which generally prevail in high latitudes, to stretch away for the coast of California. This we know hath been the practice for at least a hundred and sixty years past, as Sir Thomas Cavendish, in the year 1586, engaged off the south end of California a vessel bound from Manila to the American coast. And it was in compliance with this new plan of navigation, and to shorten the run both backwards and forwards, that the staple of this commerce to and from Manila was removed from Callao on the coast of Peru, to the port of Acapulco on the coast of Mexico, where it continues fixed to this time.

Such was the commencement, and such were the early regulations of this commerce; but its present condition being a much more interesting subject, I must beg leave to dwell longer on this head, and to be indulged in a more particular narration, beginning with a description of the island of Luconia, and of the port and bay of Manila.

The island of Luconia, though situated in the latitude of 15° north, is esteemed to be in general extremely healthy, and the water that is found upon it is said to be the best in the world. It produces all the fruits of the warm climates, and abounds in a most excellent breed of horses, supposed to be carried thither first from Spain. It is very well seated for the Indian and Chinese trade; and the bay and port of Manila, which lies on its western side, is perhaps the most remarkable on the whole globe, the bay being a large circular bason, near ten leagues in diameter, great part of it entirely land-locked. On the east side of this bay stands the city of Manila, which is large and populous, and which, at the beginning of this war, was only an open place, its principal defence consisting in a small fort, which was almost surrounded on every side by houses; but they have lately made considerable additions to its fortifications, though I have not yet learnt after what manner. The port, peculiar to the city, is called Cabite, and lies near two leagues to the southward: and in this port all the ships employed for the Acapulco trade are usually stationed.

The city of Manila itself is in a healthy situation, is well watered, and is in the neighbourhood of a very fruitful and plentiful country; but as the principal business of this place is its trade to Acapulco, it lies under some disadvantage from the difficulty there is in getting to sea to the eastward: for the passage is among islands and through channels, where the Spaniards, by reason of their unskilfulness in marine affairs, waste much time, and are often in great danger.

The trade carried on from this place to China and different parts of India is principally for such commodities as are intended to supply the kingdoms of Mexico and Peru. These are spices, all sorts of Chinese silks and manufactures, particularly silk stockings, of which I have heard that no less than fifty thousand pair were the usual number shipped in each cargoe; vast quantities of Indian stuffs, as callicoes and chints, which are much worn in America, together with other minuter articles, as goldsmiths' work, etc., which is principally wrought at the city of Manila itself by the Chinese; for it is said there are at least twenty thousand Chinese who constantly reside there, either as servants, manufacturers, or brokers. All these different commodities are collected at Manila, thence to be transported annually in one or more ships to the port of Acapulco in the kingdom of Mexico.

This trade to Acapulco is not laid open to all the inhabitants of Manila, but is confined by very particular regulations, somewhat analogous to those by which the trade of the register ships from Cadiz to the West Indies is restrained. The ships employed herein are found by the King of Spain, who pays the officers and crew; and the tunnage is divided into a certain number of bales, all of the same size: these are distributed amongst the convents at Manila, but principally to the Jesuits, as a donation to support their missions for the propagation of the Catholick faith; and the convents have thereby a right to embark such a quantity of goods on board the Manila ship as the tunnage of their bales amounts to; or if they chuse not to be concerned in trade themselves, they have the power of selling this privilege to others: nor is it uncommon, when the merchant to whom they sell their share is unprovided of a stock, for the convent to lend him considerable sums of money on bottomry.

The trade is by the royal edicts limited to a certain value, which the annual cargoe ought not to exceed. Some Spanish manuscripts I have seen, mention this limitation to be 600,000 dollars; but the annual cargoe does certainly surpass this sum, and though it may be difficult to fix its exact value, yet from many comparisons I conclude that the return cannot be much short of three millions of dollars.

As it is sufficiently obvious that the greatest share of the treasure returned from Acapulco to Manila does not remain in that place, but is again dispersed into different parts of India; and as all European nations have generally esteemed it good policy to keep their American settlements in an immediate dependence on the mother country, without permitting them to carry on directly any gainful traffick with other powers; these considerations have occasioned many remonstrances to be presented to the court of Spain against this Indian trade allowed to the kingdom of Mexico. It has been urged that the silk manufactures of Valencia and other parts of Spain are hereby greatly prejudiced, and the linens carried from Cadiz much injured in their sale: since the Chinese silks coming almost directly to Acapulco, can be afforded considerably cheaper there than any European manufactures of equal goodness, and the cotton from the Coromandel coast makes the European linens nearly useless. So that the Manila trade renders both Mexico and Peru less dependant upon Spain for a supply of their necessities than they ought to be, and exhausts those countries of a considerable quantity of silver, the greatest part of which, were this trade prohibited, would center in Spain, either in payment for Spanish commodities, or in gains to the Spanish merchant: whereas now the only advantage which arises from it is the enriching the Jesuits, and a few particular persons besides, at the other extremity of the world. These arguments did so far influence Don Joseph Patinho, who was formerly prime minister, and an enemy to the Jesuits, that about the year 1725 he had resolved to abolish this trade, and to have permitted no Indian commodities to be introduced into any of the Spanish ports in the West Indies, except such as were brought thither by the register ships from Europe. But the powerful untrigues of the Jesuits prevented this regulation from taking place.

This trade from Manila to Acapulco and back again is usually carried on in one or at most two annual ships, which set sail from Manila about July, and arrive at Acapulco in the December, January, or February following; and having there disposed of their effects, return for Manila some time in March, where they generally arrive in June; so that the whole voyage takes up very near an entire year. For this reason, though there is often no more than one ship freighted at a time, yet there is always one ready for the sea when the other arrives; and therefore the commerce at Manila is provided with three or four stout ships, that in case of any accident the trade may not be suspended. The largest of these ships, whose name I have not learnt, is described as little less than one of our first-rate men-of-war; and indeed she must be of an enormous size, as it is known that when she was employed with other ships from the same port to cruise for our China trade, she had no less than twelve hundred men on board. Their other ships, though far inferior in bulk to this, are yet stout large vessels, of the burthen of twelve hundred tun and upwards, and usually carry from three hundred and fifty to six hundred hands, passengers included, with fifty odd guns. As these are all king's ships, commissioned and paid by him, there is usually one amongst the captains stiled the general, and he carries the royal standard of Spain at the main topgallant mast-head, as we shall more particularly observe hereafter.

And now having described the city and port of Manila, and the shipping employed by its inhabitants, it is necessary to give a more circumstantial detail of the navigation from thence to Acapulco. The ship having received her cargo on board, and being fitted for the sea, generally weighs from the mole of Cabite about the middle of July, taking the advantage of the westerly monsoon, which then sets in. It appears that the getting through the channel called the Boccadero, to the eastward, must be a troublesome navigation, and in fact it is sometimes the end of August before they compleat it. When they have cleared this passage, and are disintangled from the islands, they stand to the northward of the east, till they arrive in the latitude of thirty degrees or upwards, where they expect to meet with westerly winds, before which they stretch away for the coast of California.

It is indeed most remarkable that by the concurrent testimony of all the Spanish navigators, there is not one port nor even a tolerable road as yet found out betwixt the Philippine Islands and the coast of California: so that from the time the Manila ship first loses sight of land, she never lets go her anchor till she arrives on the coast of California, and very often not till she gets to its southernmost extremity. As this voyage is rarely of less than six months' continuance, and the ship is deep laden with merchandize and crowded with people, it may appear wonderful how they can be supplied with a stock of fresh water for so long a time. The method of procuring it is indeed extremely singular, and deserves a very particular recital.

It is well known to those who are acquainted with the Spanish customs in the South Seas, that their water is preserved on shipboard, not in casks but in earthen jars, which in some sort resemble the large oil jars we often see in Europe. When the Manila ship first puts to sea, she takes on board a much greater quantity of water than can be stowed between decks, and the jars which contain it are hung all about the shrouds and stays, so as to exhibit at a distance a very odd appearance. Though it is one convenience of their jars that they are much more manageable than casks, and are liable to no leekage, unless they are broken, yet it is sufficiently obvious that a six or even a three months' store of water could never be stowed in a ship so loaded by any management whatever; and therefore without some other supply this navigation could not be performed. A supply indeed they have, but the reliance upon it seems at first sight so extremely precarious that it is wonderful such numbers should risque the perishing by the most dreadful of all deaths on the expectation of so casual a relief. In short, their only method of recruiting their water is by the rains, which they meet with between the latitudes of 30° and 40° north, and which they are always prepared to catch. For this purpose they take to sea with them a great number of mats, which, whenever the rain descends, they range slopingly against the gunwale from one end of the ship to the other, their lower edges resting on a large split bamboe; whence all the water which falls on the mats drains into the bamboe, and by this, as a trough, is conveyed into a jar. And this method of furnishing themselves with water, however accidental and extraordinary it may at first sight appear, hath never been known to fail them, but it hath been common for them, when their voyage is a little longer than usual, to fill all their water jars several times over.

However, though their distresses for fresh water are much short of what might be expected in so tedious a navigation, yet there are other inconveniences generally attendant upon a long continuance at sea from which they are not exempted. The principal of these is the scurvy, which sometimes rages with extreme violence, and destroys great numbers of the people; but at other times their passage to Acapulco (of which alone I would be here understood to speak) is performed with little loss.

The length of time employed in this passage, so much beyond what usually occurs in any other known navigation, is perhaps in part to be imputed to the indolence and unskilfulness of the Spanish sailors, and to an unnecessary degree of caution, on pretence of the great riches of the vessel: for it is said that they rarely set their main-sail in the night, and often lie by unnecessarily. Thus much is certain, that the instructions given to their captains (which I have seen) seem to have been drawn up by such as were more apprehensive of too strong a gale, though favourable, than of the inconveniences and mortality attending a lingering and tedious voyage. For the captain is particularly ordered to make his passage in the latitude of 30 degrees, if possible, and to be extremely careful to stand no farther to the northward than is absolutely necessary for the getting a westerly wind. This, according to our conceptions, appears to be a very absurd restriction, since it can scarcely be doubted but that in the higher latitudes the westerly winds are much steadier and brisker than in the latitude of 30 degrees. Indeed the whole conduct of this navigation seems liable to very great censure: since, if instead of steering E.N.E. into the latitude of 30 degrees, they at first stood N.E. or even still more northerly, into the latitude of 40 or 45 degrees, in part of which coast the trade-winds would greatly assist them, I doubt not but by this management they might considerably contract their voyage, and perhaps perform it in half the time which is now allotted for it. This may in some measure be deduced from their own journals; since in those I have seen, it appears that they are often a month or six weeks after their laying the land before they get into the latitude of 30 degrees; whereas, with a more northerly course, it might easily be done in less than a fortnight. Now when they were once well advanced to the northward, the westerly winds would soon blow them over to the coast of California, and they would be thereby freed from the other embarrassments to which they are at present subjected, only at the expence of a rough sea and a stiff gale. This is not merely matter of speculation; for I am credibly informed that about the year 1721, a French ship, by pursuing this course, ran from the coast of China to the valley of Vanderas, on the coast of Mexico, in less than fifty days: but it was said that notwithstanding the shortness of her passage, she suffered prodigiously by the scurvy, so that she had only four or five of her crew remaining alive when she arrived in America.

However, I shall descant no longer on the probability of performing this voyage in a much shorter time, but shall content myself with reciting the actual occurrences of the present navigation. The Manila ship having stood so far to the northward as to meet with a westerly wind, stretches away nearly in the same latitude for the coast of California, and when she has run into the longitude of about 100 degrees from Cape Espiritu Santo, she generally finds a plant floating on the sea, which, being called Porra by the Spaniards, is, I presume, a species of sea-leek. On the sight of this plant they esteem themselves sufficiently near the California shore, and immediately stand to the southward; and they rely so much on this circumstance, that on the first discovery of the plant, the whole ship's company chant a solemn *Te Deum*, esteeming the difficulties and hazards of their passage to be now at an end; and they constantly correct their longitude thereby, without ever coming within sight of land. After falling in with these signs, as they denominate them, they steer to the southward without endeavouring to approach the coast, till they have run into a lower latitude, for as there are many islands, and some shoals adjacent to California, the extreme caution of the Spanish navigators renders them very apprehensive of being engaged with the land. However, when they draw near its southern extremity, they venture to hale in, both for the sake of making Cape St. Lucas to ascertain their reckoning, and also to receive intelligence from the Indian inhabitants, whether or no there are any enemies on the coast; and this last circumstance, which is a particular article in the captain's instructions, obliges us to mention the late proceedings of the Jesuits among the California Indians.

Since the first discovery of California, there have been various wandering missionaries who have visited it at different times, though to little purpose. But of late years the Jesuits, encouraged and supported by a large donation from the Marquis de Valero, a most munificent bigot, have fixed themselves upon the place, and have there established a very considerable mission. Their principal settlement lies just within Cape St. Lucas, where they have collected a great number of savages, and have endeavoured to inure them to agriculture and other mechanic arts. Nor have their efforts been altogether ineffectual, for they have planted vines at their settlements with very good success, so that they already make a considerable quantity of wine, which begins to be esteemed in the neighbouring kingdom of Mexico, it resembling in flavour the inferior sorts of Madera.

The Jesuits then being thus firmly rooted on California, they have already extended their jurisdiction quite across the country from sea to sea, and are endeavouring to spread their influence farther to the northward, with which view they have made several expeditions up the gulf between California and Mexico, in order to discover the nature of the adjacent countries, all which they hope hereafter to bring under their power. And being thus occupied in advancing the interests of their society, it is no wonder if some share of attention is engaged about the security of the Manila ship, in which their convents at Manila are so deeply concerned. For this purpose there are refreshments, as fruits, wine, water, etc., constantly kept in readiness for her, and there is besides care taken at Cape St. Lucas to look out for any ship of the enemy, which might be cruising there to intercept her, this being a station where she is constantly expected, and where she has been often waited for and fought with, though generally with little success. In consequence then of the measures mutually settled between the Jesuits of Manila and their brethren at California, the captain of the galeon is ordered to fall in with the land to the northward of Cape St. Lucas, where the inhabitants are directed, on sight of the vessel, to make the proper signals with fires. On discovering these fires, the captain is to send his launch on shore with twenty men well armed, who are to carry with them the letters from the convents at Manila to the California missionaries, and are to bring back the refreshments which will be prepared for the ship, and likewise intelligence whether or no there are enemies on the coast. If the captain finds, from the account which is sent him, that he has nothing to fear, he is directed to proceed for Cape St. Lucas, and thence to Cape Corientes, after which he is to coast it along for the port of Acapulco.

The most usual time of the arrival of the galeon at Acapulco is towards the middle of January, but this navigation is so uncertain that she sometimes gets in a month sooner, and at other times has been detained at sea above a month longer. The port of Acapulco is by much the securest and finest in all the northern part of the Pacific Ocean, being, as it were, a bason surrounded by very high mountains. But the town is a most wretched place, and extremely unhealthy, for the air about it is so pent up by the hills that it has scarcely any circulation. Acapulco is besides destitute of fresh water, except what is brought from a considerable distance, and is in all respects so inconvenient, that except at the time of the mart, whilst the Manila galeon is in the port, it is almost deserted.

When the galeon arrives in this port, she is generally moored on its western side to two trees, and her cargoe is delivered with all possible expedition. And now the town of Acapulco, from almost a solitude, is immediately thronged with merchants from all parts of the kingdom of Mexico. The cargoe being landed and disposed of, the silver and the goods intended for Manila are taken on board, together with provisions and water, and the ship prepares to put to sea with the utmost expedition. There is indeed no time to be lost, for it is an express order to the captain to be out of the port of Acapulco on his return, before the first day of April, N.S.

Having mentioned the goods intended for Manila, I must observe that the principal return is always made in silver, and consequently the rest of the cargoe is but of little account; the other articles, besides the silver, being some cochineal and a few sweetmeats, the produce of the American settlements, together with European millinery ware for the women at Manila, and some Spanish wines, such as tent and sherry, which are intended for the use of their priests in the administration of the sacrament.

And this difference in the cargoe of the ship to and from Manila occasions a very remarkable variety in the manner of equipping her for these two different voyages. For the galeon when she sets sail from Manila, being deep laden with variety of bulky goods, she has not the conveniency of mounting her lower tier of guns, but carries them in her hold, till she draws near Cape St. Lucas, and is apprehensive of an enemy. Her hands too are as few as is consistent with the safety of the ship, that she may be less pestered by the stowage of provisions. But on her return from Acapulco, as her cargoe lies in less room, her lower tier is (or ought to be) always mounted before she leaves the port, and her crew is augmented with a supply of sailors, and with one or two companies of foot, which are intended to reinforce the garrison at Manila. Besides, there being many merchants who take their passage to Manila on board the galeon, her whole number of hands on her return is usually little short of six hundred, all which are easily provided for by reason of the small stowage necessary for the silver.

The galeon being thus fitted in order to her return, the captain, on leaving the port of Acapulco, steers for the latitude of 13° or 14°, and then continues on that parallel till he gets sight of the island of Guam, one of the Ladrones. In this run the captain is particularly directed to be careful of the shoals of St. Bartholomew, and of the island of Gasparico. He is also told in his instructions, that to prevent his passing the Ladrones in the dark, there are orders given that, through all the month of June, fires shall be lighted every night on the highest part of Guam and Rota, and kept in till the morning.

At Guam there is a small Spanish garrison (as will be more particularly mentioned hereafter), purposely intended to secure that place for the refreshment of the galeon, and to yield her all the assistance in their power. However, the danger of the road at Guam is so great that though the galeon is ordered to call there, yet she rarely stays above a day or two, but getting her water and refreshments on board as soon as possible, she steers away directly for Cape Espiritu Santo, on the island of Samal. Here the captain is again ordered to look out for signals, and he is told that centinels will be posted not only on that cape, but likewise in Catanduanas, Butusan, Birriborongo, and on the island of Batan. These centinels are instructed to make a fire when they discover the ship, which the captain is carefully to observe, for if, after this first fire is extinguished, he perceives that four or more are lighted up again, he is then to conclude that there are enemies on the coast, and on this he is immediately to endeavour to speak with the centinel on shore, and to procure from him more particular intelligence of their force, and of the station they cruize in; pursuant to which, he is to regulate his conduct, and to endeavour to gain some secure port amongst those islands, without coming in sight of the enemy; and in case he should be discovered when in port, and should be apprehensive of an attack, he must land his treasure, and must take some of his artillery on shore for its defence, not neglecting to send frequent and particular accounts to the city of Manila of all that passes. But if after the first fire on shore, the captain observes that two others only are made by the centinels, he is then to conclude that there is nothing to fear, and he is to pursue his course without interruption, making the best of his way to the port of Cabite, which is the port to the city of Manila, and the constant station for all ships employed in this commerce to Acapulco.

CHAPTER XI

OUR CRUISE OFF THE PORT OF ACAPULCO FOR THE MANILA SHIP

I have already mentioned, in the ninth chapter, that the return of our barge from the port of Acapulco, where she surprized three negro fishermen, gave us inexpressible satisfaction, as we learnt from our prisoners that the galeon was then preparing to put to sea, and that her departure was fixed, by an edict of the Viceroy of Mexico, to the 14th of March, N.S., that is, to the 3d of March, according to our reckoning.

What related to this Manila ship being the matter to which we were most attentive, it was necessarily the first article of our examination, but having satisfied ourselves upon this head, we then indulged our curiosity in enquiring after other news; when the prisoners informed us that they had received intelligence at Acapulco of our having plundered and burnt the town of Paita; and that, on this occasion, the Governor of Acapulco had augmented the fortifications of the place, and had taken several precautions to prevent us from forcing our way into the harbour; that in particular, he had planted a guard on the island which lies at the harbour's mouth, and that this guard had been withdrawn but two nights before the arrival of our barge. So that had the barge succeeded in her first attempt, or had she arrived at the port the second time two days sooner, she could scarcely have avoided being seized on; or if she had escaped, it must have been with the loss of the greatest part of her crew, as she would have been under the fire of the guard before she had known her danger.

The withdrawing of this guard was a circumstance that gave us much pleasure, since it seemed to demonstrate not only that the enemy had not as yet discovered us, but likewise that they had now no farther apprehensions of our visiting their coast. Indeed the prisoners assured us that they had no knowledge of our being in those seas, and that they had therefore flattered themselves that, in the long interval from our taking of Paita, we had steered another course. But we did not consider the opinion of these negro prisoners as so authentick a proof of our being hitherto concealed, as the withdrawing of the guard from the harbour's mouth; for this being the action of the governor, was of all arguments the most convincing, as he might be supposed to have intelligence with which the rest of the inhabitants were unacquainted.

Satisfied therefore that we were undiscovered, and that the day was fixed for the departure of the galeon from Acapulco, we made all necessary preparations, and waited with the utmost impatience for the important moment. As it was the 19th of February when the barge returned and brought us our intelligence, and the galeon was not to sail till the 3d of March, the commodore resolved to continue the greatest part of the intermediate time on his present station, to the westward of Acapulco, conceiving that in this situation there would be less danger of his being seen from the shore, which was the only circumstance that could deprive us of the immense treasure on which we had at present so eagerly fixed our thoughts. During this interval we were employed in scrubbing and cleansing our ships bottoms, in bringing them into their most advantageous trim, and in regulating the orders, signals, and positions, to be observed, when we should arrive off Acapulco, and the time appointed for the departure of the galeon should draw nigh.

It was on the first of March we made the high lands, usually called the paps over Acapulco, and got with all possible expedition into the situation prescribed by the commodore's orders. The distribution of our squadron on this occasion, both for the intercepting the galeon, and for avoiding a discovery from the shore, was so very judicious that it well merits to be distinctly described. The order of it was thus: the *Centurion* brought the paps over the harbour to bear N.N.E. at fifteen leagues distance, which was a sufficient offing to prevent our being seen by the enemy. To the westward of the *Centurion* there was stationed the *Carmelo*, and to the eastward the *Tryal's* prize, the *Gloucester*, and the *Carmin*; these were all ranged in a circular line, and each ship was three leagues distant from the next, so that the *Carmelo* and the *Carmin*, which were the two extremes, were twelve leagues removed from each other, and as the galeon could, without doubt, be discerned at six leagues distance from either extremity, the whole sweep of our squadron, within which nothing could pass undiscovered, was at least twenty-four leagues in extent; and yet we were so connected by our signals as to be easily and speedily informed of what was seen in any part of the line. To render this disposition still more compleat, and to prevent even the possibility of the galeon's escaping us in the night, the two cutters belonging to the *Centurion* and the *Gloucester* were both manned and sent in shore, and commanded to lie all day at the distance of four or five leagues from the entrance of the port, where, by reason of their smallness, they could not possibly be discovered, but in the night they were directed to stand nearer to the harbour's mouth, and as the light of the morning approached to come back again to their day-posts. When the cutters should first discern the Manila ship, one of them was to return to the squadron, and to make a signal, whether the galeon stood to the eastward or to the westward; whilst the other was to follow the galeon at a distance, and if it grew dark, to direct the squadron in their chace, by shewing false fires.

Besides the care we had taken to prevent the galeon from passing by us unobserved, we had not been inattentive to the means of engaging her to advantage when we came up with her, for considering the thinness of our crews, and the vaunting accounts given by the Spaniards of her size, her guns, and her strength, this was a consideration not to be neglected. As we supposed that none of our ships but the *Centurion* and *Gloucester* were capable of lying alongside of her, we took on board the *Centurion* all the hands belonging to the *Carmelo* and *Carmin*, except what were just sufficient to navigate those ships; and Captain Saunders was ordered to send from the *Tryal's* prize ten Englishmen, and as many negroes, to reinforce the crew of the *Gloucester*. At the same time, for the encouragement of our negroes, of which we had a considerable number on board, we promised them that on their good behaviour they should have their freedom. As they had been almost every day trained to the management of the great guns for the two preceding months, they were very well qualified to be of service to us; and from their hopes of liberty, and in return for the kind usage they had met with amongst us, they seemed disposed to exert themselves to the utmost of their power, whenever we should have occasion for them.

Being thus prepared for the reception of the galeon, we expected, with the utmost impatience, the often mentioned 3d of March, the day fixed for her departure. No sooner did that day dawn than we were all of us most eagerly engaged in looking out towards Acapulco, from whence neither the casual duties on board nor the calls of hunger could easily divert our eyes; and we were so strangely prepossessed with the certainty of our intelligence, and with an assurance of her coming out of port, that some or other amongst us were constantly imagining that they discovered one of our cutters returning with a signal. But, to our extreme vexation, both this day and the succeeding night passed over without any news of the galeon. However, we did not yet despair, but were all heartily disposed to flatter ourselves that some unforeseen accident had intervened, which might have put off her departure for a few days; and suggestions of this kind occurred in plenty, as we knew that the time fixed by the viceroy for her sailing was often prolonged on the petition of the merchants of Mexico. Thus we kept up our hopes, and did not abate of our vigilance, and as the 7th of March was Sunday, the beginning of Passion week, which is observed by the Papists with great strictness, and a total cessation from all kinds of labour, so that no ship is permitted to stir out of port during the whole week, this quieted our apprehensions for some time, and disposed us not to expect the galeon till the week following. On the Friday in this week our cutters returned to us, and the officers on board them were very confident that the galeon was still in port, for that she could not possibly have come out but they must have seen her. The Monday morning following, that is, on the 15th of March, the cutters were again dispatched to their old station, and our hopes were once more indulged in as sanguine prepossessions as before; but in a week's time our eagerness was greatly abated, and a general dejection and despondency took place in its room. It is true, there were some few amongst us who still kept up their spirits, and were very ingenious in finding out reasons to satisfy themselves that the disappointment we had hitherto met with had only been occasioned by a casual delay of the galeon, which a few days would remove, and not by a total suspension of her departure for the whole season. But these speculations were not adopted by the generality of our people, for they were persuaded that the enemy had, by some accident, discovered our being upon the coast, and had therefore laid an embargo on the galeon till next year. And indeed this persuasion was but too well founded, for we afterwards learnt that our barge, when sent on the discovery of the port of Acapulco, had been seen from the shore, and that this circumstance (no embarkations but canoes ever frequenting that coast) was to them a sufficient proof of the neighbourhood of our squadron; on which they stopped the galeon till the succeeding year.

The commodore himself, though he declared not his opinion, was yet in his own thoughts apprehensive that we were discovered, and that the departure of the galeon was put off; and he had, in consequence of this opinion, formed a plan for possessing himself of Acapulco, because he had no doubt but the treasure as yet remained in the town, even though the orders for dispatching of the galeon were countermanded. Indeed the place was too well defended to be carried by an open attempt, since, besides the garrison and the crew of the galeon, there were in it at least a thousand men well armed, who had marched thither as guards to the treasure, when it was brought down from the city of Mexico, for the roads thereabouts are so much infested either by independent Indians or fugitives that the Spaniards never trust the silver without an armed force to protect it. Besides, had the strength of the place been less considerable, and such as might not have appeared superior to the efforts of our squadron, yet a declared attack would have prevented us receiving any advantages from its success, for upon the first discovery of our squadron, all the treasure would have been ordered into the country, and in a few hours would have been out of our reach, so that our conquest would have been only a desolate town, where we should have found nothing that could in the least have countervailed the fatigue and hazard of the undertaking.

For these reasons, the surprisal of the place was the only method that could at all answer our purpose; and therefore the manner in which Mr. Anson proposed to conduct this enterprize was, by setting sail with the squadron in the evening, time enough to arrive at the port in the night. As there is no danger on that coast, he would have stood boldly for the harbour's mouth, where he expected to arrive, and perhaps might have entered, before the

Spaniards were acquainted with his designs. As soon as he had run into the harbour, he intended to have pushed two hundred of his men on shore in his boats, who were immediately to attempt the fort, whilst he, the commodore, with his ships, was employed in firing upon the town and the other batteries. And these different operations, which would have been executed with great regularity, could hardly have failed of succeeding against an enemy who would have been prevented by the suddenness of the attack, and by the want of daylight, from concerting any measures for their defence. So that it was extremely probable that we should have carried the fort by storm, and then the other batteries, being open behind, must have been soon abandoned, after which, the town and its inhabitants, and all the treasure, must necessarily have fallen into our hands. For the place is so cooped up with mountains that it is scarcely possible to escape out of it but by the great road which passes under the fort. This was the project which the commodore had thus far settled generally in his thoughts, but when he began to inquire into such circumstances as were necessary to be considered in order to regulate the particulars of its execution, he found there was a difficulty, which, being insuperable, occasioned the enterprize to be laid aside; as on examining the prisoners about the winds which prevail near the shore, he learnt (and it was afterwards confirmed by the officers of our cutters) that nearer in shore there was always a dead calm for the greatest part of the night, and that towards morning, when a gale sprung up, it constantly blew off the land, so that the setting sail from our present station in the evening, and arriving at Acapulco before daylight, was impossible.

This scheme, as hath been said, was formed by the commodore upon a supposition that the galeon was detained till the next year, but as this was a matter of opinion only, and not founded on intelligence, and there was a possibility that she might still put to sea in a short time, the commodore thought it prudent to continue cruising on his present station as long as the necessary attention to his stores of wood and water, and to the convenient season for his future passage to China, would give him leave. And therefore, as the cutters had been ordered to remain before Acapulco till the 23d of March, the squadron did not change its position till that day, when the cutters not appearing, we were in some pain for them, apprehending they might have suffered either from the enemy or the weather, but we were relieved from our concern the next morning, when we discovered them, though at a great distance and to the leeward of the squadron. We bore down to them and took them up, and were informed by them that, conformable to their orders, they had left their station the day before, without having seen anything of the galeon; and we found that the reason of their being so far to the leeward of us was a strong current which had driven the whole squadron to windward.

And here it is necessary to mention, that, by information which was afterwards received, it appeared that this prolongation of our cruise was a very prudent measure, and afforded us no contemptible chance of seizing the treasure on which we had so long fixed our thoughts. For after the embargo was laid on the galeon, as is before mentioned, the persons principally interested in the cargo dispatched several expresses to Mexico, to beg that she might still be permitted to depart. It seems they knew, by the accounts sent from Paita, that we had not more than three hundred men in all, whence they insisted that there was nothing to be feared, as the galeon, carrying above twice as many hands as our whole squadron, would be greatly an overmatch for us. And though the viceroy was inflexible, yet, on the account of their representation, she was kept ready for the sea near three weeks after the first order came to detain her.

When we had taken up the cutters, all the ships being joined, the commodore made a signal to speak with their commanders; and upon enquiry into the stock of fresh water remaining on board the squadron, it was found to be so very slender that we were under a necessity of quitting our station to procure a fresh supply. Consulting what place was the properest for this purpose, it was agreed that the harbour of Seguataneio or Chequetan being the nearest, was, on that account, the most eligible; so that it was immediately resolved to make the best of our way thither. But that, even while we were recruiting our water, we might not totally abandon our views upon the galeon, which, perhaps, from certain intelligence of our being employed at Chequetan, might venture to slip out to sea, our cutter, under the command of Mr. Hughes, the lieutenant of the *Tryal's* prize, was ordered to cruise off the port of Acapulco for twenty-four days, that if the galeon should set sail in that interval, we might be speedily informed of it. In pursuance of these resolutions we endeavoured to ply to the westward to gain our intended port, but were often interrupted in our progress by calms and adverse currents. At these intervals we employed ourselves in taking out the most valuable part of the cargoes of the *Carmelo* and *Carmin* prizes, which two ships we intended to destroy as soon as we had tolerably cleared them. By the 1st of April we were so far advanced towards Seguataneio that we thought it expedient to send out two boats that they might range along the coast to discover the watering-place. They were gone some days, and our water being now very short, it was a particular felicity to us that we met with daily supplies of turtle, for had we been entirely confined to salt provisions, we must have suffered extremely in so warm a climate. Indeed our present circumstances were sufficiently alarming, and gave the most considerate amongst us as much concern as any of the numerous perils we had hitherto encountered, for our boats, as we conceived by their not returning, had not as yet found a place proper to water at, and by the leakage of our casks, and other accidents, we had not ten days water on board the whole squadron, so that from the known difficulty of procuring water on this coast, and the little reliance we had on the buccaneer writers (the only guides we had to trust to), we were apprehensive of being soon exposed to a calamity the most terrible of any that occurs in the long disheartening catalogue of the distresses of a seafaring life.

But these gloomy suggestions were at length happily ended: for our boats returned on the 5th of April, having about seven miles to the westward of the rocks of Seguataneio met with a place fit for our purpose, and which, by the description they gave of it, appeared to be the port of Chequetan, mentioned by Dampier. The success of our boats was highly agreeable to us, and they were ordered out again the next day, to sound the harbour and its entrance, which they had represented as very narrow. At their return they reported the place to be free from any danger, so that on the 7th we stood for it, and that evening came to an anchor in eleven fathom. The *Gloucester* cast anchor at the same time with us, but the *Carmelo* and the *Carmin* having fallen to the leeward, the *Tryal's* prize was ordered to join them, and to bring them up, which in two or three days she effected.

Thus, after a four months' continuance at sea from the leaving of Quibo, and having but six days' water on board, we arrived in the harbour of Chequetan, the description of which, and of the adjacent coast, shall be the business of the ensuing chapter.

CHAPTER XII

DESCRIPTION OF THE HARBOUR OF CHEQUETAN, AND OF THE ADJACENT COAST AND COUNTRY

The harbour of Chequetan, which we here propose to describe, lies in the latitude of 17° 36' north, and is about thirty leagues to the westward of Acapulco. It is easy to be discovered by any ship that will keep well in with the land, especially by such as range down the coast from Acapulco, and will attend to the following particulars.

There is a beach of sand, which extends eighteen leagues from the harbour of Acapulco to the westward, against which the sea breaks so violently that with our boats it would be impossible to land on any part of it, but yet the ground is so clean that during the fair season ships may anchor in great safety at the distance of a mile or two from the shore. The land adjacent to this beach is generally low, full of villages, and planted with a great number of trees, and on the tops of some small eminencies there are several lookout towers, so that the face of the country affords a very agreeable prospect: for the cultivated part, which is the part here described, extends some leagues back from the shore, where it seems to be bounded by a chain of mountains which stretch to a considerable distance on either side of Acapulco. It is a most remarkable particularity that in this whole extent, containing in appearance the most populous and best planted district of the whole coast, there should be neither canoes, boats, nor any other embarkations, either for fishing, coasting, or for pleasure. This cannot be imputed to the difficulty of landing, because in many parts of Africa and Asia, where the same inconvenience occurs, the inhabitants have provided against it by vessels of a peculiar fabric. I therefore conceive that the government, to prevent smuggling, have prohibited the use of all kinds of small craft in that district.

The beach here described is the surest guide to those who are desirous of finding the harbour of Chequetan, for five miles to the westward of the extremity of this beach there appears a hummock, which at first makes like an island, and is in shape not very unlike the hill of Petaplan, hereafter mentioned, though much smaller. Three miles to the westward of this hummock is a white rock near the shore which cannot easily be passed by unobserved. It is about two cables'-length from the land, and lies in a large bay, about nine leagues over. The west point of this bay is the hill of Petaplan, with the view of the islands of Quicara and Quibo. This hill of Petaplan, like the forementioned hummock, may be at first mistaken for an island, though it be in reality a peninsula, which is joined to the continent by a low and

- 203 -

narrow isthmus, covered over with shrubs and small trees. The bay of Seguataneio extends from this hill a great way to the westward, and it appears by a plan of the bay of Petaplan, which is part of that of Seguataneio, that at a small distance from the hill, and opposite to the entrance of the bay, there is an assemblage of rocks which are white from the excrements of boobies and tropical birds. Four of these rocks are high and large, and together with several smaller ones, are, by the help of a little imagination, pretended to resemble the form of a cross, and are called the White Friars. These rocks, as appears by the plan, bear W. by N. from Petaplan, and about seven miles to the westward of them lies the harbour of Chequetan, which is still more minutely distinguished by a large and single rock that rises out of the water a mile and an half distant from the entrance, and bears S.½W. from the middle of it. To these directions I must add that the coast is no ways to be dreaded between the middle of October and the beginning of May, nor is there then any danger from the winds, though in the remaining part of the year there are frequent and violent tornadoes, heavy rains, and hard gales in all directions of the compass.

Such are the infallible marks by which the harbour of Chequetan may be known to those who keep well in with the land. But as to those who keep at any considerable distance from the coast, there is no other method to be taken for finding the place than that of making it by the latitude, for there are so many ranges of mountains rising one upon the back of another within land, that no drawings of the appearance of the coast can be at all depended on when off at sea, every little change of distance or variation of position bringing new mountains in view, and producing an infinity of different prospects, which render all attempts of delineating the aspect of the coast impossible.

Having discussed the methods of discovering the harbour of Chequetan, it is time to describe the harbour itself. Its entrance is but about half a mile broad; the two points which form it, and which are faced with rocks that are almost perpendicular, bear from each other S.E. and N.W. The harbour is invironed on all sides, except to the westward, with high mountains overspread with trees. The passage into it is very safe on either side of the rock that lies off the mouth of it, though we, both in coming in and going out, left it to the eastward. The ground without the harbour is gravel mixed with stones, but within it is a soft mud: and it must be remembered that in coming to an anchor a good allowance should be made for a large swell, which frequently causes a great send of the sea, as likewise for the ebbing and flowing of the tide, which we observed to be about five feet, and that it set nearly E. and W.

The watering-place is situated in that part of the harbour where there is fresh water. This, during the whole time of our stay, had the appearance of a large standing lake, without any visible outlet into the sea, from which it is separated by a part of the strand. The origin of this lake is a spring that bubbles out of the ground near half a mile within the country. We found the water a little brackish, but more considerably so towards the seaside; for the nearer we advanced towards the spring-head the softer and fresher it proved. This laid us under a necessity of filling all our casks from the furthest part of the lake, and occasioned us some trouble; and would have proved still more difficult had it not been for our particular management, which, on account of the conveniency of it, deserves to be recommended to all who shall hereafter water at this place. Our method consisted in making use of canoes which drew but little water; for, loading them with a number of small casks, they easily got up the lake to the spring-head, and the small casks being there filled, were in the same manner transported back again to the beach, where some of our hands always attended to start them into other casks of a larger size.

Though this lake, during our continuance there, appeared to have no outlet into the sea, yet there is reason to suppose that in the rainy season it overflows the strand, and communicates with the ocean; for Dampier, who was formerly here, speaks of it as a large river. Indeed it is necessary that a vast body of water should be amassed before the lake can rise high enough to overflow the strand, since the neighbouring lands are so low that great part of them must be covered with water before it can run out over the beach.

As the country hereabouts, particularly the tract of coast contiguous to Acapulco, appeared to be well peopled and cultivated, we hoped to have easily procured from thence some fresh provisions and other refreshments which we now stood greatly in need of. To facilitate these views, the commodore, the morning after we came to an anchor, ordered a party of forty men, well armed, to march into the country, and to endeavour to discover some town or village, where they were to attempt to set on foot a correspondence with the inhabitants; for when we had once begun this intercourse, we doubted not but that, by proper presents, we should allure them to bring down to us whatever fruits or fresh provisions were in their power, as our prizes abounded in various kinds of coarse merchandize, which were of little consequence to us, though to them they would be extremely valuable. Our people were directed on this occasion to proceed with the greatest circumspection, and to make as little ostentation of hostility as possible; for we were sensible we could find no wealth in these parts worth our notice, and what necessaries we really wanted, we expected would be better and more abundantly supplied by an open amicable traffic than by violence and force of arms. But this endeavour of opening a commerce with the inhabitants proved ineffectual; for towards evening, the party which had been ordered to march into the country returned greatly fatigued by their unusual exercise, and some of them so far spent that they had fainted on the road, and were obliged to be brought back upon the shoulders of their companions. They had penetrated, as they conceived, about ten miles into the country, along a beaten track, where they often saw the fresh dung of horses or mules. When they had got near five miles from the harbour, the road divided between the mountains into two branches, one running to the east and the other to the west. On deliberation concerning the course they should take, it was agreed to continue their march along the eastern road: this when they had followed it for some time led them at once into a large plain or savannah, on one side of which they discovered a centinel on horseback with a pistol in his hand. It was supposed that when they first saw him he was asleep; but his horse, startled at the glittering of their arms, and turning round suddenly, ran off with his master, who, though he was very near being unhorsed in the surprize, yet recovered his seat, and escaped with the loss only of his hat and his pistol, which he dropped on the ground. Our people pursued him in hopes of discovering the village or habitation which he would retreat to; but as he had the advantage of being on horseback, they soon lost sight of him. Notwithstanding his escape, they were unwilling to come back without making some discovery, and therefore still followed the track they were in, till the heat of the day increasing, and finding no water to quench their thirst, they were first obliged to halt, and then resolved to return; for as they saw no signs of plantations or cultivated land, they had no reason to believe that there was any village or settlement near them. However, to leave no means untried of procuring some intercourse with the people, the officers stuck up several poles in the road, to which were affixed declarations written in Spanish, encouraging the inhabitants to come down to the harbour to traffic with us, giving them the strongest assurances of a kind reception, and faithful payment for any provisions they should bring us. This was doubtless a very prudent measure, yet it produced no effect; for we never saw any of them during the whole time of our continuance at this port of Chequetan. Indeed it were to have been wished that our men, upon the division of the path, had taken the western road instead of the eastern; for then they would soon have been led to a village or town, which some Spanish manuscripts mention as being in the neighbourhood of this port, and which we afterwards learnt was not above two miles from that turning.

And on this occasion I cannot avoid mentioning another adventure which happened to some of our people in the bay of Petaplan, as it may greatly assist the reader in forming a just idea of the temper and resolution of the inhabitants of this part of the world. Some time after our arrival at Chequetan, Lieutenant Brett was sent by the commodore, with two of our boats under his command, to examine the coast to the eastward, particularly to make observations on the bay and watering-place of Petaplan. As Mr. Brett with one of the boats was preparing to go on shore towards the hill of

Petaplan, he accidentally looking across the bay, perceived on the opposite strand three small squadrons of horse parading upon the beach, and seeming to advance towards the place where he proposed to land. On sight of this he immediately put off the boat, though he had but sixteen men with him, and stood over the bay towards them: and he soon came near enough to perceive that they were mounted on very sightly horses, and were armed with carbines and lances. On seeing him make towards them, they formed upon the beach, and seemed resolved to dispute his landing, firing several distant shot at him as he drew near, till at last the boat being arrived within a reasonable distance of the most advanced squadron, Mr. Brett ordered his people to fire, upon which this resolute cavalry instantly ran with great confusion into the wood through a small opening. In this precipitate flight one of their horses fell down and threw his rider; but whether he was wounded or not we could not discern, for both man and horse soon got up again, and followed the rest into the wood. In the meantime the other two squadrons were calm spectators of the rout of their comrades, for they were drawn up at a great distance behind, out of the reach of our shot, having halted on our first approach, and never advancing a step afterwards. It was doubtless fortunate for our people that the enemy acted with so little prudence, and exerted so little spirit, since had they concealed themselves till our men had landed, it is scarcely possible but all the boat's crew must have fallen into their hands, as the Spaniards were not much short of two hundred, and the whole number with Mr. Brett only amounted to sixteen. However, the discovery of so considerable a force collected in this bay of Petaplan obliged us constantly to keep a boat or two before it: for we were apprehensive that the cutter, which we had left to cruise off Acapulco, might on her return be surprized by the enemy, if she did not receive timely information of her danger. But now to proceed with the account of the harbour of Chequetan.

After our unsuccessful attempt to engage the people of the country to furnish us with the necessaries we wanted, we desisted from any more endeavours of the same nature, and were obliged to be contented with what we could procure for ourselves in the neighbourhood of the port. We caught fish here in tolerable quantities, especially when the smoothness of the water permitted us to hale the seyne. Amongst the rest, we got cavallies, breams, mullets, soles, fiddle-fish, sea-eggs, and lobsters: and we here, and in no other place, met with that extraordinary fish called the torpedo, or numbing-fish, which is in shape very like the fiddle-fish, and is not to be known from it but by a brown circular spot about the bigness of a crown piece near the centre of its back. Perhaps its figure will be better understood when I say it is a flat fish much resembling the thorn-back. This fish, the torpedo, is indeed of a most singular nature, productive of the strangest effects on the human body: for whoever handles it, or happens even to set his foot upon it, is presently seized with a numbness all over him, but which is more distinguishable in that limb which was in immediate contact with it. The same effect too will be in some degree produced by touching the fish with anything held in the hand, since I myself had a considerable degree of numbness conveyed to my right arm, through a walking cane, which I rested on the body of the fish for a short time only; and I make no doubt but I should have been much more sensibly affected had not the fish been near expiring when I made the experiment, as it is observable that this influence acts with most vigour upon the fish's being first taken out of the water, and entirely ceases as soon as it is dead, so that it may be then handled, or even eaten, without any inconvenience. I shall only add, that the numbness of my arm upon this occasion did not go off on a sudden, as the accounts of some naturalists gave me reason to expect, but diminished gradually, so that I had some sensation of it remaining till the next day.

To the account given of the fish we met with here I must add, that though turtle now grew scarce, and we found none in this harbour of Chequetan, yet our boats, which were stationed off Petaplan, often supplied us therewith; and though this was a food that we had been long as it were confined to (since it was the only fresh provisions which we had tasted during near six months), yet we were far from being cloyed with it, or from finding that the relish we had for it at all diminished.

The animals we met with on shore were principally guanos, with which the country abounds, and which are by some reckoned delicious food. We saw no beast of prey here, except we should esteem that amphibious animal, the alligator, as such, several of which our people discovered, but none of them very large. However, we were satisfied that there were great numbers of tygers in the woods, though none of them came in sight, for we every morning found the beach near the watering-place imprinted very thick with their footsteps: but we never apprehended any mischief from them, since they are by no means so fierce as the Asiatic or African tyger, and are rarely, if ever, known to attack mankind. Birds were here in sufficient plenty; for we had abundance of pheasants of different kinds, some of them of an uncommon size, but they were all very dry and tasteless eating. And besides these we had a variety of smaller birds, particularly parrots, which we often killed for food.

The fruits and vegetable refreshments at this place were neither plentiful nor of the best kinds. There were, it is true, a few bushes scattered about the woods, which supplied us with limes, but we scarcely could procure enough for our present use: and these, with a small plum of an agreeable acid, called in Jamaica the hog-plum, together with another fruit called a papah, were the only fruits to be found in the woods. Nor is there any other useful vegetable here worth mentioning, except brook lime. This indeed grew in great quantities near the fresh-water banks; and as it was esteemed an antiscorbutic, we fed upon it frequently, though its extreme bitterness made it very unpalatable.

These are the articles most worthy of notice in this harbour of Chequetan. I shall only mention a particular of the coast lying to the westward of it, that to the eastward having been already described. As Mr. Anson was always attentive to whatever might be of consequence to those who might frequent these seas hereafter, and as we had observed that there was a double land to the westward of Chequetan, which stretched out to a considerable distance, with a kind of opening that appeared not unlike the inlet to some harbour, the commodore, soon after we came to an anchor, sent a boat to discover it more accurately, and it was found on a nearer examination that the two hills which formed the double land were joined together by a valley, and that there was no harbour nor shelter between them.

By all that hath been said it will appear that the conveniences of this port of Chequetan, particularly in the articles of refreshment, are not altogether such as might be desired: but yet, upon the whole, it must be owned to be a place of considerable consequence, and that the knowledge of it may be of great import to future cruisers, for except Acapulco, which is in the hands of the enemy, it is the only secure harbour in a vast extent of coast. It lies at a proper distance from Acapulco for the convenience of such ships as may have any designs on the Manila galeon; and it is a place where wood and water may be procured with great security in despight of the efforts of the inhabitants of the adjacent district: for there is but one narrow path which leads through the woods into the country, and this is easily to be secured by a very small party against all the strength the Spaniards in that neighbourhood can muster. After this account of Chequetan, and the coast contiguous to it, we now return to the recital of our own proceedings.

CHAPTER XIII

OUR PROCEEDINGS AT CHEQUETAN AND ON THE ADJACENT
COAST, TILL OUR SETTING SAIL FOR ASIA

The next morning after our coming to an anchor in the harbour of Chequetan, we sent about ninety of our men well armed on shore; forty of whom were ordered to march into the country, as hath been mentioned, and the remaining fifty were employed to cover the watering-place, and to prevent any interruption from the natives.

Here we compleated the unloading of the *Carmelo* and *Carmin*, which we had begun at sea; that is to say, we took out of them the indico, cacao, and cochineal, with some iron for ballast, which were all the goods we intended to preserve, though they did not amount to a tenth of their cargoes. Here too it was agreed, after a mature consultation, to destroy the *Tryal's* prize, as well as the *Carmelo* and *Carmin*, whose fate had been before resolved on. Indeed the *Tryal's* prize was in good repair, and fit for the sea; but as the whole numbers on board our squadron did not amount to the complement of a fourth-rate man-of-war, we found it was impossible to divide them into three ships without rendering each of those ships incapable of navigating in safety through the tempestuous weather we had reason to expect on the coast of China, where we supposed we should arrive about the time of the change of the monsoons. These considerations determined the commodore to destroy the *Tryal's* prize, and to reinforce the *Gloucester* with the best part of the crew. And in consequence of this resolve, all the stores on board the *Tryal's* prize were removed into the other ships, and the prize herself, with the *Carmelo* and *Carmin*, were prepared for scuttling with all the expedition we were masters of; but the great difficulties we were under in providing a store of water (which have been already touched on), together with the necessary repairs of our rigging and other unavoidable occupations, took us up so much time, and found us such unexpected employment, that it was near the end of April before we were in a condition to leave the place.

During our stay here there happened an incident which, as it proved the means of convincing our friends in England of our safety, which for some time they had despaired of, and were then in doubt about, I shall beg leave particularly to recite. I have observed, in the preceding chapter, that from this harbour of Chequetan there was but one pathway which led through the woods into the country. This we found much beaten, and were thence convinced that it was well known to the inhabitants. As it passed by the spring-head, and was the only avenue by which the Spaniards could approach us, we, at some distance beyond the spring-head, felled several large trees, and laid them one upon the other across the path; and at this barricadoe we constantly kept a guard. We besides ordered our men employed in watering to have their arms ready, and, in case of any alarm, to march instantly to this post. And though our principal intention herein was to prevent our being disturbed by any sudden attack of the enemy's horse, yet it answered another purpose which was not in itself less important: this was to hinder our own people from straggling singly into the country, where we had reason to believe they would be surprized by the Spaniards, who would doubtless be extremely solicitous to pick up some of them in hopes of getting intelligence of our future designs. To avoid this inconvenience, the strictest orders were given to the centinels to let no person whatever pass beyond their post. But notwithstanding this precaution, we missed one Lewis Leger, who was the commodore's cook. As he was a Frenchman, and was suspected to be a Papist, it was at first imagined that he had deserted, with a view of betraying all that he knew to the enemy; though this appeared, by the event, to be an ill-grounded surmise, for it was afterwards known that he had been taken by some Indians, who carried him prisoner to Acapulco, from whence he was transferred to Mexico, and then to Vera Cruz, where he was shipped on board a vessel bound to Old Spain. But the vessel being obliged by some accident to put into Lisbon, Leger escaped on shore, and was by the British consul sent from thence to England, where he brought the first authentick account of the safety of the commodore, and of his principal transactions in the South Seas. The relation he gave of his own seizure was that he rambled into the woods at some distance from the barricadoe, where he had first attempted to pass, but had been stopped and threatened to be punished; that his principal view was to get a quantity of limes for his master's store, and that in this occupation he was surprized unawares by four Indians, who stripped him naked, and carried him in that condition to Acapulco, exposed to the scorching heat of the sun, which at that time of the year shone with its greatest violence; that afterwards at Mexico his treatment in prison was sufficiently severe; so that the whole course of his captivity was a continued instance of the hatred which the Spaniards bear to all those who endeavour to disturb them in the peaceable possession of the coasts of the South Seas. Indeed Leger's fortune was, upon the whole, extremely singular, as, after the hazards he had run in the commodore's squadron, and the severities he had suffered in his long confinement amongst the enemy, a more fatal disaster attended him on his return to England: for though, when he arrived in London, some of Mr. Anson's friends interested themselves in relieving him from the poverty to which his captivity had reduced him, yet he did not long enjoy the benefit of their humanity, since he was killed in an insignificant night-brawl, the cause of which could scarcely be discovered.

And on occasion of this surprizal of Leger, I must observe, that though the enemy never appeared in sight during our stay in the harbour, yet we perceived that large parties of them were encamped in the woods about us; for we could see their smokes, and could thence determine that they were posted in a circular line surrounding us at a distance; and just before our coming away they seemed, by the increase of their fires, to have received a considerable reinforcement. But to return.

Towards the latter end of April, the unloading of our three prizes, our wooding and watering, and in short, every one of our proposed employments at the harbour of Chequetan, were compleated: so that, on the 27th of April, the *Tryal's* prize, the *Carmelo*, and the *Carmin*, all which we here intended to destroy, were towed on shore and scuttled, a quantity of combustible materials having been distributed in their upper works: and the next morning the *Centurion* with the *Gloucester* weighed anchor, though as there was but little wind, and that not in their favour, they were obliged to warp out of the harbour. When they had reached the offing, one of the boats was dispatched back again to set fire to our prizes, which was accordingly executed. After this a canoe was left fixed to a grapnel in the middle of the harbour, with a bottle in it well corked, inclosing a letter to Mr. Hughes, who commanded the cutter, which had been ordered to cruise before the port of Acapulco when we ourselves quitted that station. And on this occasion I must mention more particularly than I have yet done the views of the commodore in leaving the cutter before that port.

When we were necessitated to proceed for Chequetan to recruit our water, Mr. Anson considered that our arrival in that harbour would soon be known at Acapulco; and therefore he hoped that on the intelligence of our being employed in port, the galeon might put to sea, especially as Chequetan is so very remote from the course generally steered by the galeon. He therefore ordered the cutter to cruise twenty-four days off the port of Acapulco, and her commander was directed, on perceiving the galeon under sail, to make the best of his way to the commodore at Chequetan. As the *Centurion* was doubtless a much better sailor than the galeon, Mr. Anson, in this case, resolved to have got to sea as soon as possible, and to have pursued the galeon across the Pacifick Ocean: where supposing he should not have met with her in his passage (which, considering that he would have kept nearly the same parallel, was very improbable), yet he was certain of arriving off Cape Espiritu Santo, on the island of Samal, before her; and that being the first land she makes on her return to the Philippines, we could not have failed to have fallen in with her by cruising a few days in that station. However, the Viceroy of Mexico ruined this project by keeping the galeon in the port of Acapulco all that year.

The letter left in the canoe for Mr. Hughes, the commander of the cutter, the time of whose return was now considerably elapsed, directed him to go back immediately to his former station before Acapulco, where he would find Mr. Anson, who resolved to cruise for him there a certain number of days; after which it was added that the commodore would return to the southward to join the rest of the squadron. This last article was inserted to deceive the Spaniards, if they got possession of the canoe, as we afterwards learnt they did; but could not impose on Mr. Hughes, who well knew that the commodore had no squadron to join, nor any intention of steering back to Peru.

Being now in the offing of Chequetan, bound across the vast Pacifick Ocean in our way to China, we were impatient to run off the coast as soon as possible, since the stormy season was approaching apace. As we had no farther views in the American seas, we had hoped that nothing would have prevented us from steering to the westward the moment we got out of the harbour of Chequetan: and it was no small mortification to us that our necessary employment there had detained us so much longer than we expected, but now, when we had put to sea, we were farther detained by the absence of the cutter, and the necessity we were under of standing towards Acapulco in search of her. Indeed, as the time of her cruise had been expired for near a fortnight, we suspected that she had been discovered from the shore, and that the Governor of Acapulco had thereupon sent out a force to seize her, which, as she carried but six hands, was no very difficult enterprize. However, this being only conjecture, the commodore, as soon as he was got clear of the harbour of Chequetan, stood along the coast to the eastward in search of her: and to prevent her from passing by us in the dark, we brought to every night, and the *Gloucester*, whose station was a league within us towards the shore, carried a light, which the cutter could not but perceive if she kept along shore, as we supposed she would do; besides, as a farther security, the *Centurion* and *Gloucester* alternately shewed two false fires every half-hour. Indeed, had she escaped us, she would have found orders in the canoe to have returned immediately before Acapulco, where Mr. Anson proposed to cruise for her some days.

By Sunday, the 2d of May, we were advanced within three leagues of Acapulco, and having seen nothing of our boat, we gave her over as lost, which, besides the compassionate concern for our ship-mates, and for what it was apprehended they might have suffered, was in itself a misfortune, which, in our present scarcity of hands, we were all greatly interested in: since the crew of the cutter, consisting of six men and the lieutenant, were the very flower of our people, purposely picked out for this service, and known to be every one of them of tried and approved resolution, and as skilful seamen as ever trod a deck. However, as it was the general belief among us that they were taken and carried into Acapulco, the commodore's prudence suggested a project which we hoped would recover them. This was founded on our having many Spanish and Indian prisoners in our possession, and a number of sick negroes, who could be of no service to us in the navigating of the ship. The commodore therefore wrote a letter the same day to the Governor of Acapulco, telling him that he would release them all provided the governor returned the cutter's crew. This letter was dispatched in the afternoon by a Spanish officer, of whose honour we had a good opinion, and who was furnished with a launch belonging to one of our prizes and a crew of six other prisoners, who gave their parole for their return. The Spanish officer too, besides the commodore's letter, carried with him a joint petition, signed by all the rest of the prisoners, beseeching the governor to acquiesce in the terms proposed for their liberty. From a consideration of the number of our prisoners and the quality of some of them, we did not doubt but the governor would readily comply with Mr. Anson's proposal, and therefore we kept plying on and off the whole night, intending to keep well in with the land that we might receive an answer at the limited time, which was the next day, being Monday. But both on Monday and Tuesday we were driven so far off shore that we could not hope that any answer could reach us; and even on the Wednesday morning we found ourselves fourteen leagues from the harbour of Acapulco; however, as the wind was then favourable, we pressed forwards with all our sail, and did not doubt of getting in with the land that afternoon. Whilst we were thus standing in, the centinel called out from the mast-head that he saw a boat under sail at a considerable distance to the south-eastward. This we took for granted was the answer of the governor to the commodore's message, and we instantly edged towards her; but as we approached her we found, to our unspeakable joy, that it was our own cutter. And though, while she was still at a distance, we imagined that she had been discharged out of the port of Acapulco by the governor; yet, when she drew nearer, the wan and meagre countenances of the crew, the length of their beards, and the feeble and hollow tone of their voices, convinced us that they had suffered much greater hardships than could be expected from even the severities of a Spanish prison. They were obliged to be helped into the ship, and were immediately put to bed, where by rest and nourishing diet, which they were plentifully supplied with from the commodore's table, they recovered their health and vigour apace. And now we learnt that they had kept the sea the whole time of their absence, which was above six weeks; that when they had finished their cruise before Acapulco, and had just begun to ply to the westward, in order to join the squadron, a strong adverse current had forced them down the coast to the eastward, in spight of all their efforts to the contrary, that at length, their water being all expended, they were obliged to search the coast farther on to the eastward in quest of some convenient landing-place where they might get a fresh supply; that in this distress they ran upwards of eighty leagues to leeward, and found everywhere so large a surf that there was not the least possibility of their landing; that they passed some days in this dreadful situation without water, having no other means left them to allay their thirst than sucking the blood of the turtle which they caught; that at last, giving up all hopes of succour, the heat of the climate too augmenting their necessities, and rendering their sufferings insupportable, they abandoned themselves to despair, fully persuaded that they should perish by the most terrible of all deaths; but that soon after a most unexpected incident happily relieved them. For there fell so heavy a rain, that on spreading their sails horizontally, and putting bullets in the centers of them to draw them to a point, they caught as much water as filled all their casks; that immediately upon this fortunate supply they stood to the westward in quest of the commodore; and being now luckily favoured by a strong current, they joined us in less than fifty hours from that time, after having been absent in the whole full forty-three days. Those who have an idea of the inconsiderable size of a cutter belonging to a sixty-gun ship (being only an open boat about twenty-two feet in length), and who will reflect on the various casualties that must have attended her during a six weeks' continuance alone, in the open ocean, on so impracticable and dangerous a coast, will readily own that her return to us at last, after all the difficulties which she actually experienced, and the dangers to which she was each hour exposed, may be considered as little short of miraculous.

I cannot finish this article of the cutter without remarking how slender a reliance navigators ought to have on the accounts of the buccaneer writers; for though in this run of hers, eighty leagues to the eastward of Acapulco, she found no place where it was possible that a boat could land; yet those writers have not been ashamed to feign harbours and convenient watering-places within these limits, thereby exposing such as should confide in their relations to the risque of being destroyed by thirst.

I must farther add on this occasion that, when we stood near the port of Acapulco, in order to send our message to the governor, and to receive his answer, Mr. Brett took that opportunity of delineating a view of the entrance of the port and of the neighbouring coast, which, added to the plan of the place formerly mentioned, may be of considerable use hereafter.

Having thus recovered our cutter, the sole object of our coming a second time before Acapulco, the commodore determined not to lose a moment's time more, but to run off the coast with the utmost expedition, both as the stormy season on the coast of Mexico was now approaching apace, and as we were apprehensive of having the westerly monsoon to struggle with when we came upon the coast of China: for this reason we no longer stood towards

Acapulco, as at present we wanted no answer from the governor. However, Mr. Anson resolved not to deprive his prisoners of the liberty which he had promised them; and therefore they were all immediately embarked in two launches which belonged to our prizes, those from the *Centurion* in one launch, and those from the *Gloucester* in the other. The launches were well equipped with masts, sails, and oars; and lest the wind might prove unfavourable, they had a stock of water and provisions put on board them sufficient for fourteen days. There were discharged thirty-nine persons from on board the *Centurion*, and eighteen from the *Gloucester*, the greatest part of them Spaniards, the rest being Indians and sick negroes. Indeed, as our crews were very weak, we kept the Mulattoes and some of the stoutest of our negroes with a few Indians to assist us; but we dismissed every Spanish prisoner whatever. We have since learnt that these two launches arrived safe at Acapulco, where the prisoners could not enough extol the humanity with which they had been treated. It seems the governor, before their arrival, had returned a very obliging answer to our letter, and had at the same time ordered out two boats laden with the choicest refreshments and provisions that were to be procured at Acapulco, which he intended as a present to the commodore: but these boats not having found our ships, were at length obliged to put back again, after having thrown all their provisions overboard in a storm which threatened their destruction.

The sending away our prisoners was our last transaction on the American coast; for no sooner had we parted with them than we and the *Gloucester* made sail to the S.W., proposing to get a good offing from the land, where we hoped, in a few days, to meet with the regular trade-wind, which the accounts of former navigators had represented as much brisker and steadier in this ocean than in any other part of the world: for it has been esteemed no uncommon passage to run from hence to the eastermost isles of Asia in two months; and we flattered ourselves that we were as capable of making an expeditious voyage as any ships that had ever sailed this course before us; so that we hoped soon to gain the coast of China, for which we were now bound. As we conceived this navigation to be free from all kinds of embarrassment of bad weather, fatigue, or sickness, conformable to the general idea of it given by former travellers, we consequently undertook it with alacrity, especially as it was no contemptible step towards our arrival at our native country, for which many of us by this time began to have great longings. Thus, on the 6th of May, we, for the last time, lost sight of the mountains of Mexico, persuaded that in a few weeks we should arrive at the river of Canton in China, where we expected to meet with many English ships and with numbers of our countrymen; and hoped to enjoy the advantages of an amicable, well-frequented port, inhabited by a polished people and abounding with the conveniences and indulgencies of a civilized life; blessings which now for near twenty months had never been once in our power. But, before we take our final leave of America, there yet remains the consideration of a matter well worthy of attention, the discussion of which shall be referred to the ensuing chapter.

CHAPTER XIV

A BRIEF ACCOUNT OF WHAT MIGHT HAVE BEEN EXPECTED FROM OUR SQUADRON HAD IT ARRIVED IN THE SOUTH SEAS IN GOOD TIME

After the recital of the transactions of the commodore, and the ships under his command, on the coasts of Peru and Mexico, contained in the preceding narration, it will be no useless digression to examine what the whole squadron might have been capable of atchieving had it arrived on its destined scene of action in so good a plight as it would probably have done had the passage round Cape Horn been attempted at a more seasonable time of the year. This disquisition may be serviceable to those who shall hereafter form projects of the like nature for that part of the world, or who may be entrusted with their execution. And therefore I propose, in this chapter, to consider, as succinctly as I can, the numerous advantages which the public might have received from the operations of the squadron had it set sail from England a few months sooner than it did.

To begin then: I presume it will be granted me that in the summer time we might have got round Cape Horn with an inconsiderable loss, and without any material damage to our ships or rigging. For the *Duke* and *Duchess of Bristol*, who between them had above three hundred men, buried no more than two from the coast of Brazil to Juan Fernandez; and out of a hundred and eighty-three hands which were on board the *Duke* alone, there were only twenty-one sick of the scurvy when they arrived at that island. Whence as men-of-war are much better provided with all conveniences than privateers, we might doubtless have appeared before Baldivia in full strength, and in a condition of entering immediately on action; and therefore, as that place was in a very defenceless state, its cannon incapable of service, and its garrison in great measure unarmed, it was impossible that it could have opposed our force, or that its half-starved inhabitants, most of whom are convicts banished thither from other parts, could have had any other thoughts than that of submitting. This would have been a very important acquisition; since when Baldivia, which is an excellent port, had been once in our possession, we should immediately have been terrible to the whole kingdom of Chili, and should doubtless have awed the most distant parts of the Spanish Empire in America. Indeed it is far from improbable that, by a prudent use of this place, aided by our other advantages, we might have given a violent shock to the authority of Spain on that whole continent, and might have rendered some at least of her provinces independent. This would certainly have turned the whole attention of the Spanish ministry to that part of the world where the danger would have been so pressing, and thence Great Britain and her allies might have been rid of the numerous difficulties which the wealth of the Spanish Indies, operating in conjunction with the Gallick intrigues, have constantly thrown in their way.

But that I may not be thought to over-rate the force of this squadron by ascribing to it a power of overturning the Spanish Government in America, it is necessary to enter into a more particular discussion, and to premise a few observations on the condition of the provinces bordering near the South Seas, and on the disposition of the inhabitants, both Spaniards and Indians, at that time. For hence it will appear that the conjuncture was the most favourable we could have desired, since we shall find that the Creolian subjects were disaffected and their governors at variance, that the country was wretchedly provided with arms and stores, and they had fallen into a total neglect of all military regulations in their garrisons; and that the Indians on their frontier were universally discontented, and seemed to be watching with impatience the favourable moment when they might take a severe revenge for the barbarities they had groaned under during more than two ages: so that every circumstance concurred to facilitate the enterprizes of our squadron. Of all these articles we were amply informed by the letters we took on board our prizes; none of these vessels, as I remember, having had the precaution to throw their papers overboard.

The ill blood amongst the governors was greatly augmented by their apprehensions of our squadron; for every one being willing to have it believed that the bad condition of his government was not the effect of negligence, there were continual demands and remonstrances among them in order to throw the blame upon each other. Thus, for instance, the President of St. Jago in Chili, the President of Panama, and many other governors and military officers were perpetually soliciting the Viceroy of Peru to furnish them with the necessary sums of money for putting their provinces and places in a proper state of defence to oppose our designs: but the customary answer of the viceroy to these representations was that he was unable to comply with their requests, urging the emptiness of the royal chest at Lima, and the difficulties he was under to support the expences of his own government: he in one of his letters (which we intercepted) mentioning his apprehensions that he might soon be necessitated to stop the pay of the troops and even of the garrison of Callao, the key of the whole kingdom of Peru. Indeed he did at times remit to these governors some part of their demands; but as what he sent them was greatly short of their wants, these partial supplies rather tended to the raising jealousies and heart-burnings among them than contributed to the purposes for which they had at first been desired.

Besides these mutual janglings amongst the governors, the whole body of the people were extremely dissatisfied, they being fully persuaded that the affairs of Spain for many years before had been managed by the influence of a particular foreign interest, which was altogether detached from the advantages of the Spanish nation: so that the inhabitants of these distant provinces believed themselves to be sacrificed to an ambition which never considered their convenience or emoluments nor paid any regard to the reputation of their name or the honour of their country. That this was the temper of the Creolian Spaniards at that time might be proved from a hundred instances; but I shall content myself with one which is indeed conclusive: this is the testimony of the French mathematicians sent into America to measure the magnitude of an equatorial degree of latitude. For in the relation of the murther of a surgeon belonging to their company in one of the cities of Peru, and of the popular tumult thence occasioned, written by one of those astronomers, the author confesses that the multitude during the uproar universally joined in imprecations on their bad government, and bestowed the most abusive language upon the French, detesting them, in all probability, more particularly as being of a nation to whose influence in the Spanish counsels the Spaniards imputed all their misfortunes.

And whilst the Creolian Spaniards were thus dissatisfied, it appears by the letters we intercepted that the Indians on almost every frontier were ripe for a revolt, and would have taken up arms upon the slightest encouragement; particularly the Indians in the southern parts of Peru, as likewise the Arraucos, and the rest of the Chilian Indians, the most powerful and terrible to the Spanish name of any on that continent. For it seems in some disputes between the Spaniards and the Indians, which happened a short time before our arrival, the Spaniards had insulted the Indians with an account of the force which they expected from Old Spain under the command of Admiral Pizarro, and had vaunted that he was coming thither to compleat the great work which had been left unfinished by his ancestors. These threats alarmed the Indians, and made them believe that their extirpation was resolved on. For the Pizarros being the first conquerors of that coast, the Peruvian Indians held the name, and all that bore it, in execration; not having forgot the destruction of their monarchy, the massacre of their beloved Inca, Atapalipa, the extinction of their religion, and the slaughter of their ancestors, all perpetrated by the family of the Pizarros. The Chilian Indians too abhorred a chief who was descended of a race which, by its lieutenants, had first attempted to inslave them, and had necessitated the stoutest of their tribes for more than a century to be continually wasting their blood in defence of their independency.

Nor let it be supposed that among barbarous nations the traditions of these distant transactions could not be preserved for so long an interval; since those who have been acquainted with that part of the world agree that the Indians, in their publick feasts and annual solemnities, constantly revive the memory of these tragick incidents; and such as have been present at these spectacles have constantly observed that all the recital and representations of this kind were received with emotions so vehement, and with so enthusiastick a rage, as plainly demonstrated how strongly the memory of their former wrongs was implanted in them, and how acceptable the means of revenge would at all times prove. To this I must add too, that the Spanish governors themselves were so fully informed of the disposition of the Indians at this conjuncture, and were so apprehensive of a general defection among them, that they employed all their industry to reconcile the most dangerous tribes, and to prevent them from immediately taking up arms. Among the rest, the President of Chili in particular made large concessions to the Arraucos and the other Chilian Indians, by which, and by distributing considerable presents to their leading men, he at last got them to consent to a prolongation of the truce between the two nations. But these negociations were not concluded at the time when we might have been in the South Seas; and had they been compleated, yet the hatred of these Indians to the Spaniards was so great that it would have been impossible for their chiefs, how deeply soever corrupted, to have kept them from joining us against their old detested enemy.

Thus then it appears that on our arrival in the South Seas we might have found the whole coast unprovided with troops and destitute even of arms: for we well know, from very particular intelligence, that there were not three hundred fire-arms, of which too the greatest part were matchlocks, in all the province of Chili. Whilst at the same time, the Indians were ripe for a revolt, the Spaniards disposed to mutiny, and the governors enraged with one another, and each prepared to rejoice in the disgrace of his antagonist. At this fortunate crisis we, on the other hand, might have consisted of near two thousand men, the greatest part in health and vigour, all well armed, and united under a chief whose enterprising genius (as we have seen) could not be depressed by a continued series of the most sinister events, and whose equable and prudent turn of temper would have remained unvaried in the midst of the greatest degree of good success; and who besides possessed, in a distinguished manner, the two qualities the most necessary for these uncommon undertakings—I mean that of maintaining his authority and preserving, at the same time, the affections of his people. Our other officers too, of every rank, appear, by the experience the public hath since had of them, to have been equal to any attempt they might have been charged with by their commander: and our men (at all times brave if well conducted) in such a cause, where treasure was the object, and under such leaders, would doubtless have been prepared to rival the most celebrated achievements hitherto performed by British mariners.

It cannot then be contested but that Baldivia must have surrendered on the appearance of our squadron: after which, it may be presumed, that the Arraucos, the Pulches, and Penguinches, inhabiting the banks of the river Imperial, about twenty-five leagues to the northward of this place, would have immediately taken up arms, being disposed thereto, as hath been already related, and encouraged by the arrival of so considerable a force in their neighbourhood. As these Indians can bring into the field near thirty thousand men, the greatest part of them horse, their first step would have been the invading the province of Chili, which they would have found totally unprovided both of ammunition and weapons; and as its inhabitants are a luxurious and effeminate race, they would have been incapable, on such an emergency, of giving any opposition to this rugged enemy: so that it is no strained conjecture to imagine that the Indians would have been soon masters of the whole country. Moreover, the other Indians, on the frontiers of Peru, being equally disposed with the Arraucos to shake off the Spanish yoke, it is highly probable that they likewise would have embraced this favourable occasion, and that a general insurrection would have taken place through all the Spanish territories of South America; in which case, the only resource left to the Creolians (dissatisfied as they were with the Spanish government) would have been to have made the best terms they could with their Indian neighbours, and to have withdrawn themselves from the obedience of a master who had shown so little regard to their security. This last supposition may perhaps appear chimerical to those who measure the possibility of all events by the scanty standard of their own experience; but the temper of the times, and the strong dislike of the natives to the measures then pursued by the Spanish court, sufficiently evince at least its possibility. However, not to insist on the presumption of a general revolt, it is sufficient for our purpose to conclude that the Arraucos would scarcely have failed of taking arms on our appearance: since this alone would so far have terrified the enemy that they would no longer have employed their thoughts on the means of opposing us, but would have turned all their care to the Indian affairs; as they still remember, with the utmost horror, the sacking of their cities, the rifling of their convents, the captivity of their wives and daughters, and the desolation of their country by these resolute savages in the last war between the two nations. For it must be observed that the Chilian Indians have been frequently successful against the Spaniards, and possess at this time a large tract of country which was formerly full of Spanish towns and villages, whose inhabitants were all either destroyed or carried into captivity by the Arraucos and the other neighbouring Indians, who in a war against the Spaniards never fail to join their forces.

But even, independent of an Indian revolt, there were two places only, on all the coast of the South Sea, which could be supposed capable of resisting our squadron; these were the cities of Panama and Callao: as to the first of these, its fortifications were so decayed, and it was so much in want of powder, that the president himself, in an intercepted letter, acknowledged it was incapable of being defended; whence I take it for granted it would have given us but little trouble, especially if we had opened a communication across the isthmus with our fleet on the other side. And with regard to the city and port of Callao, its condition was not much better than that of Panama; since its walls are built upon the plain ground, without either out-work or ditch before them, and consist only of very slender feeble masonry, without any earth behind them; so that a battery of five or six pieces of cannon, raised anywhere within four or five hundred paces of the place, would have had a full view of the whole rampart, and would have opened it in a short time; and the breach hereby formed, as the walls are so extremely thin, could not have been difficult of ascent; for the ruins would have been but little higher than the surface of the ground; and it would have yielded this particular advantage to the assailants, that the bullets, which grazed upon it, would have driven before them such shivers of brick and stone as would have prevented the garrison from forming behind it, supposing that the troops employed in defence of the place should have so far surpassed the usual limits of Creolian bravery as to resolve to stand a general assault. Indeed, such a resolution cannot be imputed to them; for the garrison and people were in general dissatisfied with the viceroy's behaviour, and were never expected to act a vigorous part. On the contrary, the viceroy himself greatly apprehended that the commodore would make him a visit at Lima, the capital of the kingdom of Peru; to prevent which, if possible, he had ordered twelve gallies to be built at Guaiaquil and other places, which were intended to oppose the landing of our boats, and to hinder us from pushing our men on shore. But this was an impracticable project of defence, and proceeded on the supposition that our ships, when we should land our men, would keep at such a distance that these gallies, by drawing little water, would have been out of the reach of our guns; whereas the commodore, before he had made such an attempt, would doubtless have been possessed of several prize ships, which he would not have hesitated to have run on shore for the protection of his boats; and besides, there were many places on that coast, and one particularly in the neighbourhood of Callao, where there was good anchoring, though a great depth of water, within a cable's length of the shore; consequently the cannon of the man-of-war would have swept all the coast to above a mile's distance from the water's edge, and would have effectually prevented any force from assembling to oppose the landing and forming of our men. And this landing-place had the additional advantage that it was but two leagues distant from Lima; so that we might have been at that city within four hours after we should have been first discovered from the shore. The place I have in view is about two leagues south of Callao, and just to the northward of the headland called, in Frezier's draught of that coast, Morro

Solar. Here there is seventy or eighty fathom of water within two cables' length of the shore; and here the Spaniards themselves were so apprehensive of our attempting to land, that they had projected to build a fort close to the water; but as there was no money in the royal chests, they could not compleat so considerable a work, and therefore they contented themselves with keeping a guard of a hundred horse there, that they might be sure to receive early notice of our appearance on that coast. Indeed some of them (as we were told), conceiving our management at sea to be as pusillanimous as their own, pretended that this was a road where the commodore would never dare to hazard his ships, for fear that in so great a depth of water their anchors could not hold them.

And let it not be imagined that I am proceeding upon groundless and extravagant presumptions, when I conclude that fifteen hundred or a thousand of our people, well conducted, should have been an over-match for any numbers the Spaniards could muster in South America. Since, not to mention the experience we had of them at Paita and Petaplan, it must be remembered that our commodore was extremely solicitous to have all his men trained to the dexterous use of their fire-arms; whereas the Spaniards, in this part of the world, were wretched provided with arms, and were very awkward in the management of the few they had: and though on their repeated representations the court of Spain had ordered several thousand firelocks to be put on board Pizarro's squadron, yet those, it is evident, could not have been in America time enough to have been employed against us. Hence then by our arms, and our readiness in the use of them (not to insist on the timidity and softness of our enemy), we should in some degree have had the same advantages which the Spaniards themselves had on the first discovery of this country against its naked and unarmed inhabitants.

Now let it in the next place be considered what were the events which we had to fear, or what were the circumstances which could have prevented us from giving law to all the coast of South America, and thereby cutting off from Spain the resources which she drew from those immense provinces. By sea there was no force capable of opposing us; for how soon soever we had sailed, Pizarro's squadron could not have sailed sooner than it did, and therefore could not have avoided the fate it met with. As we should have been masters of the ports of Chili, we could thereby have supplied ourselves with the provisions we wanted in the greatest plenty; and from Baldivia to the equinoctial we ran no risque of losing our men by sickness (that being of all climates the most temperate and healthy), nor of having our ships disabled by bad weather. And had we wanted sailors to assist in the navigating of our squadron whilst a considerable proportion of our men were employed on shore, we could not have failed of getting whatever numbers we pleased in the ports we should have taken, and from the prizes which would have fallen into our hands. For I must observe that the Indians, who are the principal mariners in that part of the world, are extremely docile and dexterous; and though they are not fit to struggle with the inclemencies of a cold climate, yet in temperate seas they are most useful and laborious seamen.

Thus then it appears what important revolutions might have been brought about by our squadron had it departed from England as early as it ought to have done: and from hence it is easy to conclude what immense advantages might have thence accrued to the public. For, as on our success it would have been impossible that the kingdom of Spain should have received any treasure from the provinces bordering on the South Seas, or should even have had any communication with them, it is certain that the whole attention of that monarchy would have been immediately employed in endeavouring to regain these inestimable territories, either by force of arms or compact. By the first of these methods it was scarcely possible they could succeed; for it must have been at least a twelvemonth after our arrival before any ships from Spain could have got into the South Seas, and when they had been there, they would have found themselves without resource, since they would probably have been separated, disabled, and sickly, and would then have had no port remaining in their possession where they could either rendezvous or refit. Whilst we might have been supplied across the isthmus with whatever necessaries, stores, or even men we wanted; and might thereby have supported our squadron in as good a plight as when it first set sail from St. Helens. In short, it required but little prudence so to have conducted this business as to have rendered all the efforts of Spain, seconded by the power of France, ineffectual, and to have maintained our conquest in defiance of them both. Whence they must either have resolved to have left Great Britain mistress of the wealth of South America (the principal support of all their destructive projects), or they must have submitted to her terms, and have been contented to receive these provinces back again, as an equivalent for such restrictions to their future ambition as she in her prudence should have dictated to them. Having thus discussed the prodigious weight which the operations of our squadron might have added to the national influence of this kingdom, I shall here end this second book, referring to the next the passage of the shattered remains of our force across the Pacific Ocean, and all their subsequent transactions till the commodore's arrival in England.

BOOK III

CHAPTER I

THE RUN FROM THE COAST OF MEXICO TO THE LADRONES OR MARIAN ISLANDS

When, on the 6th of May 1642, we left the coast of America, we stood to the S.W. with a view of meeting the N.E. tradewind, which the accounts of former writers taught us to expect at seventy or eighty leagues from the land. We had besides another reason for standing to the southward, which was the getting into the latitude of 13° or 14° north, that being the parallel where the Pacific Ocean is most usually crossed, and consequently where the navigation is esteemed the safest: this last purpose we had soon answered, being in a day or two sufficiently advanced to the south. But though we were at the same time more distant from the shore than we had presumed was necessary for the falling in with the trade-wind, yet in this particular we were most grievously disappointed, the wind still continuing to the westward, or at best variable. As the getting into the N.E. trade was to us a matter of the last consequence, we stood yet more to the southward, and made many experiments to meet with it; but all our efforts were for a long time unsuccessful; so that it was seven weeks from our leaving the coast before we got into the true trade-wind. This was an interval in which we had at first believed we should well-nigh have reached the eastermost parts of Asia; but we were so baffled with the contrary and variable winds, which for all that time perplexed us, that we were not as yet advanced above a fourth of the way. The delay alone would have been a sufficient mortification; but there were other circumstances attending it which rendered this situation not less terrible, and our apprehensions perhaps still greater, than in any of our past calamities. For our two ships were by this time extremely crazy; and many days had not passed before we discovered a spring in the fore-mast of the *Centurion*, which rounded about twenty-six inches of its circumference, and which was judged to be at least four inches deep. And no sooner had the carpenters secured this mast with fishing it, than the *Gloucester* made a signal of distress to inform us that she had a spring in her main-mast, twelve feet below the trussel trees; which appeared so dangerous that she could not carry any sail upon it. Our carpenters on a strict examination of this mast found it excessively rotten and decayed; and it being judged necessary to cut it down as low as it was defective, it was by this means reduced to nothing but a stump, which served only as a step to the top-mast. These accidents augmented our delay, and being added to our other distresses occasioned us great anxiety about our future safety. For though after our departure from Juan Fernandes we had enjoyed a most uninterrupted state of health, till our leaving the coast of Mexico, yet the scurvy now began to make fresh havock amongst our people: and we too well knew the effects of this disease by our former fatal experience to suppose that anything except a speedy passage could secure the greater part of our crew from being destroyed thereby. But as, after being seven weeks at sea, there did not appear any reasons that could persuade us we were nearer the trade-wind than when we set out, there was no ground for us to imagine that our passage would not prove at least three times as long as we at first expected; and consequently we had the melancholy prospect either of dying by the scurvy or of perishing with the ship for want of hands to navigate her. Indeed, several amongst us were willing to believe that in this warm climate, so different from what we felt in passing round Cape Horn, the violence of this disease, and its fatality, might be in some degree mitigated; as it had not been unusual to suppose that its particular virulence during that passage was in a great measure owing to the severity of the weather: but the ravage of the distemper, in our present circumstances, soon convinced us of the falsity of this speculation; as it likewise exploded certain other opinions which usually pass current about the cause and nature of this disease.

For it has been generally presumed that sufficient supplies of water and of fresh provisions are effectual preventives of this malady; but it happened that in the present case we had a considerable stock of fresh provisions on board, being the hogs and fowls which were taken at Paita; we besides almost daily caught great abundance of bonitos, dolphins, and albicores; and the unsettled season, which deprived us of the benefit of the trade-wind, proved extremely rainy; so that we were enabled to fill up our water-casks almost as fast as they were empty; and each man had five pints of water allowed him every day during the passage. But notwithstanding this plenty of water, notwithstanding that the fresh provisions were distributed amongst the sick, and the whole crew often fed upon fish; yet neither were the sick hereby relieved or the progress or malignity of the disease at all abated. Nor was it in these instances only that we found the general maxims upon this head defective: for tho' it has been usually esteemed a necessary piece of management to keep all ships where the crews are large as clean and airy between decks as possible; and it hath been believed by many that this particular alone, if well attended to, would prevent the appearance of the scurvy, or at least mitigate its virulence; yet we observed during the latter part of our run that, though we kept all our ports open and took uncommon pains in cleansing and sweetning the ships, the disease still raged with as much violence as ever; nor did its advancement seem to be thereby sensibly retarded.

However, I would not be understood to assert that fresh provisions, plenty of water, and a constant supply of sweet air between decks are matters of no moment: I am, on the contrary, well satisfied that they are all of them articles of great importance, and are doubtless extremely conducive to the health and vigour of a crew, and may in many cases prevent this fatal malady from taking place. All I have aimed at in what I have advanced is only to evince that, in some instances, both the cure and prevention of this malady is impossible to be effected by any management, or by the application of any remedies which can be made use of at sea. Indeed, I am myself fully persuaded that, when it has got to a certain head, there are no other means in nature for relieving the sick but carrying them on shore, or at least bringing them into the neighbourhood of the land. Perhaps a distinct and adequate knowledge of the source of this disease may never be discovered; but, in general, there is no difficulty in conceiving that, as a continued supply of fresh air is necessary to all animal life, and as this air is so particular a fluid that, without losing its elasticity, or any of its obvious properties, it may be rendered unfit for this purpose by the mixing with it some very subtle and otherwise imperceptible effluvia; it may be easily conceived, I say, that the steams arising from the ocean may have a tendency to render the air they are spread through less properly adapted to the support of the life of terrestrial animals, unless these steams are corrected by effluvia of another kind, which perhaps the land alone can afford.

To what hath been already said in relation to this disease, I shall add that our surgeon (who during our passage round Cape Horn had ascribed the mortality we suffered to the severity of the climate) exerted himself in the present run to the utmost: but he at last declared that all his measures were totally ineffectual, and did not in the least avail his patients. On this it was resolved by the commodore to try the success of two medicines which, just before his departure from England, were the subject of much discourse, I mean the pill and drop of Mr. Ward. For however violent the operations of these medicines are said to have sometimes proved, yet in the present instance, where, without some remedy, destruction seemed inevitable, the experiment at least was thought adviseable: and, therefore, one or both of them at different times were administred to persons in every stage of the distemper. Out of the numbers who took them, one, soon after swallowing the pill, was seized with a violent bleeding at the nose. He was before given over by the surgeon and lay almost at the point of death; but he immediately found himself much better, and continued to recover, tho' slowly, till we arrived on shore, which was near a fortnight after. A few others too were relieved for some days, but the disease returned again with as much virulence as ever. Though neither did these, nor the rest, who received no benefit, appear to be reduced to a worse condition than they would have been if they had taken nothing. The most remarkable property of these medicines, and what was obvious in almost every one that took them, was that they acted in proportion to the vigour of the patient; so that those who were within two or three days of dying were scarcely affected; and as the patient was differently advanced in the disease, the operation was either a gentle perspiration, an easy vomit, or a moderate purge: but if they were taken by one in full strength, they then produced all the forementioned effects with considerable violence, which sometimes continued for six or eight hours together with little intermission. However, let us return to the prosecution of our voyage.

I have already observed that a few days after our running off the coast of Mexico the *Gloucester* had her main-mast cut down to a stump, and we were obliged to fish our foremast; and that these misfortunes were greatly aggravated by our meeting with contrary and variable winds for near seven weeks. I shall now add that when we reached the trade-wind, and it settled between the north and the east, yet it seldom blew with so much strength that the *Centurion* might not have carried all her small sails abroad without the least danger; so that, had we been a single ship, we might have run down our longitude apace, and have arrived at the Ladrones soon enough to have recovered great numbers of our men who afterwards perished. But the *Gloucester*, by the loss of her main-mast, sailed so very heavily that we had seldom any more than our top-sails set, and yet were frequently obliged to lie to for her: and, I conceive, that on the whole we lost little less than a month by our attendance upon her, in consequence of the various mischances she encountered. During all this run it was remarkable that we were rarely many days together without seeing great numbers of birds; which is a proof that there are several islands, or at least rocks, scattered all along, at no very considerable distance from our track: but the frequency of these birds seem to ascertain that there are many more than have been hitherto discovered; for the most part of the birds we observed were such as are known to roost on shore; and the manner of their appearance sufficiently evinced that they came from some distant haunt every morning, and returned thither again in the evening, since we never saw them early or late; and the hour of their arrival and departure gradually varied, which we supposed was occasioned by our running nearer their haunts or getting farther from them.

The trade-wind continued to favour us, without any fluctuation, from the end of June till towards the end of July. But on the 26th of July, being then, as we esteemed, about three hundred leagues from the Ladrones, we met with a westerly wind, which did not come about again to the eastward in four days' time. This was a most dispiriting incident, as it at once damped all our hopes of speedy relief, especially too as it was attended with a vexatious accident to the *Gloucester*: for in one part of these four days the wind flatted to a calm, and the ships rolled very deep; by which means the *Gloucester's* forecap splitting, her fore top-mast came by the board, and broke her fore-yard directly in the slings. As she was hereby rendered incapable of making any sail for some time, we were under a necessity, as soon as a gale sprung up, to take her in tow; and near twenty of the healthiest and ablest of our seamen were removed from the duty of our own ship, and were continued eight or ten days together on board the *Gloucester* to assist in repairing her damages. But these things, mortifying as we thought them, were only the commencement of our disasters; for scarce had our people finished their business in the *Gloucester* before we met with a most violent storm from the western board, which obliged us to lie to. At the beginning of this storm our ship sprung a leak, and let in so much water that all our people, officers included, were constantly employed about the pumps: and the next day we had the vexation so see the *Gloucester* with her fore top-mast once more by the board. Nor was that the whole of her calamity, since whilst we were viewing her with great concern for this new distress, we saw her main top-mast, which had hitherto served her as a jury main-mast, share the same fate. This compleated our misfortunes, and rendered them without resource: for we knew the *Gloucester's* crew were so few and feeble that without our assistance they could not be relieved; whilst at the same time our sick were now so far increased, and those who remained in health so continually fatigued with the additional duty of our pumps, that it was impossible for us to lend them any aid. Indeed we were not as yet fully apprized of the deplorable situation of the *Gloucester's* crew; for when the storm abated, which during its continuance prevented all communication with them, the *Gloucester* bore up under our stern, and Captain Mitchel informed the commodore that besides the loss of his masts, which was all that was visible to us, the ship had then no less than seven feet of water in her hold, although his officers and men had been kept constantly at the pumps for the last twenty-four hours.

This new circumstance was indeed a most terrible accumulation to the other extraordinary distresses of the *Gloucester*, and required if possible the most speedy and vigorous assistance, which Captain Mitchel begged the commodore to afford him. But the debility of our people, and our own immediate preservation, rendered it impracticable for the commodore to comply with his request. All that could be done was to send our boat on board for a more particular account of the ship's condition, as it was soon suspected that the taking her people on board us, and then destroying her, was the only measure that could be prosecuted in the present emergency, both for the security of their lives and of our own.

Our boat soon returned with a representation of the state of the *Gloucester*, and of her several defects, signed by Captain Mitchel and all his officers; whence it appeared that she had sprung a leak by the stern post being loose, and working with every roll of the ship, and by two beams amidships being broken in the orlope, no part of which, as the carpenters reported, could possibly be repaired at sea; that both officers and men had wrought twenty-four hours at the pump without intermission, and were at length so fatigued that they could continue their labour no longer, but had been forced to desist, with seven feet of water in the hold, which covered all their casks, so that they could neither come at fresh water nor provision: that they had no mast standing, except the foremast, the mizen-mast, and the mizen top-mast, nor had they any spare masts to get up in the room of those they had lost: that the ship was, besides, extremely decayed in every part; for her knees and clamps were all become quite loose, and her upper works in general were so crazy that the quarter-deck was ready to drop down: that her crew was greatly reduced, as there remained alive on board her, officers included, no more than seventy-seven men, eighteen boys, and two prisoners, and that of this whole number only sixteen men and eleven boys were capable of keeping the deck, several of these too being very infirm.

The commodore, on the perusal of this melancholy representation, presently ordered them a supply of water and provisions, of which they seemed to be in the most pressing want, and at the same time sent his own carpenter on board them to examine into the truth of every particular; and it being found on the strictest enquiry that the preceding account was in no instance exaggerated, it plainly appeared there was no possibility of preserving the *Gloucester* any longer, as her leaks were irreparable, and the united hands on board both ships would not be able to free her, could we have spared the whole of our crew to her relief. What then could be resolved on, when it was the utmost we ourselves could do to manage our own pumps? Indeed there was no room for deliberation; the only step to be taken was the saving the lives of the few that remained on board the *Gloucester*, and the getting out of her as much as we could before she was destroyed. The commodore therefore immediately sent an order to Captain Mitchel to put his people on board the *Centurion* as expeditiously as he could, now the weather was calm and favourable, and to take out such stores as he could get at whilst the ship could be kept above water. And as our leak required less attention whilst the present easy weather continued, we sent our boats with as many men as we could spare to Captain Mitchel's assistance.

The removing the *Gloucester's* people on board us, and the getting out such stores as could most easily be come at, gave us full employment for two days. Mr. Anson was extremely desirous to have saved two of her cables and an anchor, but the ship rolled so much, and the men were so excessively fatigued, that they were incapable of effecting it; nay, it was even with the greatest difficulty that the prize-money which the *Gloucester* had taken in the South Seas was secured and sent on board the *Centurion*. However, the prize goods in the *Gloucester*, which amounted to several thousand pounds in value, and were principally the *Centurion's* property, were entirely lost; nor could any more provision be got out than five casks of flour, three of which were spoiled by the salt water. Their sick men, amounting to near seventy, were conveyed into the boats with as much care as the circumstances of that time would permit; but three or four of them expired as they were hoisting them into the *Centurion*.

It was the 15th of August, in the evening, before the *Gloucester* was cleared of everything that was proposed to be removed; and though the hold was now almost full of water, yet, as the carpenters were of opinion that she might still swim for some time, if the calm should continue and the water become smooth, it was resolved she should be burnt, as we knew not how little distant we might be at present from the island of Guam, which was in the possession of our enemies, to whom the wreck of such a ship would have been no contemptible acquisition. When she was set on fire, Captain Mitchel and his officers left her, and came on board the *Centurion*: and we immediately stood from the wreck, not without some apprehensions (as we had only a light breeze) that if she blew up soon the concussion of the air might damage our rigging; but she fortunately continued burning the whole night, so that though her guns fired successively as the flames reached them, yet it was six in the morning, when we were about four leagues distant, before she blew up. The report she made upon this occasion was but small, although the blast produced an exceeding black pillar of smoke, which shot up into the air to a very considerable height.

Thus perished his Majesty's ship the *Gloucester*. And now it might have been expected that, being freed from the embarrassments which her frequent disasters had involved us in, we should have proceeded on our way much brisker than we had hitherto done, especially as we had received some small addition to our strength by the taking on board the *Gloucester's* crew. However, we were soon taught that our anxieties were not yet to be relieved, and that, notwithstanding all we had already suffered, there remained much greater distresses which we were still to struggle with. For the late storm, which had proved so fatal to the *Gloucester*, had driven us to the northward of our intended course; and the current setting the same way, after the weather abated, had forced us yet a degree or two farther, so that we were now in $17\text{-}1/4°$ of north latitude, instead of being in $13\text{-}1/2°$, which was the parallel we proposed to keep, in order to reach the island of Guam. As it had been a perfect calm for some days since the cessation of the storm, and we were ignorant how near we were to the meridian of the Ladrones, though we supposed ourselves not to be far from it, we apprehended that we might be driven to the leeward of them by the current without discovering them. On this supposition, the only land we could make would be some of the eastern parts of Asia, where, if we could arrive, we should find the western monsoon in its full force, so that it would be impossible for the stoutest, best-manned ship to get in. Besides, this coast being between four and five hundred leagues distant from us, we, in our languishing circumstances, could expect no other than to be destroyed by the scurvy long before the most favourable gale could enable us to compleat so extensive a navigation. For our deaths were by this time extremely alarming, no day passing in which we did not bury eight or ten, and sometimes twelve, of our men; and those who had as yet continued healthy began to fall down apace. Indeed we made the best use we could of our present calm, by employing our carpenters in searching after the leak, which, notwithstanding the little wind we had, was now considerable. The carpenters at length discovered it to be in the gunner's fore store-room, where the water rushed in under the breast-hook on each side of the stern: but though they found where it was, they agreed it was impossible to stop it till they could come at it on the outside, which was evidently a matter not to be attempted till we should arrive in port. However, they did the best they could within board, and were fortunate enough to reduce it, which was a considerable relief to us.

We hitherto considered the calm which succeeded the storm, and which had now continued for some days, as a very great misfortune, since the currents were all the time driving us to the northward of our parallel, and we thereby risqued the missing of the Ladrones, which we at present conceived ourselves to be very near. But when a gale sprung up our condition was still worse; for it blew from the S.W., and consequently was directly opposed to the course we wanted to steer: and though it soon veered to the N.E., yet this served only to tantalize us, as it returned back again in a very short time to its old quarter. However, on the 22d of August we had the satisfaction to find that the current was shifted, and had set us to the southward; and the 23d, at daybreak, we were cheered with the discovery of two islands in the western board. This gave us all great joy, and raised our drooping spirits, for till then an universal dejection had seized us, and we almost despaired of ever seeing land again. The nearest of these islands, as we learnt afterwards, was Anatacan; this we judged to be full fifteen leagues from us; it seemed to be high land, though of an indifferent length. The other was the island of Serigan, which had rather the appearance of a rock than of a place we could hope to anchor at. We were extremely impatient to get in with the nearest island, where we expected to find anchoring ground and an opportunity of refreshing our sick. But the wind proved so variable all day, and there was so little of it that we advanced towards it but slowly; however, by the next morning we were got so far to the westward that we were in sight of a third island, which was that of Paxaros, and which is marked in the chart only as a rock. This was very small, and the land low, so that we had passed within less than a mile of it in the night without observing it. At noon, being then not four miles from the island of Anatacan, the boat was sent away to examine the anchoring ground and the produce of the place, and we were not a little solicitous for her return, as we conceived our fate to depend upon the report we should receive; for the other two islands were obviously enough incapable of furnishing us with any assistance, and we knew not that there were any besides which we could reach. In the evening the boat came back, and the crew informed us that there was no road for a ship to anchor in, the bottom being everywhere foul ground, and all except one small spot not less than fifty fathom in depth; that on that spot there was thirty fathom, though not above half a mile from the shore; and that the bank was steep too, and could not be depended on. They farther told us that they had landed on the island, not without some difficulty on account of the greatness of the swell; that they found the ground was everywhere covered with a kind of wild cane or rush; but that they met with no water, and did not believe the place to be inhabited, though the soil was good and abounded with groves of coconut trees.

The account of the impossibility of anchoring at this island occasioned a general melancholy on board, for we considered it as little less than the prelude to our destruction; and our despondency was increased by a disappointment we met with the succeeding night, when, as we were plying under top-sails, with an intention of getting nearer to the island, and of sending our boat on shore to load with coconuts for the refreshment of our sick, the wind proved squally, and blew so strong off shore, that we were driven too far to the southward to venture to send off our boat. And now the only possible circumstance that could secure the few which remained alive from perishing, was the accidental falling in with some other of the Ladrone Islands better prepared for our accommodation; but as our knowledge of these islands was extremely imperfect, we were to trust entirely to chance for our guidance; only as they are all of them usually laid down near the same meridian, and we conceived those we had already seen to be part of them, we concluded to stand to the southward, as the most probable means of discovering the rest. Thus, with the most gloomy persuasion of our approaching destruction, we stood from the island of Anatacan, having all of us the strongest apprehensions (and those not ill grounded) either of dying by the scurvy, or of being destroyed with the ship, which, for want of hands to work her pumps, might in a short time be expected to founder.

CHAPTER II

OUR ARRIVAL AT TINIAN, AND AN ACCOUNT OF THE ISLAND AND OF OUR PROCEEDINGS THERE TILL THE "CENTURION" DROVE OUT TO SEA

It was the 26th of August, 1742, in the morning, when we lost sight of the island of Anatacan, dreading that it was the last land we should ever fix our eyes on. But the next morning we discovered three other islands to the eastward, which were between ten and fourteen leagues distant from us. These were, as we afterwards learnt, the island of Saypan, Tinian, and Aguigan. We immediately steered towards Tinian, which was the middlemost of the three; but we had so much of calms and light airs, that though we were helped forwards by the currents, yet on the morrow, at daybreak, we had not advanced nearer than within five leagues of it. However, we kept on our course, and about ten o'clock we perceived a proa under sail to the southward between Tinian and Aguigan. As we imagined from hence that these islands were inhabited, and knew that the Spaniards had always a force at Guam, we took the necessary precautions for our own security: and endeavoured to prevent the enemy as much as possible from making an advantage of our present wretched circumstances, of which we feared they would be sufficiently informed by the manner of our working the ship. We therefore mustered all our hands who were capable of standing to their arms, and loaded our upper and quarter-deck guns with grape shot; and that we might the more readily procure some intelligence of the state of these islands, we showed Spanish colours, and hoisted a red flag at the fore top-mast-head, hoping thereby to give our ship the appearance of the Manila galeon, and to decoy some of the inhabitants on board us. Thus preparing ourselves, and standing towards the land, we were near enough, at three in the afternoon, to send the cutter on shore to find out a proper birth for the ship; and we soon perceived that a proa put off from the island to meet the cutter, fully persuaded, as we afterwards found, that we were the Manila ship. As we saw the cutter returning with the proa in tow, we instantly sent the pinnace to receive the proa and the prisoners, and to bring them on board, that the cutter might proceed on her errand. The pinnace came back with a Spaniard and four Indians, which were the people taken in the proa: and the Spaniard being immediately examined as to the produce and circumstances of this island of Tinian, his account of it surpassed even our most sanguine hopes. For he informed us that though it was uninhabited (which in itself, considering our present defenceless condition, was a convenience not to be despised), yet it wanted but few of the accommodations that could be expected in the most cultivated country. In particular, he assured us that there was plenty of very good water; that there were an incredible number of cattle, hogs, and poultry running wild on the island, all of them excellent in their kind; that the woods afforded sweet and sour oranges, limes, lemons, and coconuts in great abundance, besides a fruit peculiar to these islands, which served instead of bread; that from the quantity and goodness of the provisions produced here, the Spaniards at Guam made use of it as a store for supplying the garrison; and that he himself was a serjeant of that garrison, who was sent hither with twenty-two Indians to jerk beef, which he was to load for Guam on board a small bark of about fifteen tun, which lay at anchor near the shore.

This relation was received by us with inexpressible joy. Part of it we were ourselves able to verify on the spot, as we were by this time near enough to discover several numerous herds of cattle feeding in different places of the island; and we did not any ways doubt the rest of his narration, since the appearance of the shore prejudiced us greatly in its favour, and made us hope that not only our necessities might be there fully relieved, and our diseased recovered, but that, amidst those pleasing scenes which were then in view, we might procure ourselves some amusement and relaxation, after the numerous fatigues we had undergone. For the prospect of the country did by no means resemble that of an uninhabited and uncultivated place; but had much more the air of a magnificent plantation where large lawns and stately woods had been laid out together with great skill, and where the whole had been so artfully combined, and so judiciously adapted to the slopes of the hills, and the inequalities of the ground, as to produce a most striking effect, and to do honour to the invention of the contriver. Thus (an event not unlike what we had already seen) we were forced upon the most desirable and salutary measures by accidents which at first sight we considered as the greatest of misfortunes; for had we not been driven by the contrary winds and currents to the northward of our course (a circumstance which at that time gave us the most terrible apprehensions), we should, in all probability, never have arrived at this delightful island, and consequently we should have missed of that place where alone all our wants could be most amply relieved, our sick recovered, and our enfeebled crew once more refreshed, and enabled to put again to sea.

The Spanish serjeant, from whom we received the account of the island, having informed us that there were some Indians on shore under his command, employed in jerking beef, and that there was a bark at anchor to take it on board, we were desirous, if possible, to prevent the Indians from escaping, since they would certainly have given the Governor of Guam intelligence of our arrival: we therefore immediately dispatched the pinnace to secure the bark, as the serjeant told us that was the only embarkation on the place; and then about eight in the evening we let go our anchor in twenty-two fathom. But though it was almost calm, and whatever vigour and spirit was to be found on board was doubtless exerted to the utmost on this pleasing occasion, when, after having kept the sea for some months, we were going to take possession of this little paradise, yet we were full five hours in furling our sails. It is true we were somewhat weakened by the crews of the cutter and pinnace which were sent on shore; but it is not less true that, including those absent with the boats and some negroes and Indians prisoners, all the hands we could muster capable of standing at a gun amounted to no more than seventy-one, most of which too were incapable of duty except on the greatest emergencies. This, inconsiderable as it may appear, was the whole force we could collect in our present enfeebled condition from the united crews of the *Centurion*, the *Gloucester*, and the *Tryal*, which, when we departed from England, consisted all together of near a thousand hands.

When we had furled our sails, our people were allowed to repose themselves during the remainder of the night, to recover them from the fatigue they had undergone. But in the morning a party was sent on shore well armed, of which I myself was one, to make ourselves masters of the landing-place, since we were not certain what opposition might be made by the Indians on the island. We landed, however, without difficulty, for the Indians having perceived, by our seizure of the bark the night before, that we were enemies, they immediately fled into the woody parts of the island. We found on shore many huts which they had inhabited, and which saved us both the time and trouble of erecting tents. One of these huts, which the Indians made use of for a store-house, was very large, being twenty yards long and fifteen broad: this we immediately cleared of some bales of jerked beef which had been left in it, and converted it into an hospital for our sick, who as soon as the place was ready to receive them, were brought on shore, being in all a hundred and twenty-eight. Numbers of these were so very helpless that we were obliged to carry them from the boats to the hospital upon our shoulders, in which humane employment (as before at Juan Fernandes) the commodore himself, and every one of his officers, were engaged without distinction; and notwithstanding the extreme debility and the dying aspects of the greatest part of our sick, it is almost incredible how soon they began to feel the salutary influence of the land: for, though we buried twenty-one men on this and the preceding day, yet we did not lose above ten men more during the whole two months we staid here; but our diseased in general reaped so much benefit from the fruits of the island, particularly those of the acid kind, that in a week's time there were but few of them who were not so far recovered as to be able to move about without help.

Being now in some sort established at this place, we were enabled more distinctly to examine its qualities and productions; and that the reader may the better judge of our manner of life here, and future navigators be better apprized of the conveniencies we met with, I shall, before I proceed any farther in the history of our own adventures, throw together the most interesting particulars that came to our knowledge relating to the situation, soil, produce, and accommodations of this island of Tinian.

This island lies in the latitude of 15° 8' north, and longitude from Acapulco 114° 50' west. Its length is about twelve miles, and its breadth about half as much, it extending from the S.S.W. to N.N.E. The soil is everywhere dry and healthy, and being withal somewhat sandy, it is thereby the less disposed to a rank and over-luxuriant vegetation; and hence the meadows and the bottoms of the woods are much neater and smoother than is customary in hot climates. The land rose in gentle slopes from the very beach where we watered to the middle of the island, though the general course of its ascent was often interrupted by vallies of an easy descent, many of which wind irregularly through the country. These vallies and the gradual swellings of the ground which their different combinations gave rise to were most beautifully diversified by the mutual encroachments of woods and lawns, which coasted each other and traversed the island in large tracts. The woods consisted of tall and well-spread trees, the greatest part of them celebrated either for their aspect or their fruit: whilst the lawns were usually of a considerable breadth, their turf quite clean and uniform, it being composed of a very fine trefoil, which was intermixed with a variety of flowers. The woods too were in many places open, and free from all bushes and underwood, so that they terminated on the lawns with a well-defined outline, where neither shrubs nor weeds were to be seen; but the neatness of the adjacent turf was frequently extended to a considerable distance under the hollow shade formed by the trees. Hence arose a great number of the most elegant and entertaining prospects, according to the different blendings of these woods and lawns, and their various intersections with each other, as they spread themselves differently through the vallies, and over the slopes and declivities in which the place abounded. Nor were the allurements of Tinian confined to the excellency of its landskips only; since the fortunate animals, which during the greatest part of the year are the sole lords of this happy soil, partake in some measure of the romantic cast of the island, and are no small addition to its wonderful scenery; for the cattle, of which it is not uncommon to see herds of some thousands feeding together in a large meadow, are certainly the most remarkable in the world, as they are all of them milk-white, except their ears, which are generally brown or black. And though there are no inhabitants here, yet the clamour and frequent parading of domestic poultry, which range the woods in great numbers, perpetually excite the idea of the neighbourhood of farms and villages, and greatly contribute to the chearfulness and beauty of the place. The cattle on Tinian we computed were at least ten thousand; we had no difficulty in getting near them, for they were not at all shy of us. Our first method of killing them was shooting them; but at last, when by accidents to be hereafter recited we were obliged to husband our ammunition, our men ran them down with ease. Their flesh was extremely well tasted, and was believed by us to be much more easily digested than any we had ever met with. The fowls too were exceeding good, and were likewise run down with little trouble; for they could scarce fly further than an hundred yards at a flight, and even that fatigued them to such a degree that they could not readily rise again, so that, aided by the openness of the woods, we could at all times furnish ourselves with whatever number we wanted. Besides the cattle and the poultry we found here abundance of wild hogs. These were most excellent food, but as they were a very fierce animal, we were obliged either to shoot them, or to hunt them with large dogs, which we found upon the place at our landing, and which belonged to the detachment which was then upon the island amassing provisions for the garrison of Guam. As these dogs had been purposely trained to the killing of the wild hogs, they followed us very readily and hunted for us; but though they were a large bold breed, the hogs fought with so much fury that they frequently destroyed them, whence we by degrees lost the greatest part of them.

This place was not only extremely grateful to us, from the plenty and excellency of its fresh provisions, but was as much perhaps to be admired on account of its fruits and vegetable productions, which were most fortunately adapted to the cure of the sea scurvy, the disease which had so terribly reduced us. For in the woods there were inconceivable quantities of coco-nuts, with the cabbages growing on the same tree. There were besides, guavoes, limes, sweet and sour oranges, and a kind of fruit peculiar to these islands, called by the Indians Rhymay, but by us the Bread Fruit, for it was constantly eaten by us during our stay upon the island instead of bread, and so universally preferred to it that no ship's bread was expended in that whole interval. It grew upon a tree which is somewhat lofty, and which towards the top divides into large and spreading branches. The leaves of this tree are of a remarkable deep green, are notched about the edges, and are generally from a foot to eighteen inches in length. The fruit itself is found indifferently on all parts of the branches; it is in shape rather elliptical than round; it is covered with a rough rind, and is usually seven or eight inches long; each of them grows singly and not in clusters. This fruit is fittest to be used when it is full grown but still green, in which state, after it is properly prepared by being roasted in the embers, its taste has some distant resemblance to that of an artichoke's bottom, and its texture is not very different, for it is soft and spongy. As it ripens it becomes softer and of a yellow colour, when it contracts a luscious taste and an agreeable smell, not unlike a ripe peach; but then it is esteemed unwholsome and is said to produce fluxes. I shall only add that it is described both by Dampier and in Ray's *History of Plants*. Besides the fruits already enumerated, there were many other vegetables extremely conducive to the cure of the malady we had long laboured under, such as water melons, dandelion, creeping purslan, mint, scurvy grass, and sorrel; all which, together with the fresh meats of the place, we devoured with great eagerness, prompted thereto by the strong inclination which, in scorbutic disorders, nature never fails of exciting for those powerful specifics.

It will easily be conceived from what hath been already said that our chear upon this island was in some degree luxurious; but I have not yet recited all the varieties of provision which we here indulged in. Indeed we thought it prudent totally to abstain from fish, the few we caught at our first arrival having surfeited those who eat of them; but considering how much we had been inured to that species of food we did not regard this circumstance as a disadvantage, especially as the defect was so amply supplied by the beef, pork, and fowls already mentioned, and by great plenty of wild fowl; for it is to be remembered that near the centre of the island there were two considerable pieces of fresh water, which abounded with duck, teal, and curlew; not to mention the whistling plover, which we found there in prodigious plenty.

It may now perhaps be wondered at that an island so exquisitely furnished with the conveniencies of life, and so well adapted not only to the subsistence but likewise to the enjoyment of mankind, should be entirely destitute of inhabitants, especially as it is in the neighbourhood of other islands, which in some measure depend upon this for their support. To obviate this difficulty, I must observe that it is not fifty years since the island was depopulated. The Indians we had in our custody assured us that formerly the three islands of Tinian, Rota, and Guam were all full of inhabitants; and that Tinian alone contained thirty thousand souls: but a sickness raging amongst these islands which destroyed multitudes of the people, the Spaniards, to recruit their numbers at Guam, which were extremely diminished by the mortality, ordered all the inhabitants of Tinian thither; where, languishing for their former habitations and their customary method of life, the greatest part of them in a few years died of grief. Indeed, independent of that attachment which all mankind have ever shown to the places of their birth and bringing up, it should seem from what has been already said that there were few countries more worthy to be regretted than this of Tinian.

These poor Indians might reasonably have expected, at the great distance from Spain where they were placed, to have escaped the violence and cruelty of that haughty nation, so fatal to a large proportion of the whole human race: but it seems their remote situation could not protect them from sharing in the common destruction of the western world; all the advantage they received from their distance being only to perish an age or two later. It may perhaps be doubted if the number of the inhabitants of Tinian, who were banished to Guam, and who died there pining for their native home, was so considerable as what we have related above; but not to mention the concurrent assertion of our prisoners and the commodiousness of the island and its great fertility, there are still remains to be met with on the place which show it to have been once extremely populous. For there are in all parts of the island many ruins of a very particular kind. These usually consist of two rows of square pyramidal pillars, each pillar being about six feet from the next, and the distance between the rows being about twelve feet; the pillars themselves are about five feet square at the base, and about thirteen feet high; and on the top of each of them there is a semi-globe with the flat surface upwards; the whole of the pillars and semi-globe is solid, being composed of sand and stone cemented together and plaistered over. If the account our prisoners gave us of these structures was true, the island must indeed have been most extraordinary well peopled; since they assured us that they were the foundations of particular buildings set apart for those Indians only who had engaged in some religious vow; monastic institutions being often to be met with in many Pagan nations. However, if these ruins were originally the basis of the common dwelling-houses of the natives, their numbers must have been considerable; for in many parts of the island they are extremely thick planted, and sufficiently evince the great plenty of its former inhabitants. But to return to the present state of the island.

Having briefly recounted the conveniencies of this place, the excellency and quantity of its fruits and provisions, the neatness of its lawns, the stateliness, freshness, and fragrance of its woods, the happy inequality of its surface, and the variety and elegance of the views it afforded, I must now observe that all these advantages were greatly enhanced by the healthiness of its climate, by the almost constant breezes which prevail there, and by the frequent showers which fell there; for these, instead of the heavy continued rains which in some countries render great part of the year so unpleasing, were usually of a very short and almost momentary duration. Hence they were extremely grateful and refreshing, and were perhaps one cause of the salubrity of the air, and of the extraordinary influence it was observed to have upon us in increasing and invigorating our appetites and digestion. This effect was indeed remarkable, since those amongst our officers who were at all other times spare and temperate eaters, who, besides a slight breakfast, used to make but one moderate repast a day, were here, in appearance, transformed into gluttons; for instead of one reasonable flesh meal, they were now scarcely satisfied with three, each of them too so prodigious in quantity as would at another time have produced a fever or a surfeit. And yet our digestion so well corresponded to the keenness of our appetites that we were neither disordered nor even loaded by this uncommon repletion; for after having, according to the custom of the island, made a large beef breakfast, it was not long before we began to consider the approach of dinner as a very desirable, though somewhat tardy, incident.

After giving these large encomiums to this island, in which, however, I conceive I have not done it justice, it is necessary I should speak of those circumstances in which it is defective, whether in point of beauty or utility. And, first, with respect to its water. I must own that, before I had seen this spot, I did not conceive that the absence of running water, of which it is entirely destitute, could have been so well replaced by any other means as it is in this island; since though there are no streams, yet the water of the wells and springs, which are to be met with everywhere near the surface, is extremely good; and in the midst of the island there are two or three considerable pieces of excellent water, the turf of whose banks was as clean, as even, and as regularly disposed as if they had been basons purposely made for the decoration of the place. It must, however, be confessed that with regard to the beauty of the prospects, the want of rills and streams is a very great defect, not to be compensated either by large pieces of standing water or by the neighbourhood of the sea, though that, from the smallness of the island generally, makes a part of every extensive landskip.

As to the residence upon the island, the principal inconvenience attending it is the vast numbers of muscatos, and various other species of flies, together with an insect called a tick; this, though principally attached to the cattle, would yet frequently fasten upon our limbs and bodies, and if not perceived and removed in time would bury its head under the skin and raise a painful inflammation. We found here too centipedes and scorpions, which we supposed were venomous, though none of us ever received any injury from them.

But the most important and formidable exception to this place remains still to be told. This is the inconvenience of the road and the little security there is in some seasons for a ship at anchor. The only proper anchoring place for ships of burthen is at the S.W. end of the island. Here the *Centurion* anchored in twenty and twenty-two fathom water about a mile and an half distant from the shore opposite to a sandy bay. The bottom of this road is full of sharp-pointed coral rocks, which, during four months of the year, that is from the middle of June to the middle of October, render it a very unsafe anchorage. This is the season of the western monsoons, when near the full and change of the moon, but more particularly at the change, the wind is usually variable all round the compass, and seldom fails to blow with such fury that the stoutest cables are not to be confided in. What adds to the danger at these times is the excessive rapidity of the tide of flood which sets to the S.E. between this island and that of Aguiguan, a small islet near the southern extremity of Tinian. This tide runs at first with a vast head and overfall of water, occasioning such a hollow and overgrown sea as is scarcely to be conceived; so that (as will be more particularly recited in the sequel) we were under the dreadful apprehensions of being pooped by it, though we were in a sixty-gun ship. In the remaining eight months of the year, that is from the middle of October to the middle of June, there is a constant season of settled weather, when, if the cables are but well armed, there is scarcely any danger of their being even rubbed, so that during all that interval it is as secure a road as could be wished for. I shall only add that the anchoring bank is very shelving, and stretches along the S.W. end of the island, and is entirely free from shoals, except a reef of rocks which is visible, and lies about half a mile from the shore, affording a narrow passage into a small sandy bay, which is the only place where boats can possibly land. Having given this account of the island and its produce, it is necessary to return to our own history.

Our first undertaking after our arrival was the removal of our sick on shore, as hath been related. Whilst we were thus employed, four of the Indians on the island, being part of the Spanish Serjeant's detachment, came and surrendered themselves to us, so that with those we took in the proa, we had now eight of them in our custody. One of the four who submitted undertook to show us the most convenient places for killing cattle, and two of our men were ordered to attend him on that service; but one of them unwarily trusting the Indian with his firelock and pistol, the Indian escaped with them into the woods. His countrymen, who remained behind, were apprehensive of suffering for this perfidy of their comrade, and therefore begged leave to send one of their own party into the country, who they engaged should both bring back the arms and persuade the whole detachment from Guam to submit to us. The commodore granted their request, and one of them was dispatched on this errand, who returned next day and brought back the firelock and pistol, but assured us he had found them in a pathway in the wood, and protested that he had not been able to meet with any one of his countrymen. This report had so little the air of truth that we suspected there was some treachery carrying on, and therefore to prevent any future communication amongst them, we immediately ordered all the Indians who were in our power on board the ship, and did not permit them to go any more on shore.

When our sick were well settled on the island, we employed all the hands that could be spared from attending them in arming the cables with a good rounding, several fathom from the anchor, to secure them from being rubbed by the coral rocks which here abounded. This being compleated, our next occupation was our leak, and in order to raise it out of water, we, on the first of September, began to get the guns aft to bring the ship by the stern; and now the carpenters, being able to come at it on the outside, they ripped off what was left of the old sheathing, caulked all the seams on both sides the cut-water, and leaded them over, and then new sheathed the bows to the surface of the water. By this means we conceived the defect was sufficiently secured, but upon our beginning to return the guns to their ports, we had the mortification to perceive that the water rushed into the ship in the old place with as much violence as ever. Hereupon we were necessitated to begin again, and that our second attempt might be more successful, we cleared the fore store-room and sent a hundred and thirty barrels of powder on board the small Spanish bark we had seized here, by which means we raised the ship about three feet out of the water forwards. The carpenters now ripped off the sheathing lower down, new caulked all the seams, and afterwards laid on new sheathing; and then, supposing the leak to be effectually stopped, we began to move the guns forwards; but the upper deck guns were scarcely replaced when, to our amazement, it burst out again. As we durst not cut away the lining within board, lest a but end or a plank might start, and we might go down immediately, we had no other resource left than chincing and caulking within board. Indeed by this means the leak was stopped for some time; but when our guns were all fixed in their ports, and our stores were taken on board, the water again forced its way through a hole in the stem where one of the bolts was driven in. We on this desisted from all farther efforts, being at last well assured that the defect was in the stem itself, and that it was not to be remedied till we should have an opportunity of heaving down.

In the first part of the month of September, several of our sick were tolerably recovered by their residence on shore; and on the 12th of September all those who were so far relieved since their arrival as to be capable of doing duty were sent on board the ship: and then the commodore, who was himself ill of the scurvy, had a tent erected for him on shore, where he went with the view of staying a few days to establish his health, being convinced by the general experience of his people that no other method but living on the land was to be trusted to for the removal of this dreadful malady. The place where his tent was pitched on this occasion was near the well whence we got all our water, and was indeed a most elegant spot.

As the crew on board were now reinforced by the recovered hands returned from the island, we began to send our casks on shore to be fitted up, which till this time could not be done, for the coopers were not well enough to work. We likewise weighed our anchors, that we might examine our cables, which we suspected had by this time received considerable damage. And as the new moon was now approaching, when we apprehended violent gales, the commodore, for our greater security, ordered that part of the cables next to the anchors to be armed with the chains of the fire-grapnels; besides which, they were cackled twenty fathom from the anchors and seven fathom from the service with a good rounding of a 4-½-inch hauser; and being persuaded that the dangers of this road demanded our utmost foresight, we to all these precautions added that of lowering the main and fore-yard close down, that in case of blowing weather the wind might have less power upon the ship to make her ride a strain.

Thus effectually prepared, as we conceived, we waited till the new moon, which was the 18th of September, when riding safe that and the three succeeding days (though the weather proved very squally and uncertain), we flattered ourselves (for I was then on board) that the prudence of our measures had secured us from all accidents; but on the 22d, the wind blew from the eastward with such fury that we soon despaired of riding out the storm. In this conjuncture we should have been extremely glad that the commodore and the rest of our people on shore, which were the greatest part of our hands, had been on board us, since our only hopes of safety seemed to depend on our putting immediately to sea; but all communication with the shore was now absolutely cut off, for there was no possibility that a boat could live, so that we were necessitated to ride it out till our cables parted. Indeed we were not long expecting this dreadful event, for the small bower parted at five in the afternoon, and the ship swung off to the best bower; and as the night came on the violence of the wind still increased, tho' notwithstanding its inexpressible fury, the tide ran with so much rapidity as to prevail over it: for the tide which set to the northward at the beginning of the hurricane, turning suddenly to the southward about six in the evening, forced the ship before it, in despight of the storm which blew upon the beam. The sea now broke most surprizingly all round us, and a large tumbling swell threatened to poop us, by which the long-boat at this time, moored astern, was on a sudden canted so high that it broke the transon of the commodore's gallery, whose cabin was on the quarter-deck, and would doubtless have risen as high as the trafferel had it not been for the stroke, which stove the boat all to pieces; and yet the poor boat-keeper, though extremely bruised, was saved almost by miracle. About eight the tide slackened, but the wind not abating, the best bower cable, by which alone we rode, parted at eleven. Our sheet anchor, which was the only one we had left, was instantly cut from the bow; but before it could reach the bottom, we were driven from twenty-two into thirty-five fathom; and after we had veered away one whole cable and two-thirds of another, we could not find ground with sixty fathom of line. This was a plain indication that the anchor lay near the edge of the bank, and could not hold us long. In this pressing danger, Mr. Saumarez, our first lieutenant, who now commanded on board, ordered several guns to be fired and lights to be shown as a signal to the commodore of our distress; and in a short time after, it being then about one o'clock and the night excessively dark, a strong gust, attended with rain and lightning, drove us off the bank, and forced us out to sea, leaving behind us on the island Mr. Anson with many more of our officers and great part of our crew, amounting in the whole to a hundred and thirteen persons. Thus were we all, both at sea and on shore, reduced to the utmost despair by this catastrophe; those on shore conceiving they had no means left them ever to depart from the island, whilst we on board, being utterly unprepared to struggle with the fury of the seas and winds we were now exposed to, expected each moment to be our last.

CHAPTER III

TRANSACTIONS AT TINIAN AFTER THE DEPARTURE OF THE "CENTURION"

The storm which drove the *Centurion* to sea blew with too much turbulence to permit either the commodore or any of the people on shore to hear the guns which she fired as signals of distress, and the frequent glare of the lightning had prevented the explosions from being observed: so that when at daybreak it was perceived from the shore that the ship was missing, there was the utmost consternation amongst them: for much the greatest part of them immediately concluded that she was lost, and intreated the commodore that the boat might be sent round the island to look after the wreck; and those who believed her safe had scarcely any expectation that she would ever be able to make the island again, since the wind continued to blow strong at east, and they well knew how poorly she was manned and provided for struggling with so tempestuous a gale. In either of these views their situation was indeed most deplorable: for if the *Centurion* was lost, or should be incapable of returning, there appeared no possibility of their ever getting off the island, as they were at least six hundred leagues from Macao, which was their nearest port, and they were masters of no other vessel than the small Spanish bark of about fifteen tun seized at their first arrival, which would not even hold a fourth part of their number. And the chance of their being taken off the island by the casual arrival of any other ship was altogether desperate, as perhaps no European ship had ever anchored here before, and it were madness to expect that like incidents should send another here in an hundred ages to come: so that their desponding thoughts could only suggest to them the melancholy prospect of spending the remainder of their days on this island, and bidding adieu for ever to their country, their friends, their families, and all their domestic endearments.

Nor was this the worst they had to fear: for they had reason to apprehend that the Governor of Guam, when he should be informed of their circumstances, might send a force sufficient to overpower them, and to remove them to that island; and then the most favourable treatment they could expect would be to be detained prisoners during life; since from the known policy and cruelty of the Spaniards in their distant settlements, it was rather to be supposed that the governor, if he once had them in his power, would make their want of commissions (all of them being on board the *Centurion*) a pretext for treating them as pirates, and for depriving them of their lives with infamy.

In the midst of these gloomy reflections, Mr. Anson, though he always kept up his usual composure and steadiness, had doubtless his share of disquietude. However, having soon projected a scheme for extricating himself and his men from their present anxious situation, he first communicated it to some of the most intelligent persons about him; and having satisfied himself that it was practicable, he then endeavoured to animate his people to a speedy and vigorous prosecution of it. With this view he represented to them how little foundation there was for their apprehensions of the *Centurion's* being lost: that he should have presumed they had been all of them better acquainted with sea affairs than to give way to the impression of so chimerical a fright: that he doubted not but if they would seriously consider what such a ship was capable of enduring, they would confess there was not the least probability of her having perished: that he was not without hopes that she might return in a few days; but if she did not, the worst that could be imagined was, that she was driven so far to the leeward of the island that she could not regain it, and that she would consequently be obliged to bear away for Macao on the coast of China: that as it was necessary to be prepared against all events, he had, in this case, considered of a method of carrying them off the island, and of joining their old ship the *Centurion* again at Macao: that this method was to hale the Spanish bark on shore, to saw her asunder, and to lengthen her twelve feet, which would enlarge her to near forty tun burthen, and would enable her to carry them all to China: that he had consulted the carpenters, and they had agreed that this proposal was very feasible, and that nothing was wanting to execute it but the united resolution and industry of the whole body: and having added that for his own part he would share the fatigue and labour with them, and would expect no more from any man than what he, the commodore himself, was ready to submit to, he concluded with representing to them the importance of saving time, urging that, in order to be the better secured at all events, it was expedient to set about the work immediately, and to take it for granted that the *Centurion* would not be able to put back (which was indeed the commodore's secret opinion), since if she did return, they should only throw away a few days' application; but if she did not, their situation and the season of the year required their utmost dispatch.

These remonstrances, though not without effect, did not at first operate so powerfully as Mr. Anson could have wished. He indeed raised their spirits by showing them the possibility of their getting away, of which they had before despaired; but then from their confidence in this resource they grew less apprehensive of their situation, gave a greater scope to their hopes, and flattered themselves that the *Centurion* would be able to regain the island, and prevent the execution of the commodore's scheme, which they could easily foresee would be a work of considerable labour. Hence it was some days before they were all of them heartily engaged in the project; but at last being convinced of the impossibility of the ship's return, they betook themselves zealously to the different tasks allotted them, and were as industrious and as eager as their commander could desire, punctually assembling by daybreak at the rendezvous, whence they were distributed to their different employments, which they followed with unusual vigour till night came on.

And here I must interrupt the course of this transaction to relate an incident which for a short time gave Mr. Anson more concern than all the preceding disasters. A few days after the ship was driven off, some of the people on shore cried out, "A sail!" This spread a general joy, every one supposing that it was the ship returning; but presently, a second sail was descried, which quite destroyed their first conjecture, and made it difficult to guess what they were. The commodore eagerly turned his glass towards them, and saw they were two boats, on which it immediately occurred to him that the *Centurion* was gone to the bottom, and that these were her two boats coming back with the remains of her people; and this sudden and unexpected suggestion wrought on him so powerfully that to conceal his emotion he was obliged (without speaking to any one) instantly to retire to his tent, where he passed some bitter moments, in the firm belief that the ship was lost, and that now all his views of farther distressing the enemy, and of still signalizing his expedition by some important exploit, were at an end.

However, he was soon relieved from these disturbing thoughts by discovering that the two boats in the offing were Indian proas; and perceiving that they made towards the shore, he directed every appearance that could give them any suspicion to be removed, concealing his people in the adjacent thickets, ready to secure the Indians when they should land: but after the proas had stood in within a quarter of a mile of the beach, they suddenly stopt short, and remaining there motionless for near two hours, they then got under sail again, and steered to the southward. Let us now return to the projected enlargement of the bark.

If we examine how they were prepared for going through with this undertaking, on which their safety depended, we shall find that, independent of other matters which were of as much consequence, the lengthning of the bark alone was attended with great difficulty. Indeed, in a proper place, where all the necessary materials and tools were to be had, the embarrassment would have been much less; but some of these tools were to be made, and many of the materials were wanting, and it required no small degree of invention to supply all these deficiencies. And when the hull of the bark should be compleated, this was but one article, and there were others of equal weight which were to be well considered: these were the rigging it, the victualling it, and lastly the navigating it, for the space of six or seven hundred leagues, through unknown seas where no one of the company had ever passed before. And in these particulars such obstacles occurred, that without the intervention of very extraordinary and unexpected accidents, the possibility of the whole enterprize would have fallen to the ground, and their utmost industry and efforts must have been fruitless. Of all these circumstances I shall make a short recital.

It fortunately happened that the carpenters, both of the *Gloucester* and of the *Tryal*, with their chests of tools, were on shore when the ship drove out to sea; the smith too was on shore, and had with him his forge and several of his tools, but unhappily his bellows had not been brought from on board, so that he was incapable of working, and without his assistance they could not hope to proceed with their design. Their first attention, therefore, was to make him a pair of bellows, but in this they were for some time puzzled by their want of leather; however, as they had hides in sufficient plenty, and they had found a hogshead of lime, which the Indians or Spaniards had prepared for their own use, they tanned a few hides with this lime; and though we may suppose the workmanship to be but indifferent, yet the leather they thus procured answered the intention tolerably well, and the bellows, to which a gun-barrel served for a pipe, had no other inconvenience than that of being somewhat strong scented from the imperfection of the tanner's work.

Whilst the smith was preparing the necessary iron-work, others were employed in cutting down trees and sawing them into planks; and this being the most laborious task, the commodore wrought at it himself for the encouragement of his people. But there being neither blocks nor cordage sufficient for tackles to haul the bark on shore, this occasioned a new difficulty; however, it was at length resolved to get her up on rollers, since for these the body of the coconut tree was extremely well fitted, as its smoothness and circular turn prevented much labour, and suited it to the purpose with very little workmanship. A number of these trees were therefore felled, and the ends of them properly opened for the insertion of hand-spikes; and in the meantime a dry dock was dug to receive the bark, and ways were laid from thence quite into the sea to facilitate the bringing her up. Neither were these the whole of their occupations, since, besides those who were thus busied in preparing measures towards the future enlargement of the bark, a party was constantly ordered to kill and provide provisions for the rest. And though in these various employments, some of which demanded considerable dexterity, it might have been expected there would have been great confusion and delay, yet good order being once established and all hands engaged, their preparations advanced apace. Indeed, the common men, I presume, were not the less tractable for their want of spirituous liquors: for there being neither wine nor brandy on shore, the juice of the coconut was their constant drink; and this, though extremely pleasant, was not at all intoxicating, but kept them very temperate and orderly.

The main work now proceeding successfully, the officers began to consider of all the articles which would be necessary to the fitting out the bark for the sea. On this consultation it was found that the tents on shore and the spare cordage accidentally left there by the *Centurion*, together with the sails and rigging already belonging to the bark, would serve to rig her indifferently well when she was lengthened. And as they had tallow in plenty, they proposed to pay her bottom with a mixture of tallow and lime, which it was known was not ill adapted to that purpose: so that with respect to her equipment she would not have been very defective. There was, however, one exception, which would have proved extremely inconvenient, and that was her size: for as they could not make her quite forty tun burthen, she would have been incapable of containing half the crew below the deck, and she would have been so top-heavy that if they were all at the same time ordered upon deck, there would be no small hazard of her oversetting; but this was a difficulty not to be removed, as they could not augment her beyond the size already proposed. After the manner of rigging and fitting up the bark was considered and regulated, the next essential point to be thought on was how to procure a sufficient stock of provisions for their voyage; and here they were greatly at a loss what expedient to have recourse to, as they had neither grain nor bread of any kind on shore, their bread-fruit, which would not keep at sea, having all along supplied its place; and though they had live cattle enough, yet they had no salt to cure beef for a sea-store, nor would meat take salt in that climate. Indeed, they had preserved a small quantity of jerked beef, which they found upon the place at their landing; but this was greatly disproportioned to the run of near six hundred leagues which they were to engage in, and to the number of hands they should have on board. It was at last, however, resolved to put on board as many coconuts as they possibly could, to prolong to the utmost their jerked beef by a very sparing distribution of it, and to endeavour to supply their want of bread by rice; to furnish themselves with which, it was proposed, when the bark was fitted up, to make an expedition to the island of Rota, where they were told that the Spaniards had large plantations of rice under the care of the Indian inhabitants. But as this last measure was to be executed by force, it became necessary to examine what ammunition had been left on shore, and to preserve it carefully; and on this enquiry, they had the mortification to find that their firelocks would be of little service to them, since all the powder that could be collected, by the strictest search, did not amount to more than ninety charges, which was considerably short of one apiece to each of the company, and was indeed a very slender stock of ammunition for such as were to eat no grain or bread during a whole month, except what they were to procure by force of arms.

But the most alarming circumstance, and which, without the providential interposition of very improbable events, would have rendered all their schemes abortive, remains yet to be related. The general idea of the fabric and equipment of the vessel was settled in a few days; and this being done, it was not difficult to frame some estimation of the time necessary to compleat her. After this, it was natural to expect that the officers would consider the course they were to steer, and the land they were to make. These reflections led them to the disheartning discovery that there was neither compass nor quadrant on the island. Indeed the commodore had brought a pocket-compass on shore for his own use, but Lieutenant Brett had borrowed it to determine the position of the neighbouring islands, and he had been driven to sea in the *Centurion* without returning it. And as to a quadrant, that could not be expected to be found on shore, since as it was of no use at land, there could be no reason for bringing it from on board the ship. There were now eight days elapsed since the departure of the *Centurion*, and yet they were not in any degree relieved from this terrible perplexity. At last, in rumaging a chest belonging to the Spanish bark, they discovered a small compass, which, though little better than the toys usually made for the amusement of schoolboys, was to them an invaluable treasure. And a few days after, by a similar piece of good fortune, they met with a quadrant on the sea-shore, which had been thrown overboard amongst other lumber belonging to the dead. The quadrant was eagerly seized, but on examination it unluckily wanted vanes, and therefore in this present state was altogether useless; however, fortune still continuing in a favourable mood, it was not long before a person, through curiosity pulling out the drawer of an old table which had been driven on shore, found therein some vanes which fitted the quadrant very well; and it being thus compleated, it was examined by the known latitude of the place, and upon trial answered to a sufficient degree of exactness.

When now all these obstacles were in some degree removed (which were always as much as possible concealed from the vulgar, that they might not grow remiss with the apprehension of labouring to no purpose), the business proceeded very successfully and vigorously. The necessary iron-work was in great forwardness, and the timbers and planks (which, tho' not the most exquisite performances of the sawyer's art, were yet sufficient for the purpose) were all prepared; so that on the 6th of October, being the 14th day from the departure of the ship, they hauled the bark on shore, and on the two succeeding days she was sawn asunder, though with the caution not to cut her planks: and her two parts being separated the proper distance from each other, and the materials being all ready beforehand, they, the next day, being the 9th of October, went on with no small dispatch in their proposed enlargement of her; whence by this time they had all their future operations so fairly in view, and were so much masters of them, that they were able to determine when the whole would be finished, and had accordingly fixed the 5th of November for the day of their putting to sea. But their projects and labours were now drawing to a speedier and happier conclusion; for on the 11th of October, in the afternoon, one of the *Gloucester's* men being upon a hill in the middle of the island, perceived the *Centurion* at a distance, and running down with his utmost speed towards the landing-place, he, in the way, saw some of his comrades, to whom he hallooed out with great extasy, "The ship, the ship!" This being heard by Mr. Gordon, a lieutenant of marines, who was convinced by the fellow's transport that his report was true, Mr. Gordon directly hastened towards the place where the commodore and his people were at work, and being fresh and in breath easily outstripped the *Gloucester's* man, and got before him to the commodore, who, on hearing this pleasing and unexpected news, threw down his axe, with which he was then at work, and by his joy broke through, for the first time, the equable and unvaried character which he had hitherto preserved: whilst the others who were present instantly ran down to the seaside in a kind of frenzy, eager to feast themselves with a sight they had so ardently longed after, and of which they had now for a considerable time despaired. By five in the evening the *Centurion* was visible in the offing to them all; and, a boat being sent off with eighteen men to reinforce her, and with fresh meat and fruits for the refreshment of her crew, she, the next afternoon, happily cast anchor in the road, where the commodore immediately came on board her, and was received by us with the sincerest and heartiest acclamations: for, by the following short recital of the fears, the dangers, and fatigues we in the ship underwent during our nineteen days' absence from Tinian, it may be easily conceived that a harbour, refreshments, repose, and the joining of our commander and shipmates were not less pleasing to us than our return was to them.

CHAPTER IV

PROCEEDINGS ON BOARD THE "CENTURION" WHEN DRIVEN OUT TO SEA

The *Centurion* being now once more safely arrived at Tinian, to the mutual respite of the labours of our divided crew, it is high time that the reader, after the relation already given of the projects and employment of those left on shore, should be apprized of the fatigues and distresses to which we, whom the *Centurion* carried off to sea, were exposed during the long interval of nineteen days that we were absent from the island.

It has been already mentioned that it was the 22d of September, about one o'clock, in an extreme dark night, when by the united violence of a prodigious storm and an exceeding rapid tide, we were driven from our anchors and forced to sea. Our condition then was truly deplorable; we were in a leaky ship with three cables in our hawses, to one of which hung our only remaining anchor: we had not a gun on board lashed, nor a port barred in; our shrouds were loose, and our top-masts unrigged, and we had struck our fore and main-yards close down before the hurricane came on, so that there were no sails we could set, except our mizen. In this dreadful extremity we could muster no more strength on board to navigate the ship than an hundred and eight hands, several negroes and Indians included: this was scarcely the fourth part of our complement, and of these the greater number were either boys, or such as, being but lately recovered from the scurvy, had not yet arrived at half their former vigour. No sooner were we at sea, but by the violence of the storm and the working of the ship we made a great quantity of water through our hawse-holes, ports, and scuppers, which, added to the constant effect of our leak, rendered our pumps alone a sufficient employment for us all. But though we knew that this leakage, by being a short time neglected, would inevitably end in our destruction, yet we had other dangers then hanging over us which occasioned this to be regarded as a secondary consideration only. For we all imagined that we were driving directly on the neighbouring island of Aguiguan, which was about two leagues distant; and as we had lowered our main and fore-yards close down, we had no sails we could set but the mizen, which was altogether insufficient to carry us clear of this imminent peril. Urged therefore by this pressing emergency, we immediately applied ourselves to work, endeavouring with the utmost of our efforts to heave up the main and fore-yards, in hopes that if we could but be enabled to make use of our lower canvass, we might possibly weather the island, and thereby save ourselves from this impending shipwreck. But after full three hours' ineffectual labour, the jeers broke, and the men being quite jaded, we were obliged, by mere debility, to desist, and quietly to expect our fate, which we then conceived to be unavoidable. For we soon esteemed ourselves to be driven just upon the shore, and the night was so extremely dark that we expected to discover the island no otherwise than by striking upon it; so that the belief of our destruction, and the uncertainty of the point of time when it should take place, occasioned us to pass several hours under the most serious apprehensions that each succeeding moment would send us to the bottom. Nor did these continued terrors of instantly striking and sinking end but with the daybreak, when we with great transport perceived that the island we had thus dreaded was at a considerable distance, and that a strong northern current had been the cause of our preservation.

The turbulent weather which forced us from Tinian did not abate till three days after, and then we swayed up the fore-yard, and began to heave up the main-yard, but the jeers broke again and killed one of our people, and prevented us at that time from proceeding. The next day, being the 26th of September, was a day of most severe fatigue to us all, for it must be remembered that in these exigences no rank or office exempted any person from the manual application and bodily labour of a common sailor. The business of this day was no less than an endeavour to heave up the sheet-anchor, which we had hitherto dragged at our bows with two cables an end. This was a work of great importance to our future preservation: for not to mention the impediment it would be to our navigation, and hazard to our ship, if we attempted to make sail with the anchor in its present situation, we had this most interesting consideration to animate us, that it was the only anchor we had left, and without securing it we should be under the utmost difficulties and hazards whenever we fell in with the land again; and therefore, being all of us fully apprized of the consequence of this enterprize, we laboured at it with the severest application for twelve hours, when we had indeed made a considerable progress, having brought the anchor in sight; but it growing dark, and we being excessively fatigued, we were obliged to desist, and to leave our work unfinished till the next morning, and then, refreshed by the benefit of a night's rest, we compleated it, and hung the anchor at our bow.

It was the 27th of September, that is, five days after our departure, before we had thus secured our anchor. However, we the same day got up our main-yard, so that having now conquered, in some degree, the distress and disorder which we were necessarily involved in at our first driving out to sea, and being enabled to make use of our canvass, we set our courses, and for the first time stood to the eastward in hopes of regaining the island of Tinian, and joining our commodore in a few days, since, by our accounts, we were only forty-seven leagues distant to the south-west. Hence, on the first day of

October, having then run the distance necessary for making the island according to our reckoning, we were in full expectation of seeing it: but here we were unhappily disappointed, and were thereby convinced that a current had driven us considerably to the westward. This discovery threw us into a new perplexity; for as we could not judge how much we might hereby have deviated, and consequently how long we might still expect to be at sea, we had great apprehensions that our stock of water would prove deficient, since we were doubtful about the quantity we had on board, finding many of our casks so decayed as to be half leaked out. However, we were delivered from our uncertainty the next day, having then a sight of the island of Guam, and hence we computed that the currents had driven us forty-four leagues to the westward of our accounts. Being now satisfied of our situation by this sight of land, we kept plying to the eastward, though with excessive labour; for the wind continuing fixed in the eastern board, we were obliged to tack often, and our crew was so weak that, without the assistance of every man on board, it was not in our power to put the ship about. This severe employment lasted till the 11th of October, being the nineteenth day from our departure, when arriving in the offing of Tinian, we were reinforced from the shore, as hath been already related; and on the evening of the same day we, to our inexpressible joy, came to an anchor in the road, thereby procuring to our shipmates on shore, as well as to ourselves, a cessation from the fatigues and apprehensions which this disastrous incident had given rise to.

CHAPTER V

EMPLOYMENT AT TINIAN TILL THE FINAL DEPARTURE OF THE "CENTURION" FROM THENCE; WITH A DESCRIPTION OF THE LADRONES

When the commodore came on board the *Centurion* after her return to Tinian, he resolved to stay no longer at the island than was absolutely necessary to compleat our stock of water, a work which we immediately set ourselves about. But the loss of our long-boat, which was staved against our poop before we were driven out to sea, put us to great inconveniences in getting our water on board, for we were obliged to raft off all our cask, and the tide ran so strong, that besides the frequent delays and difficulties it occasioned, we more than once lost the whole raft. Nor was this our only misfortune; for on the 14th of October, being but the third day after our arrival, a sudden gust of wind brought home our anchor, forced us off the bank, and drove the ship out to sea a second time. The commodore, it is true, and the principal officers were now on board; but we had near seventy men on shore, who had been employed in filling our water and procuring provisions. These had with them our two cutters: but as they were too many for the cutters to bring off at once, we sent the eighteen-oared barge to assist them, and at the same time made a signal for all that could to embark. The two cutters soon came off to us full of men; but forty of the company, who were busied in killing cattle in the woods, and in bringing them down to the landing-place, remained behind; and though the eighteen-oared barge was left for their conveyance, yet as the ship soon drove to a considerable distance, it was not in their power to join us. However, as the weather was favourable, and our crew was now stronger than when we were first driven out, we in about five days' time returned again to an anchor at Tinian, and relieved those we had left behind us from their second fears of being deserted by their ship.

On our arrival, we found that the Spanish bark, the old object of their hopes, had undergone a new metamorphosis: for those on shore despairing of our return, and conceiving that the lengthening the bark, as formerly proposed, was both a toilsome and unnecessary measure, considering the small number they consisted of, they had resolved to join her again and to restore her to her first state; and in this scheme they had made some progress, for they had brought the two parts together, and would have soon compleated her, had not our coming back put a period to their labours and disquietudes.

These people we had left behind informed us that just before we were seen in the offing two proas had stood in very near the shore, and had continued there for some time; but on the appearance of our ship they crowded away, and were presently out of sight. And on this occasion I must mention an incident, which though it happened during the first absence of the ship, was then omitted, to avoid interrupting the course of the narration.

It hath been already observed that a part of the detachment sent to this island under the command of the Spanish Serjeant lay concealed in the woods. Indeed we were the less solicitous to find them out, as our prisoners all assured us that it was impossible for them to get off, and consequently that it was impossible for them to send any intelligence about us to Guam. But when the *Centurion* drove out to sea and left the commodore on shore, he one day, attended by some of his officers, endeavoured to make the tour of the island. In this expedition, being on a rising ground, they observed in the valley beneath them the appearance of a small thicket, which by attending to more nicely they found had a progressive motion. This at first surprized them; but they soon perceived that it was no more than several large coco bushes, which were dragged along the ground by persons concealed beneath them. They immediately concluded that these were some of the Serjeant's party, which was indeed true; and therefore the commodore and his people made after them, in hopes of tracing out their retreat. The Indians, remarking that they were discovered, hurried away with precipitation; but Mr. Anson was so near them that he did not lose sight of them till they arrived at their cell, which he and his officers entering, found to be abandoned, there being a passage from it which had been contrived for the conveniency of flight, and which led down a precipice. They here met with an old firelock or two, but no other arms. However, there was a great quantity of provisions, particularly salted sparibs of pork, which were excellent; and from what our people saw, they concluded that the extraordinary appetite which they had acquired at this island was not confined to themselves alone; for it being about noon, the Indians laid out a very plentiful repast, considering their numbers, and had their bread-fruit and coconuts prepared ready for eating, in a manner too which plainly evinced that with them a good meal was neither an uncommon nor an unheeded article. The commodore having in vain searched after the path by which the Indians had escaped, he and his officers contented themselves with sitting down to the dinner which was thus luckily fitted to their present hunger; after which they returned back to their old habitation, displeased at missing the Indians, as they hoped to have engaged them in our service, if they could have had any conference with them. I must add, that notwithstanding what our prisoners had asserted, we were afterwards assured that these Indians were carried off to Guam long before we left the place. But to return to our history.

On our coming to an anchor again, after our second driving off to sea, we laboured indefatigably at getting in our water; and having, by the 20th of October, compleated it to fifty tun, which we supposed would be sufficient during our passage to Macao, we on the next day sent one of each mess on shore to gather as large a quantity of oranges, lemons, coconuts, and other fruits of the island as they possibly could, for the use of themselves and their messmates when at sea. And these purveyors returning on the evening of the same day, we then set fire to the bark and proa, hoisted in our boats and got under sail, steering away towards the south end of the island of Formosa, and taking our leaves, for the third and last time, of the island of Tinian: an island which, whether we consider the excellence of its productions, the beauty of its appearance, the elegance of its woods and lawns, the healthiness of its air, or the adventures it gave rise to, may in all these views be justly stiled romantic.

And now, postponing for a short time our run to Formosa, and thence to Canton, I shall interrupt the narration with a description of that range of islands usually called the Ladrones, or Marian Islands, of which this of Tinian is one.

These islands were discovered by Magellan in the year 1521; and from the account given of the two he first fell in with, it should seem that they were those of Saypan and Tinian, for they are described as very beautiful islands, and as lying between 15 and 16 degrees of north latitude. These characteristics are particularly applicable to the two above-mentioned places; for the pleasing appearance of Tinian hath occasioned the Spaniards to give it the additional name of Buenavista; and Saypan, which is in the latitude of 15° 22' north, affords no contemptible prospect when seen at sea, as is sufficiently evident from a view of its north-west side.

There are usually reckoned twelve of these islands; but it will appear that if the small islets and rocks are counted, that their whole number will amount to above twenty. They were formerly most of them well inhabited; and even not sixty years ago, the three principal islands, Guam, Rota, and Tinian together, are asserted to have contained above fifty thousand people: but since that time Tinian had been entirely depopulated; and no more than two or three hundred Indians have been left at Rota to cultivate rice for the island of Guam; so that now Guam alone can properly be said to be inhabited. This island of Guam is the only settlement of the Spaniards; here they keep a governor and garrison, and here the Manila ship generally touches for refreshment in her passage from Acapulco to the Philippines. It is esteemed to be about thirty leagues in circumference, and contains, by the Spanish accounts, near four thousand inhabitants, of which a thousand are supposed to live in the city of San Ignatio de Agana, where the governor generally resides, and where the houses are represented as considerable, being built with stone and timber, and covered with tiles, a very uncommon fabric for these warm climates and savage countries. Besides this city, there are upon the island thirteen or fourteen villages. As Guam is a post of some consequence, on account of the refreshment it yields to the Manila ship, there are two castles on the seashore; one is the castle of St. Angelo, which lies near the road where the Manila ship usually anchors, and is but an insignificant fortress, mounting only five guns, eight-pounders; the other is the castle of St. Lewis, which is N.E. from St. Angelo, and four leagues distant, and is intended to protect a road where a small vessel anchors which arrives here every other year from Manila. This fort mounts the same number of guns as the former: and besides these forts, there is a battery of five pieces of cannon on an eminence near the seashore. The Spanish troops employed at this island consist of three companies of foot, betwixt forty and fifty men each, and this is the principal strength the governor has to depend on; for he cannot rely on any assistance from the Indian inhabitants, being generally upon ill terms with them, and so apprehensive of them that he has debarred them the use both of firearms and lances.

The rest of these islands, though not inhabited, do yet abound with many kinds of refreshment and provision; but there is no good harbour or road amongst them all. Of that of Tinian we have treated largely already; nor is the road of Guam much better, since it is not uncommon for the Manila ship, though she proposes to stay there but twenty-four hours, to be forced to sea, and to leave her boat behind her. This is an inconvenience so sensibly felt by the commerce at Manila, that it is always recommended to the Governor of Guam to use his best endeavours for the discovery of some secure port in the neighbouring ocean. How industrious he may be to comply with his instructions I know not; but this is certain, that notwithstanding the many islands already found out between the coast of Mexico and the Philippines, there is not any one safe port to be met with in that whole track, though in other parts of the world it is not uncommon for very small islands to furnish most excellent harbours.

From what has been said, it appears that the Spaniards on the island of Guam are extremely few compared to the Indian inhabitants; and formerly the disproportion was still greater, as may be easily conceived from the account given in another chapter of the numbers heretofore on Tinian alone. These Indians are a bold, strong, well-limbed people, and, as it should seem from some of their practices, are no ways defective in understanding, for their flying proas in particular, which during ages past have been the only vessels employed by them, are so singular and extraordinary an invention that it would do honour to any nation, however dextrous and acute. Since, if we consider the aptitude of this proa to the navigation of these islands, which lying all of them nearly under the same meridian, and within the limits of the trade-wind, require the vessels made use of in passing from one to the other to be peculiarly fitted for sailing with the wind upon the beam; or if we examine the uncommon simplicity and ingenuity of its fabric and contrivance, or the extraordinary velocity with which it moves, we shall, in each of these articles, find it worthy of our admiration, and deserving a place amongst the mechanical productions of the most civilized nations where arts and sciences have most eminently flourished. As former navigators, though they have mentioned these vessels, have yet treated of them imperfectly, and, as I conceive that besides their curiosity they may furnish both the shipwright and seaman with no contemptible observations, I shall here insert a very exact description of the build, rigging, and working of these vessels, which I am the better enabled to perform, as one of them fell into our hands on our first arrival at Tinian, and Mr. Brett took it to pieces that he might delineate its fabric and dimensions with greater accuracy: so that the following account may be relied on.

The name of flying proa, appropriated to these vessels, is owing to the swiftness with which they sail. Of this the Spaniards assert such stories as must appear altogether incredible to one who has never seen these vessels move; nor are they the only people who recount these extraordinary tales of their celerity, for those who shall have the curiosity to enquire at Portsmouth dock about an experiment tried there some years since with a very imperfect one built at that place, will meet with accounts not less wonderful than any the Spaniards have related. However, from some rude estimations made by us of the velocity with which they crossed the horizon at a distance while we lay at Tinian, I cannot help believing that with a brisk trade-wind they will run near twenty miles an hour; which, though greatly short of what the Spaniards report of them, is yet a prodigious degree of swiftness. But let us give a distinct idea of its figure.

The construction of this proa is a direct contradiction to the practice of all the rest of mankind: for as it is customary to make the head of the vessel different from the stern, but the two sides alike, the proa, on the contrary, has her head and stern exactly alike, but her two sides very different; the side intended to be always the lee side being flat, whilst the windward side is built rounding, in the manner of other vessels: and to prevent her oversetting, which from her small breadth and the strait run of her leeward side, would, without this precaution, infallibly happen, there is a frame laid out from her to windward, to the end of which is fastened a log fashioned into the shape of a small boat, and made hollow. The weight of the frame is intended to balance the proa, and the small boat is by its buoyancy (as it is always in the water) to prevent her oversetting to windward; and this frame is usually called an outrigger. The body of the proa (at least of that we took) is formed of two pieces joined endways and sewed together with bark, for there is no iron used in her construction. She is about two inches thick at the bottom, which at the gunwale is reduced to less than one. The proa generally carries six or seven Indians, two of which are placed in the head and stern, who steer the vessel alternately with a paddle according to the tack she goes on, he in the stern being the steersman; the other Indians are employed either in baling out the water which she accidentally ships, or in setting and trimming the sail. From the description of these vessels it is sufficiently obvious how dexterously they are fitted for ranging this collection of islands called the Ladrones: since as these islands bear nearly N. and S. of each other, and are all within the limits of the trade-wind, the proas, by sailing most excellently on a wind, and with either end foremost, can run from one of these islands to the other and back again only by shifting the sail, without ever putting about; and by the flatness of their lee side, and their small breadth, they are capable of lying much nearer the wind than any other vessel hitherto known, and thereby have an advantage which no vessels that go large can ever pretend to.

The advantage I mean is that of running with a velocity nearly as great, and perhaps sometimes greater, than what the wind blows with. This, however paradoxical it may appear, is evident enough in similar instances on shore, since it is well known that the sails of a windmill often move faster than the wind; and one great superiority of common windmills over all others that ever were, or ever will be, contrived to move with an horizontal motion, is analogous to the case we have mentioned of a vessel upon a wind and before the wind: for the sails of an horizontal windmill, the faster they move the more they detract from the impulse of the wind upon them; whereas the common windmills, by moving perpendicular to the torrent of air, are nearly as forcibly acted on by the wind when they are in motion as when they are at rest.

Thus much may suffice as to the description and nature of these singular embarkations. I must add that vessels bearing some obscure resemblance to these are to be met with in various parts of the East Indies, but none of them, that I can learn, to be compared with those of the Ladrones, either for their construction or celerity; which should induce one to believe that this was originally the invention of some genius of these islands, and was afterwards imperfectly copied by the neighbouring nations: for though the Ladrones have no immediate intercourse with any other people, yet there lie to the S. and S.W. of them a great number of islands, which are imagined to extend to the coast of New Guinea. These islands are so near the Ladrones that canoes from them have sometimes by distress been driven to Guam, and the Spaniards did once dispatch a bark for their discovery, which left two Jesuits amongst them, who were afterwards murthered. Whence it may be presumed that the inhabitants of the Ladrones, with their proas, may by storms or casualties have been driven amongst those islands. Indeed, I should conceive that the same range of islands stretches to the S.E. as well as the S.W., and to a prodigious distance, for Schouten, who traversed the south part of the Pacific Ocean in the year 1615, met with a large double canoe full of people above a thousand leagues from the Ladrones, towards the S.E. If that double canoe was any distant imitation of the flying proa, which is no very improbable conjecture, it must then be supposed that a range of islands, near enough to each other to be capable of an accidental communication, is continued thither from the Ladrones. This seems to be farther evinced from hence, that all those who have crossed from America to the East Indies in a southern latitude have never failed of discovering several very small islands scattered over that immense ocean.

And as there may be hence some reason to conclude that there is a chain of islands spreading themselves southward towards the unknown boundaries of the Pacific Ocean of which the Ladrones are only a part, so it appears that the same chain is extended from the northward of the Ladrones to Japan: whence in this light the Ladrones will be only one small portion of a range of islands reaching from Japan perhaps to the unknown southern continent. After this short account of these places, I shall now return to the prosecution of our voyage.

CHAPTER VI

FROM TINIAN TO MACAO

On the 21st of October, in the evening, we took our leave of the Island of Tinian, steering the proper course for Macao in China. The eastern monsoon was now, we reckoned, fairly settled; and we had a constant gale blowing right astern, so that we generally ran from forty to fifty leagues a day. But we had a large hollow sea pursuing us, which occasioned the ship to labour much; whence our leak was augmented, and we received great damage in our rigging, which by this time was grown very rotten. However, our people were now happily in full health, so that there were no complaints of fatigue, but all went through their attendance on the pumps, and every other duty of the ship, with ease and chearfulness.

Before we left Tinian we swept for our best and small bower, and employed the Indians to dive in search of them; but all to no purpose. Hence, except our prize anchors, which were stowed in the hold, and were too light to be depended on, we had only our sheet-anchor left: and that being obviously much too heavy for a coasting-anchor, we were under great concern how we should manage on the coast of China, where we were entire strangers, and where we should doubtless be frequently under the necessity of coming to an anchor. But we at length removed the difficulty by fixing two of our largest prize anchors into one stock and placing between their shanks two guns, four pounders. This we intended to serve as a best bower: and a third prize anchor being in like manner joined to our stream-anchor, with guns between them, made us a small bower; so that, besides our sheet-anchor, we had again two others at our bows, one of which weighed 3900, and the other 2900 pounds.

The 3d of November, about three in the afternoon, we saw an island, which at first we imagined to be Botel Tobago Xima, but on our nearer approach we found it to be much smaller than that is usually represented; and about an hour after we saw another island, five or six miles farther to the westward. As no chart or journal we had seen took notice of any island to the eastward of Formosa but Botel Tobago Xima, and as we had no observation of our latitude at noon, we were in some perplexity, apprehending that an extraordinary current had driven us into the neighbourhood of the Bashee Islands. We therefore, when night came on, brought to, and continued in that posture till the next morning, which proving dark and cloudy, for some time prolonged our uncertainty; but it clearing up about nine o'clock, we again discerned the two islands abovementioned; and having now the day before us, we pressed forwards to the westward, and by eleven got a sight of the southern part of the island of Formosa. This satisfied us that the second island we saw was Botel Tobago Xima, and the first a small islet or rock, lying five or six miles due east of it, which, not being mentioned in any of our books or charts, had been the occasion of all our doubts.

When we had made the Island of Formosa we steered W. by S. in order to double its extremity, and kept a good look-out for the rocks of Vele Rete, which we did not discover till two in the afternoon. They then bore from us W.N.W. three miles distant, the south end of Formosa at the same time bearing N. by W.½W. about five leagues distant. To give these rocks a good birth we immediately haled up S. by W. and so left them between us and the land. Indeed we had reason to be careful of them; for though they appeared as high out of the water as a ship's hull, yet they are environed with breakers on all sides, and there is a shoal stretching from them at least a mile and a half to the southward, whence they may be truly called dangerous. The course from Botel Tobago Xima to these rocks is S.W. by W. and the distance about twelve or thirteen leagues: and the south end of Formosa, off which they lie, is in the latitude of 21° 50' north, and according to our most approved reckonings in 23° 50' west longitude from Tinian; though some of our accounts made its longitude above a degree more.

While we were passing by these rocks of Vele Rete there was an outcry of fire on the forecastle; this occasioned a general alarm, and the whole crew instantly flocked together in the utmost confusion; so that the officers found it difficult for some time to appease the uproar: but having at last reduced the people to order, it was perceived that the fire proceeded from the furnace, where the bricks being overheated, had begun to communicate the fire to the adjacent woodwork: hence by pulling down the brickwork it was extinguished with great facility. In the evening we were surprized with a view of what we at first sight conceived to have been breakers, but on a stricter examination we discerned them to be only a great number of fires on the Island of Formosa. These we imagined were intended by the inhabitants of that island as signals to invite us to touch there, but that suited not our views, we being impatient to reach the port of Macao as soon as possible. From Formosa we steered W.N.W. and sometimes still more northerly, proposing to fall in with the coast of China to the eastward of Pedro Blanco, as the rock so called is usually esteemed an excellent direction for ships bound to Macao. We continued this course till the following night, and then frequently brought to, to try if we were in soundings: but it was the 5th of November, at nine in the morning, before we struck ground, and then we had forty-two fathom, and a bottom of grey sand mixed with shells. When we had run about twenty miles farther W.N.W. we had thirty-five fathom and the same bottom; then our soundings gradually decreased from thirty-five to twenty-five fathom; but soon after, to our great surprize, they jumped back again to thirty fathom. This was an alteration we could not very well account for, since all the charts laid down regular soundings everywhere to the northward of Pedro Blanco. We for this reason kept a careful look out, and altered our course to N.N.W., and having run thirty-five miles in that direction, our soundings again gradually diminished to twenty-two fathom, and we at last, about midnight, got sight of the main land of China, bearing N. by W. four leagues distant. We then brought the ship to, with her head to the sea, proposing to wait for the morning; and before sunrise we were surprized to find ourselves in the midst of an incredible number of fishing boats, which seemed to cover the surface of the sea as far as the eye could reach. I may well style their number incredible, since I cannot believe, upon the lowest estimate, that there were so few as six thousand, most of them manned with five hands, and none of those we saw with less than three. Nor was this swarm of fishing vessels peculiar to that spot: for, as we ran on to the westward, we found them as abundant on every part of the coast. We at first doubted not but we should procure a pilot from them to carry us to Macao; but though many of them came close to the ship, and we endeavoured to tempt them by showing them a number of dollars, a most alluring bait for Chinese of all ranks and professions, yet we could not entice them on board us, nor procure any directions from them; though, I presume, the only difficulty was their not comprehending what we wanted them to do, as we could have no communication with them but by signs. Indeed we often pronounced the word Macao; but this we had reason to suppose they understood in a different sense, since in return they sometimes held up fish to us; and we afterwards learnt that the Chinese name for fish is of a somewhat similar sound. But what surprized us most was the inattention and want of curiosity which we observed in this herd of fishermen. A ship like ours had doubtless never been in these seas before; and perhaps there might not be one amongst all the Chinese, employed in that fishery, who had ever seen any European vessel; so that we might reasonably have expected to have been considered by them as a very uncommon and extraordinary object. But though many of their boats came close to the ship, yet they did not appear to be at all interested about us, nor did they deviate in the least from their course to regard us. Which insensibility, especially of maritime persons, in a matter relating to their own profession, is scarcely to be credited, did not the general behaviour of the Chinese in other instances furnish us with continual proofs of a similar turn of mind. It may perhaps be doubted whether this cast of temper be the effect of nature or education; but, in either case, it is an incontestable symptom of a mean and contemptible disposition, and is alone a sufficient confutation of the extravagant praises which many prejudiced writers have bestowed on the ingenuity and capacity of this nation. But to return.

Not being able to procure any information from the Chinese fishermen about our proper course to Macao, it was necessary for us to rely entirely on our own judgment: and concluding from our latitude, which was 22° 42' north, and from our soundings, which were only seventeen or eighteen fathoms, that we were yet to the eastward of Pedro Blanco, we still stood on to the westward. And for the assistance of future navigators, who may hereafter doubt what part of the coast they are upon, I must observe that besides the latitude of Pedro Blanco, which is 22° 18', and the depth of water, which to the westward of that rock is almost everywhere twenty fathoms, there is another circumstance which will be greatly assistant in judging of the position of the ship: this is the kind of ground; for, till we came within thirty miles of Pedro Blanco, we had constantly a sandy bottom; but there the bottom changed to soft and muddy, and continued so quite to the Island of Macao; only while we were in sight of Pedro Blanco, and very near it, we had for a short space a bottom of greenish mud, intermixed with sand.

It was on the 5th of November, at midnight, when we first made the coast of China. The next day, about two o'clock, as we were standing to the westward, within two leagues of the coast, still surrounded by fishing vessels in as great numbers as at first, we perceived that a boat ahead of us waved a red flag and blew a horn. This we considered as a signal made to us either to warn us of some shoal, or to inform us that they would supply us with a pilot. We therefore immediately sent our cutter to the boat to know their intentions, when we were soon convinced of our mistake, and found that this boat was the commodore of the whole fishery, and that the signal she had made was to order them all to leave off fishing and to return in shore, which we saw them instantly obey. Being thus disappointed we kept on our course, and shortly after passed by two very small rocks, which lay four or five miles distant from the shore. We were now in hourly expectation of descrying Pedro Blanco, but night came on before we got sight of it, and we therefore brought to till the morning, when we had the satisfaction to discover it. Pedro Blanco is a rock of a small circumference, but of a moderate height, resembling a sugar-loaf, both in shape and colour, and is about seven or eight miles distant from the shore. We passed within a mile and an half of it, and left it between us and the land, still keeping on to the westward; and the next day, being the 7th, we were abreast of a chain of islands which stretched from east to west. These, as we afterwards found, were called the islands of Lema; they are rocky and barren, and are, in all, small and great, fifteen or sixteen; but there are, besides, many more between them and the main land of China. We left these islands on the starboard side, passing within four miles of them, where we had twenty-four fathom water. Being still surrounded by fishing boats, we once more sent the cutter on board some of them to endeavour to procure a pilot, but we could not prevail; however, one of the Chinese directed us by signs to sail round the westermost of the islands or rocks of Lema, and then to hale up. We followed this direction, and in the evening came to an anchor in eighteen fathom; at which time the rock bore S.S.E. five miles distant, and the grand Ladrone W. by S. about two leagues distant. The rock is a most excellent direction for ships coming from the eastward: its latitude is 21° 52' north, and it bears from Pedro Blanco S. 64° W. distant twenty-one leagues. You are to leave it on the starboard side, and you may come within half a mile of it in eighteen fathom water: and then you must steer N. by W.½W. for the channel, between the island of Cabouce and Bamboo, which are to the northward of the grand Ladrone.

After having continued at anchor all night, we, on the 9th, at four in the morning, sent our cutter to sound the channel where we proposed to pass; but before the return of the cutter, a Chinese pilot put on board the *Centurion*, and told us, in broken Portuguese, he would carry the ship to Macao for thirty dollars: these were immediately paid him, and we then weighed and made sail. Soon after several other pilots came on board, who, to recommend themselves, produced certificates from the captains of many European ships they had pilotted in, but we still continued under the management of the Chinese whom we at first engaged. By this time we learnt that we were not far distant from Macao, and that there were in the river of Canton, at the mouth of which Macao lies, eleven European ships of which four were English. Our pilot carried us between the islands of Bamboo and Cabouce; but the winds hanging in the northern board, and the tides often setting strongly against us, we were obliged to come frequently to an anchor; so that we did not get through between the two islands till the 12th of November, at two in the morning. In passing through, our depth of water was from twelve to fourteen fathom; and as we steered on N. by W.½W. between a number of other islands, our soundings underwent little or no variation till towards the evening, when they encreased to seventeen fathom, in which depth, the wind dying away, we anchored not far from the Island of Lantoon, the largest of all this range of islands. At seven in the morning we weighed again, and steering W.S.W. and S.W. by W. we at ten o'clock happily anchored in Macao road, in five fathom water, the city of Macao bearing W. by N. three leagues distant; the peak of Lantoon E. by N. and the grand Ladrone S. by E., each of them about five leagues distant. Thus, after a fatiguing cruise of above two years' continuance, we once more arrived at an amicable port and a civilized country, where the conveniencies of life were in great plenty; where the naval stores, which we now extremely wanted, could be in some degree procured; where we expected the inexpressible satisfaction of receiving letters from our relations and friends; and where our countrymen, who were lately arrived from England, would be capable of answering the numerous enquiries we were prepared to make, both about public and private occurrences, and to relate to us many particulars which, whether of importance or not, would be listened to by us with the utmost attention, after the long suspension of our correspondence with our country, to which the nature of our undertaking had hitherto subjected us.

CHAPTER VII

PROCEEDINGS AT MACAO

The city of Macao, in the road of which we came to an anchor on the 12th of November, is a Portuguese settlement, situated in an island at the entrance of the river of Canton. It was formerly very rich and populous, and capable of defending itself against the power of the adjacent Chinese governors: but at present it is much fallen from its antient splendor; for though it is inhabited by Portuguese, and hath a governor nominated by the King of Portugal, yet it subsists merely by the sufferance of the Chinese, who can starve the place and dispossess the Portuguese whenever they please. This obliges the Governor of Macao to behave with great circumspection, and carefully to avoid every circumstance that may give offence to the Chinese. The river of Canton, off the mouth of which this city lies, is the only Chinese port frequented by European ships; and is, on many accounts, a more commodious harbour than Macao: but the peculiar customs of the Chinese, solely adapted to the entertainment of trading ships, and the apprehensions of the commodore, lest he should embroil the East India Company with the Regency of Canton if he should insist on being treated upon a different footing than the merchantmen, made him resolve rather to go to Macao than to venture into the river of Canton. Indeed, had not this reason prevailed with him, he himself had nothing to fear. For it is certain that he might have entered the port of Canton, and might have continued there as long as he pleased, and afterwards have left it again, although the whole power of the Chinese empire had been brought together to oppose him.

The commodore, not to depart from his usual prudence, no sooner came to an anchor in Macao road than he dispatched an officer with his compliments to the Portuguese Governor of Macao, requesting his excellency, by the same officer, to advise him in what manner it would be proper to act to avoid offending the Chinese, which, as there were then four of our ships in their power at Canton, was a matter worthy of attention. The difficulty which the commodore principally apprehended related to the duty usually paid by ships in the river of Canton, according to their tunnage. For, as men-of-war are exempted in every foreign harbour from all manner of port charges, the commodore thought it would be derogatory to the honour of his country to submit to this duty in China: and therefore he desired the advice of the Governor of Macao, who, being an European, could not be ignorant of the privileges claimed by a British man-of-war, and consequently might be expected to give us the best lights for obviating this perplexity. Our boat returned in the evening with two officers sent by the governor, who informed the commodore that it was the governor's opinion that if the *Centurion* ventured into the river of Canton the duty would certainly be expected; and therefore, if the commodore approved of it, he would send him a pilot, who should conduct us into another safe harbour, called the Typa, which was every way commodious for careening the ship (an operation we were resolved to begin upon as soon as possible) and where, in all probability, the above-mentioned duty would never be demanded.

This proposal the commodore agreed to, and in the morning weighed anchor, under the direction of the Portuguese pilot, and steered for the intended harbour. As we entered between two islands, which form the eastern passage to it, we found our soundings decreased to three fathom and a half. However, the pilot assuring us that this was the least depth we should meet with, we continued our course, till at length the ship stuck fast in the mud, with only eighteen foot water abaft; and, the tide of ebb making, the water sewed to sixteen feet, but the ship remained perfectly upright; we then sounded all round us, and discovering that the water deepened to the northward, we carried out our small bower with two hawsers an end, and at the return of the tide of flood hove the ship afloat, and a breeze springing up at the same instant, we set the fore-top sail and, slipping the hawser, ran into the harbour, where we moored in about five fathom water. This harbour of the Typa is formed by a number of islands, and is about six miles distant from Macao. Here we saluted the castle of Macao with eleven guns, which were returned by an equal number.

The next day the commodore paid a visit in person to the governor, and was saluted at his landing by eleven guns, which were returned by the *Centurion*. Mr. Alison's business in this visit was to solicit the governor to grant us a supply both of provisions and of such naval stores as were necessary to refit the ship. The governor seemed really inclined to do us all the service he could, and assured the commodore, in a friendly manner, that he would privately give us all the assistance in his power; but he at the same time frankly owned that he dared not openly to furnish us with anything we demanded unless we first produced an order for it from the Viceroy of Canton, since he himself neither received provisions for his garrison nor any other necessaries but by permission from the Chinese Government; and as they took care only to victual him from day to day, he was indeed no other than their vassal, whom they could at all times compel to submit to their own terms by laying an embargo on his provisions.

On this declaration of the governor, Mr. Anson resolved himself to go to Canton to procure a licence from the viceroy, and he accordingly hired a Chinese boat for himself and his attendants; but just as he was ready to embark, the hoppo, or Chinese custom-house officer of Macao, refused to grant a permit to the boat, and ordered the watermen not to proceed at their peril. The commodore at first endeavoured to prevail with the hoppo to withdraw his injunction and to grant a permit; and the governor of Macao employed his interest with the hoppo to the same purpose. But the officer continuing inflexible, Mr. Anson told him the next day that if the permit was any longer refused he would man and arm the *Centurion's* boats, asking the hoppo at the same time who he imagined would dare to oppose them in their passage. This threat immediately brought about what his intreaties had endeavoured at in vain; the permit was granted, and Mr. Anson went to Canton. On his arrival there, he consulted with the supercargoes and officers of the English ships how to procure an order from the viceroy for the necessaries he wanted: but in this he had reason to suppose that the advice they gave him, though well intended, was yet not the most prudent; for as it is the custom with these gentlemen never to apply to the supreme magistrate himself, whatever difficulties they labour under, but to transact all matters relating to the government by the mediation of the principal Chinese merchants, Mr. Anson was persuaded to follow the same method upon this occasion, the English promising, in which they were doubtless sincere, to exert all their interest to engage the merchants in his favour. Indeed, when the Chinese merchants were spoke to, they readily undertook the management of this business, and promised to answer for its success; but after near a month's delay, and reiterated excuses, during which interval they pretended to be often upon the point of compleating it, they at last, when they were pressed, and measures were taken for delivering a letter to the viceroy, threw off the mask, and declared they neither had made application to the viceroy, nor could they, as he was too great a man, they said, for them to approach on any occasion: and not contented with having themselves thus grossly deceived the commodore, they now used all their persuasion with the English at Canton to prevent them from intermeddling with anything that regarded him; representing to them that it would in all probability embroil them with the government, and occasion them a great deal of unnecessary trouble; which groundless insinuations had unluckily but too much weight with those they were intended to influence.

It may be difficult to assign a reason for this perfidious conduct of the Chinese merchants. Interest indeed is known to exert a boundless influence over the inhabitants of that empire; but how their interest could be affected in the present case is not easy to discover, unless they apprehended that the presence of a ship of force might damp their Manila trade, and therefore acted in this manner with a view of forcing the commodore to Batavia: though it might be as natural in this light to suppose that they would have been eager to have got him dispatched. I therefore rather impute their behaviour to the unparalleled pusillanimity of the nation, and to the awe they are under of the government, since such a ship as the *Centurion*, fitted for war only, having never been seen in those parts before, she was the horror of these dastards, and the merchants were in some degree terrified even with the idea of her, and could not think of applying to the viceroy, who is doubtless fond of all opportunities of fleecing them, without representing to themselves the occasion which a hungry and tyrannical magistrate might possibly find for censuring their intermeddling with so unusual a transaction, in which he might pretend the interest of the state was immediately concerned. However, be this as it may, the commodore was satisfied that nothing was to be done by the interposition of the merchants, as it was on his pressing them to deliver a letter to the viceroy that they had declared they durst not interfere in the affair, and had confessed that, notwithstanding all their pretences of serving him, they had not yet taken one step towards it. Mr. Anson therefore told them that he would proceed to Batavia and refit his ship there, but informed them at the same time that this was impossible to be done unless he was supplied with a stock of provisions sufficient for his passage. The merchants, on this, undertook to procure him provisions, though they assured him that it was what they durst not engage in openly, but they proposed to manage it in a clandestine manner by putting a quantity of bread, flour, and other provisions on board the English ships, which were now ready to sail, and these were to stop at the mouth of the Typa, where the *Centurion's* boats were to receive it. This article, which the merchants represented as a matter of great favour, being settled, the commodore, on the 16th of December, came back from Canton to the ship, seemingly resolved to proceed to Batavia to refit as soon as he should get his supplies of provisions on board.

But Mr. Anson (who never intended going to Batavia) found on his return to the *Centurion* that her main-mast was sprung in two places and that the leak was considerably increased; so that, upon the whole, he was fully satisfied that though he should lay in a sufficient stock of provisions, yet it would be impossible for him to put to sea without refitting. Since, if he left the port with his ship in her present condition she would be in the utmost danger of foundring; and therefore, notwithstanding the difficulties he had met with, he resolved at all events to have her hove down before he departed from Macao. He was fully convinced, by what he had observed at Canton, that his great caution not to injure the East India Company's affairs, and the regard he had shown to the advice of their officers, had occasioned all his perplexity. For he now saw clearly that if he had at first carried his ship into the river of Canton, and had immediately addressed himself to the mandarines, who are the chief officers of state, instead of employing the merchants to apply on his behalf, he would, in all probability, have had all his requests granted and would have been soon dispatched. He had already lost a month by the wrong measures he had pursued, but he resolved to lose as little more time as possible; and therefore, the 17th of December, being the next day after his return from Canton, he wrote a letter to the viceroy of that place acquainting him that he was commander-in-chief of a squadron of his Britannick Majesty's ships of war, which had been cruising for two years past in the South Seas against the Spaniards, who were at enmity with the king his master; that on his way back to England he had put into the port of Macao, having a considerable leak in his ship and being in great want of provisions, so that it was impossible for him to proceed on his voyage till his ship was repaired and he was supplied with the necessaries he wanted; that he had been at Canton in hopes of being admitted to a personal audience of his excellency; but being a stranger to the customs of the country, he had not been able to inform himself what steps were necessary to be taken to procure such an audience, and therefore was obliged to apply in this manner, to desire his excellency to give orders for his being permitted to employ carpenters and proper workmen to refit his ship, and to furnish himself with provisions and stores, that he might be enabled to pursue his voyage to Great Britain. Hoping, at the same time, that these orders would be issued with as little delay as possible lest it might occasion his loss of the season, and he might be prevented from departing till the next winter.

This letter was translated into the Chinese language, and the commodore delivered it himself to the hoppo or chief officer of the emperor's customs at Macao, desiring him to forward it to the Viceroy of Canton with as much expedition as he could. The officer at first seemed unwilling to take charge of it, and raised many difficulties about it; so that Mr. Anson suspected him of being in league with the merchants of Canton, who had always shewn a great apprehension of the commodore's having any immediate intercourse with the viceroy or mandarines; and therefore the commodore, not without some resentment, took back his letter from the hoppo and told him he would immediately send it to Canton in his own boat, and would give his officer positive orders not to return without an answer from the viceroy. The hoppo perceiving the commodore to be in earnest, and fearing to be called to an account for his refusal, begged to be entrusted with the letter, and promised to deliver it, and to procure an answer as soon as possible. And now it was presently seen how justly Mr. Anson had at last judged of the proper manner of dealing with the Chinese; for this letter was written but the 17th of December, as hath been already observed; and on the 19th in the morning a mandarine of the first rank, who was governor of the city of Janson, together with two mandarines of an inferior class and a considerable retinue of officers and servants, having with them eighteen half gallies furnished with music, and decorated with a great number of streamers, and full of men, came to grapnel ahead of the *Centurion*; whence the mandarine sent a message to the commodore, telling him that he (the mandarine) was ordered by the Viceroy of Canton to examine the condition of the ship; therefore desiring the ship's boat might be sent to fetch him on board. The *Centurion's* boat was immediately dispatched, and preparations were made for receiving him; in particular a hundred of the most sightly of the crew were uniformly dressed in the regimentals of the marines, and were drawn up under arms on the main-deck against his arrival. When he entered the ship he was saluted by the drums and what other military music there was on board, and passing by the new-formed guard, he was met by the commodore on the quarter-deck, who conducted him to the great cabin. Here the mandarine explained his commission, declaring that he was directed to examine all the articles mentioned in the commodore's letter to the viceroy, and to confront them with the representation that had been given of them: that he was in the first place instructed to inspect the leak, and had for that purpose brought with him two Chinese carpenters; and that for the more regular dispatch of his business he had every head of enquiry separately wrote down on a sheet of paper, with a void space opposite to it, where he was to insert such information and remarks thereon as he could procure by his own observation.

This mandarine appeared to be a person of very considerable parts, and endowed with more frankness and honesty than is to be found in the generality of the Chinese. After the necessary inspections had been made, particularly about the leak, which the Chinese carpenters reported to be to the full as dangerous as it had been described, and consequently that it was impossible for the *Centurion* to proceed to sea without being refitted, the mandarine expressed himself satisfied with the account given in the commodore's letter. And this magistrate, as he was more intelligent than any other person of his nation that came to our knowledge, so likewise was he more curious and inquisitive, viewing each part of the ship with extraordinary attention, and appearing greatly surprized at the largeness of the lower deck guns and at the weight and size of the shot. The commodore, observing his astonishment, thought this a proper opportunity to convince the Chinese of the prudence of granting him all his demands in the most speedy and ample manner: he therefore told the mandarine and those who were with him that besides the request he made for a general licence to furnish himself with whatever his present situation required, he had a particular complaint to prefer against the proceedings of the custom-house of Macao; that at his first arrival the Chinese boats had brought on board him plenty of greens and variety of fresh provisions for daily use: that though they had always been paid to their full satisfaction, yet the custom-house officers at Macao had soon forbid them; by which means he was deprived of those refreshments which were of the utmost consequence to the health of his men after their long and sickly voyage; that as they, the mandarines, had informed themselves of his wants and were eye-witnesses of the force and strength of his ship, they might be satisfied it was not because he had no power to supply himself that he desired the permission of the government to purchase what provisions he stood in need of, since he presumed they were convinced that the *Centurion* alone was capable of destroying the whole navigation of the port of Canton, or of any other port in China, without running the least risque from all the force the Chinese could collect; that it was true this was not the manner of proceeding between nations in friendship with each other; but it was likewise true that it was not customary for any nation to permit the ships of their friends to starve and sink in their ports, when those friends had money to purchase necessaries, and only desired liberty to lay it out; that they must confess he and his people had hitherto behaved with great modesty and reserve; but that, as his distresses were each day increasing, famine would at last prove too strong for any restraint, and necessity was acknowledged in all countries to be superior to every other law; and therefore it could not be expected that his crew would long continue to starve in the midst of that plenty to which their eyes were every day witnesses. To this the commodore added (though perhaps with a less serious air) that if, by the delay of supplying him with provisions, his men should, from the impulses of hunger, be obliged to turn cannibals, and to prey upon their own species, it was easy to be foreseen that, independent of their friendship to their comrades, they would in point of luxury prefer the plump well-fed Chinese to their own emaciated ship-mates. The first mandarine acquiesced in the justness of this reasoning, and told the commodore that he should that night proceed for Canton; that on his arrival a council of mandarines would be summoned, of which he was a member, and that, by being employed in the present commission, he was of course the commodore's advocate; that as he was himself fully convinced of the urgency of Mr. Anson's necessity, he did not doubt but on the representation he should make of what he had seen, the council would be of the same opinion, and that all which was demanded would be amply and speedily granted; that with regard to the commodore's complaint of the custom-house of Macao, this he would undertake to rectify immediately by his own authority. And then desiring a list to be given him of the quantity of provision necessary for the expence of the ship during one day, he wrote a permit under it, and delivered it to one of his attendants, directing him to see that quantity sent on board early every morning; which order from that time forwards was punctually complied with.

When this weighty affair was thus in some degree regulated, the commodore invited him and his two attendant mandarines to dinner, telling them at the same time that if his provision, either in kind or quantity, was not what they might expect, they must thank themselves for having confined him to so hard an allowance. One of his dishes was beef, which the Chinese all dislike, tho' Mr. Anson was not apprized of it. This seems to be derived from the Indian superstition, which for some ages past has made a great progress in China. However, his guests did not entirely fast, for the three mandarines completely finished the white part of four large fowls. They were indeed extremely embarrassed with their knives and forks, and were quite incapable of making use of them: so that after some fruitless attempts to help themselves, which were sufficiently aukward, one of the attendants was obliged to cut their meat in small pieces for them. But whatever difficulty they might have in complying with the European manner of eating, they seemed not to be novices at drinking. In this part of the entertainment the commodore excused himself under the pretence of illness; but there being another gentleman present, of a florid and jovial complexion, the chief mandarine clapped him on the shoulder and told him by the interpreter that certainly he could not plead sickness, and therefore insisted on his bearing him company; and that gentleman perceiving that after they had dispatched four or five bottles of Frontiniac the mandarine still continued unruffled, he ordered a bottle of citron-water to be brought up, which the Chinese seemed much to relish; and this being near finished, they arose from table in appearance cool and uninfluenced by what they had drank; and the commodore having, according to custom, made the mandarine a present, they all departed in the same vessels that brought them.

After their departure the commodore with great impatience expected the resolution of the council, and the proper licences to enable him to refit the ship. For it must be observed, as hath already appeared from the preceding narration, that the Chinese were forbid to have any dealings with him, so that he could neither purchase stores nor necessaries, nor did any kind of workmen dare to engage themselves in his service until the permission of the government was first obtained. And in the execution of these particular injunctions the magistrates never fail of exercising great severity, since, notwithstanding the fustian elogiums bestowed upon them by the Romish missionaries residing in the East, and their European copiers, they are composed of the same fragil materials with the rest of mankind, and often make use of the authority of the law, not to suppress crimes, but to enrich themselves by the pillage of those who commit them. This is the more easily effected in China, because capital punishments are rare in that country, the effeminate genius of the nation, and their strong attachment to lucre, disposing them rather to make use of fines. And as from these there arises no inconsiderable profit to those who compose their tribunals, it is obvious enough that prohibitions of all kinds, particularly such as the alluring prospect of great profit may often tempt the subject to infringe, cannot but be favourite institutions in such a government.

A short time before this, Captain Saunders took his passage to England on board a Swedish ship, and was charged with dispatches from the commodore; and in the month of December, Captain Mitchel, Colonel Cracherode, and Mr. Taswel, one of the agent victuallers, with his nephew

Mr. Charles Herriot, embarked on board some of our company's ships; and I, having obtained the commodore's leave to return home, embarked with them. I must observe, too, having omitted it before, that whilst we lay at Macao, we were informed by the officers of our Indiamen that the *Severn* and *Pearl*, the two ships of our squadron which had separated from us off Cape Noir, were safely arrived at Rio Janeiro on the coast of Brazil. I have formerly taken notice that at the time of their separation we suspected them to be lost: and there were many reasons which greatly favoured this suspicion, for we knew that the *Severn* in particular was extremely sickly; which was the more obvious to the rest of the ships, as in the preceding part of the voyage her commander, Captain Legg, had been remarkable for his exemplary punctuality in keeping his station, and yet, during the last ten days before his separation, his crew was so diminished and enfeebled, that with his utmost efforts he could not possibly maintain his proper position with his wonted exactness. The extraordinary sickness on board him was by many imputed to the ship, which was new, and on that account was believed to be the more unhealthy; but whatever was the cause of it, the *Severn* was by much the most sickly of the squadron, since before her departure from St. Catherine's she buryed more men than any of them, insomuch that the commodore was obliged to recruit her with a number of fresh hands; and, the mortality still continuing on board her, she was supplied with men a second time at sea, after our setting sail from St. Julians; yet, notwithstanding these different reinforcements, she was at last reduced to the distressed condition I have already mentioned. Hence the commodore himself firmly believed she was lost, and therefore it was with great joy we received the news of her and the *Pearl's* safety, after the strong persuasion, which had so long prevailed amongst us, of their having both perished. But to proceed with the transactions between Mr. Anson and the Chinese.

Notwithstanding the favourable disposition of the mandarine Governor of Janson at his leaving Mr. Anson, several days were elapsed before there was any advice from him; and Mr. Anson was privately informed there were great debates in council upon his affair, partly perhaps owing to its being so unusual a case, and in part to the influence, as I suppose, of the intrigues of the French at Canton: for they had a countryman and fast friend residing on the spot who spoke the language well, and was not unacquainted with the venality of the government, nor with the persons of several of the magistrates, and consequently could not be at a loss for means of traversing the assistance desired by Mr. Anson. Indeed this opposition of the French was not merely the effect of national prejudice, or a contrariety of political interests; but was in good measure owing to vanity, a motive of much more weight with the generality of mankind than any attachment to the public service of the community. For the French pretending their Indiamen to be men-of-war, their officers were apprehensive that any distinction granted to

Mr. Anson on account of his bearing the king's commission would render them less considerable in the eyes of the Chinese, and would establish a prepossession at Canton in favour of ships of war, by which they, as trading vessels, would suffer in their importance. And I wish the affectation of endeavouring to pass for men-of-war, and the fear of sinking in the estimation of the Chinese, if the *Centurion* was treated in a different manner from themselves, had been confined to the officers of the French ships only. However, notwithstanding all these obstacles, it should seem that the representation of the commodore to the mandarines, of the facility with which he could right himself if justice were denied him, had at last its effect, since, on the 6th of January, in the morning, the Governor of Janson, the commodore's advocate, sent down the Viceroy of Canton's warrant for the refitment of the *Centurion*, and for supplying her people with all they wanted. Having now the necessary licences, a number of Chinese smiths and carpenters went on board the next day to treat about the work they were to do, all which they proposed to undertake by the great. They demanded at first to the amount of a thousand pounds sterling for the repairs of the ship, the boats, and the masts. This the commodore seemed to think an unreasonable sum, and endeavoured to persuade them to work by the day; but that was a method they would not hearken to; so it was at last agreed that the carpenters should have to the amount of about six hundred pounds for their work, and that the smiths should be paid for their iron-work by weight, allowing them at the rate of three pounds a hundred nearly for the small work, and forty-six shillings for the large.

This being regulated, the commodore next exerted himself to get the most important business of the whole compleated; I mean the heaving down the *Centurion* and examining the state of her bottom. The first lieutenant therefore was dispatched to Canton to hire two country vessels, called in their language junks, one of them being intended to heave down by, and the other to serve as a magazine for the powder and ammunition: whilst at the same time the ground was smoothed on one of the neighbouring islands, and a large tent was pitched for lodging the lumber and provisions, and near a hundred Chinese caulkers were soon set to work on the decks and sides of the ship. But all these preparations, and the getting ready the careening gear, took up a great deal of time, for the Chinese caulkers, though they worked very well, were far from being expeditious. Besides, it was the 26th of January before the junks arrived, and the necessary materials, which were to be purchased at Canton, came down very slowly, partly from the distance of the place, and partly from the delays and backwardness of the Chinese merchants. And in this interval Mr. Anson had the additional perplexity to discover that his fore-mast was broken asunder above the upper-deck partners, and was only kept together by the fishes which had been formerly clapt upon it.

However, the *Centurion's* people made the most of their time, and exerted themselves the best they could; and as by clearing the ship the carpenters were enabled to come at the leak, they took care to secure that effectually whilst the other preparations were going forwards. The leak was found to be below the fifteen-foot mark, and was principally occasioned by one of the bolts being wore away and loose in the joining of the stern, where it was scarfed.

At last, all things being prepared, they, on the 22d of February, in the morning, hove out the first course of the *Centurion's* starboard-side, and had the satisfaction to find that her bottom appeared sound and good; and the next day (having by that time compleated the new sheathing of the first course) they righted her again, to set up anew the careening gear, which had stretched much. Thus they continued heaving down and often righting the ship, from a suspicion of their careening tackle, till the 3d of March, when, having compleated the paying and sheathing the bottom, which proved to be everywhere very sound, they for the last time righted the ship, to their great joy, since not only the fatigue of careening had been considerable, but they had been apprehensive of being attacked by the Spaniards whilst the ship was thus incapacitated for defence. Nor were their fears altogether groundless, for they learnt afterwards, by a Portuguese vessel, that the Spaniards at Manila had been informed that the *Centurion* was in the Typa, and intended to careen there, and that thereupon the governor had summoned his council, and had proposed to them to endeavour to burn her whilst she was careening, which was an enterprize which, if properly conducted, might have put them in great danger. It was farther reported that this scheme was not only proposed, but resolved on, and that a captain of a vessel had actually undertaken to perform the business for forty thousand dollars, which he was not to receive unless he succeeded; but the governor pretending that there was no treasure in the royal chest, and insisting that the merchants should advance the money, and they refusing to comply with the demand, the affair was dropped. Perhaps the merchants suspected that the whole was only a pretext to get forty thousand dollars from them, and indeed this was affirmed by some who bore the governor no good-will, but with what truth it is difficult to ascertain.

As soon as the *Centurion* was righted, they took on board her powder and gunners' stores, and proceeded with getting in their guns as fast as possible, and then used their utmost expedition in repairing the fore-mast, and in compleating the other articles of her refitment. But whilst they were thus employed, they were alarmed on the 10th of March by a Chinese fisherman, who brought them intelligence that he had been on board a large Spanish ship off the Grand Ladrone, and that there were two more in company with her. He added several particulars to his relation, as that he had brought one of their officers to Macao, and that, on this, boats went off early in the morning from Macao to them: and the better to establish the belief of his veracity, he said he desired no money if his information should not prove true. This was presently believed to be the fore-mentioned expedition from Manila, and the commodore immediately fitted his cannon and small arms in the best manner he could for defence; and having then his pinnace and cutter in the offing, who had been ordered to examine a Portuguese vessel which was getting under sail, he sent them the advice he had received, and directed them to look out strictly. Indeed, no Spanish ships ever appeared, and they were soon satisfied the whole of the story was a fiction, though it was difficult to conceive what reason could induce the fellow to be at such extraordinary pains to impose on them.

It was the beginning of April when they had new rigged the ship, stowed their provisions and water on board, and had fitted her for the sea; and before this time the Chinese grew very uneasy, and extremely desirous that she should be gone, either not knowing, or pretending not to believe, that this was a point the commodore was as eagerly set on as they could be. At length, about the 3d of April, two mandarine boats came on board from Macao to press him to leave their port, and this having been often urged before, though there had been no pretence to suspect Mr. Anson of any affected delays, he at this last message answered them in a determined tone, desiring them to give him no further trouble, for he would go when he thought proper, and not sooner. After this rebuke the Chinese (though it was not in their power to compel him to depart) immediately prohibited all provisions from being carried on board him, and took such care their injunctions should be complied with, that from thence forwards nothing could be purchased at any rate whatever.

The 6th of April the *Centurion* weighed from the Typa, and warped to the southward; and by the 15th she was got into Macao road, compleating her water as she past along, so that there remained now very few articles more to attend to, and her whole business being finished by the 19th, she, at three in the afternoon of that day, weighed and made sail, and stood to sea.

CHAPTER VIII

FROM MACAO TO CAPE ESPIRITU SANTO—THE TAKING OF THE
MANILA GALEON, AND RETURNING BACK AGAIN

The commodore was now got to sea, with his ship well refitted, his stores replenished, and an additional stock of provisions on board. His crew too was somewhat reinforced, for he had entered twenty-three men during his stay at Macao, the greatest part of them Lascars or Indian sailors, and the rest Dutch. He gave out at Macao that he was bound to Batavia, and thence to England, and though the westerly monsoon was now set in, when that passage is considered as impracticable, yet, by the confidence he had expressed in the strength of his ship, and the dexterity of his hands, he had persuaded not only his own crew, but the people at Macao likewise, that he proposed to try this unusual experiment, so that there were many letters sent on board him by the inhabitants of Canton and Macao for their friends at Batavia.

But his real design was of a very different nature: for he supposed that instead of one annual ship from Acapulco to Manila, there would be this year, in all probability, two, since, by being before Acapulco, he had prevented one of them from putting to sea the preceding season. He therefore, not discouraged by his former disasters, resolved again to risque the casualties of the Pacific Ocean, and to cruise for these returning vessels off Cape Espiritu Santo on the island of Samal, which is the first land they always make at the Philippine Islands: and as June is generally the month in which they arrive there, he doubted not but he should get to his intended station time enough to intercept them. It is true they were said to be stout vessels, mounting forty-four guns apiece, and carrying above five hundred hands, and might be expected to return in company, and he himself had but two hundred and twenty-seven hands on board, of which near thirty were boys. But this disproportion of strength did not deter him, as he knew his ship to be much better fitted for a sea engagement than theirs, and as he had reason to expect that his men would exert themselves after a most extraordinary manner when they had in view the immense wealth of these Manila galeons.

This project the commodore had resolved on in his own thoughts ever since his leaving the coast of Mexico, and the greatest mortification which he received from the various delays he had met with in China, was his apprehension lest he might be thereby so long retarded as to let the galeons escape him. Indeed, at Macao it was incumbent on him to keep these views extremely secret, for there being a great intercourse and a mutual connection of interests between that port and Manila, he had reason to fear that if his designs were discovered, intelligence would be immediately sent to Manila, and measures would be taken to prevent the galeons from falling into his hands. But being now at sea, and entirely clear of the coast, he summoned all his people on the quarter-deck and informed them of his resolution to cruise for the two Manila ships, of whose wealth they were not ignorant. He told them he should chuse a station where he could not fail of meeting with them, and though they were stout ships, and full manned, yet, if his own people behaved with their accustomed spirit, he was certain he should prove too hard for them both, and that one of them at least could not fail of becoming his prize. He further added that many ridiculous tales had been propagated about the strength of the sides of these ships, and their being impenetrable to cannon shot; that these fictions had been principally invented to palliate the cowardice of those who had formerly engaged them, but he hoped there were none of those present weak enough to give credit to so absurd a story. For his own part, he did assure them upon his word that, whenever he fell in with them, he would fight them so near that they should find his bullets, instead of being stopped by one of their sides, should go through them both.

This speech of the commodore was received by his people with great joy, since no sooner had he ended than they expressed their approbation, according to naval custom, by three strenuous cheers, and declared their determination to succeed, or perish, whenever the opportunity presented itself. Immediately too their hopes, which on their departure from the coast of Mexico had entirely subsided, were again revived, and they persuaded themselves that notwithstanding the various casualties and disappointments they had hitherto met with, they should yet be repaid the price of their fatigues, and should at last return home enriched with the spoils of the enemy. For, firmly relying on the assurances of the commodore, that they should certainly meet with the galeons, they were all of them too sanguine to doubt a moment of mastering them, so that they considered themselves as having them already in their possession. And this confidence was so universally spread through the whole ship's company, that the commodore, who had taken some Chinese sheep to sea with him for his own provision, enquiring one day of his butcher why he had lately seen no mutton at his table, and asking him if all the sheep were killed, the fellow very seriously replied that there were indeed two sheep left, but that if his honour would give him leave, he proposed to keep those for the entertainment of the general of the galeons.

When the *Centurion* left the port of Macao, she stood for some days to the westward, and, on the 1st of May, they saw part of the island of Formosa; and steering thence to the southward, they, on the 4th of May, were in the latitude of the Bashee Islands, as laid down by Dampier; but they suspected his account of inaccuracy, as they knew that he had been considerably mistaken in the latitude of the south end of Formosa, and therefore they kept a good look-out, and about seven in the evening discovered from the masthead five small islands, which were judged to be the Bashees. As they afterwards saw Botel Tobago Xima, they by this means found an opportunity of correcting the position of the Bashee Islands, which had been hitherto laid down twenty-five leagues too far to the westward: for by their observations they esteemed the middle of these islands to be in 21° 4' north, and to bear from Botel Tobago Xima S.S.E. twenty leagues distant, that island itself being in 21° 57' north.

After getting a sight of the Bashee Islands, they stood between the S. and S.W. for Cape Espiritu Santo, and the 20th of May at noon they first discovered that cape, which about four o'clock they brought to bear S.S.W. near eleven leagues distant. It appeared to be of a moderate height, with several round hummocks on it. As it was known that there were centinels placed upon this cape to make signals to the Acapulco ship when she first falls in with the land, the commodore immediately tacked, and ordered the top-gallant sails to be taken in, to prevent being discovered. And this being the station where it was resolved to cruise for the galeons, they kept the cape between the south and the west, and endeavoured to confine themselves between the latitude of 12° 50' and 13° 5', the cape itself lying, by their observations, in 12° 40' north, and in 4° of east longitude from Botel Tobago Xima.

It was the last of May, by the foreign stile, when they arrived off this cape, and the month of June, by the same stile, being that in which the Manila ships are usually expected, the *Centurion's* people were now waiting each hour with the utmost impatience for the happy crisis which was to balance the account of all their past calamities. As from this time there was but small employment for the crew, the commodore ordered them almost every day to be exercised in the working of the great guns, and in the use of their small arms. This had been his practice, more or less, at every convenient season during the whole course of his voyage, and the advantages which he received from it in his engagement with the galeon were an ample recompence for all his care and attention. Indeed, it should seem that there are few particulars of a commander's duty of more importance, how much soever it may have been sometimes overlooked or misunderstood: since it will, I suppose, be confessed that in two ships of war equal in the number of their men and guns, the disproportion of strength arising from a greater or less dexterity in the use of their great guns and small arms is what can scarcely be ballanced by any other circumstances whatever. For, as these are the weapons with which they are to engage, what greater inequality can there be betwixt two contending parties than that one side should perfectly understand the management of them, and should have the skill to employ them in the most effectual manner for the annoyance of their enemy; while the other side should, by their awkward handling of their arms, render them rather terrible to themselves than mischievous to their antagonist? This seems so obvious and natural a conclusion, that a person unacquainted with these matters would suppose the first care of a commander to be the training his people to the ready use of their arms.

But human affairs are not always conducted by the plain dictates of common sense. There are many other principles which influence our transactions, and there is one in particular, which tho' of a very erroneous complexion, is scarcely ever excluded from our most serious deliberations; I mean custom, or the practice of those who have preceded us. This is usually a power too mighty for reason to grapple with, and is often extremely troublesome to those who oppose it, since it has much of superstition in its nature, and pursues all those who question its authority with unrelenting vehemence. However, in these latter ages of the world, some lucky encroachments have been made upon its prerogative, and it may surely be expected that the gentlemen of the navy, whose particular profession hath within a few years been considerably improved by a number of new inventions, will of all others be the readiest to give up any usage which has nothing to plead in its behalf but prescription, and will not suppose that every branch of their business hath already received all the perfection of which it is capable. Indeed, it must be owned that if a dexterity in the use of small arms, for instance, hath been sometimes less attended to on board our ships of war than might have been wished for, it hath been rather owing to unskilful methods of teaching it than to negligence: since the common sailors, how strongly soever attached to their own prejudices, are very quick-sighted in finding out the defects of others, and have ever shewn a great contempt for the formalities practised in the training of land troops to the use of their arms. But when those who have undertaken to instruct the seamen have contented themselves with inculcating only what was useful, in the simplest manner, they have constantly found their people sufficiently docile, and the success hath even exceeded their expectation. Thus on board Mr. Anson's ship, where they were taught no more of the manual exercise than the shortest method of loading with cartridges, and were constantly trained to fire at a mark, which was usually hung at the yard-arm, and where some little reward was given to the most expert, the whole crew, by this management, were rendered extremely skilful, for besides an uncommon readiness in loading, they were all of them good marksmen, and some of them most extraordinary ones. Whence I doubt not but, in the use of small arms, they were more than a match for double their number who had not been habituated to the same kind of exercise. But to return.

It was the last of May, N.S., as hath been already said, when the *Centurion* arrived off Cape Espiritu Santo, and consequently the next day the month began in which the galeons were to be expected. The commodore therefore made all necessary preparations for receiving them, hoisting out his long-boat and lashing her alongside, that the ship might be ready for engaging if they fell in with the galeons during the night. All this time too he was very solicitous to keep at such a distance from the cape as not to be discovered. But it hath been since learnt, that notwithstanding his care, he was seen from the land, and advice of him was sent to Manila, where, tho' it was at first disbelieved, yet, on reiterated intelligence (for it seems he was seen more than once), the merchants were alarmed, and the governor was applied to, who undertook (the commerce supplying the necessary sums) to fit out a force consisting of two ships of thirty-two guns, one of twenty guns, and two sloops of ten guns each, to attack the *Centurion* on her station. With this view some of these vessels actually weighed, but the principal ship not being ready, and the monsoon being against them, the commerce and the governor disagreed, so that the enterprize was laid aside. This frequent discovery of the *Centurion* from the shore was somewhat extraordinary, since the pitch of the cape is not high, and she usually kept from ten to fifteen leagues distant, though once indeed, by an indraught of the tide, as was supposed, they found themselves in the morning within seven leagues of the land.

As the month of June advanced, the expectancy and impatience of the commodore's people each day increased. And I think no better idea can be given of their great eagerness on this occasion than by copying a few paragraphs from the journal of an officer who was then on board, as it will, I presume, be a more natural picture of the full attachment of their thoughts to the business of their cruise than can be given by any other means. The paragraphs I have selected, as they occur in order of time, are as follow:—

"May 31. Exercising our men at their quarters, in great expectation of meeting with the galeons very soon, this being the eleventh of June, their stile."

"June 3. Keeping in our stations, and looking out for the galeons."

"June 5. Begin now to be in great expectation, this being the middle of June, their stile."

"June 11. Begin to grow impatient at not seeing the galeons."

"June 13. The wind having blown fresh easterly for the forty-eight hours past, gives us great expectations of seeing the galeons soon."

"June 15. Cruising on and off, and looking out strictly."

"June 19. This being the last day of June, N.S., the galeons, if they arrive at all, must appear soon."

From these samples it is sufficiently evident how completely the treasure of the galeons had engrossed their imagination, and how anxiously they passed the latter part of their cruise, when the certainty of the arrival of those vessels was dwindled down to probability only, and that probability became each hour more and more doubtful. However, on the 20th of June, O.S., being just a month after their gaining their station, they were relieved out of this state of uncertainty, for at sunrise they discovered a sail from the mast-head, in the S.E. quarter. On this, a general joy spread through the whole ship, for they had no doubt but this was one of the galeons, and they expected soon to descry the other. The commodore instantly stood towards her, and at half an hour after seven they were near enough to see her from the *Centurion's* deck, at which time the galeon fired a gun, and took in her top-gallant sails. This was supposed to be a signal to her consort to hasten her up, and therefore the *Centurion* fired a gun to leeward to amuse her. The commodore was surprized to find that during all this interval the galeon did not change her course, but continued to bear down upon him; for he hardly believed, what afterwards appeared to be the case, that she knew his ship to be the *Centurion*, and resolved to fight him.

About noon the commodore was little more than a league distant from the galeon, and could fetch her wake, so that she could not now escape; and, no second ship appearing, it was concluded that she had been separated from her consort. Soon after, the galeon haled up her fore-sail and brought to under top-sails, with her head to the northward, hoisting Spanish colours, and having the standard of Spain flying at the top-gallant mast-head. Mr. Anson in the meantime had prepared all things for an engagement on board the *Centurion*, and had taken every possible measure, both for the most effectual exertion of his small strength, and for the avoiding the confusion and tumult too frequent in actions of this kind. He picked out about thirty of his choicest hands and best marksmen, whom he distributed into his tops, and who fully answered his expectation by the signal services they performed. As he had not hands enough remaining to quarter a sufficient number to each great gun in the customary manner, he therefore, on his lower tire, fixed only two men to each gun, who were to be solely employed in loading it, whilst the rest of his people were divided into different gangs of ten or twelve men each, who were continually moving about the decks to run out and fire such guns as were loaded. By this management he was enabled to make use of all his guns, and instead of whole broadsides, with intervals between them, he kept up a constant fire without intermission, whence he doubted not to procure very signal advantages. For it is common with the Spaniards to fall down upon the decks when they see a broadside preparing, and to continue in that posture till it is given, after which they rise again, and, presuming the danger to be for some time over, work their guns, and fire with great briskness, till another broadside is ready. But the firing gun by gun, in the manner directed by the commodore, rendered this practice of theirs impossible.

The *Centurion* being thus prepared, and nearing the galeon apace, there happened, a little after noon, several squalls of wind and rain, which often obscured the galeon from their sight; but whenever it cleared up they observed her resolutely lying to. Towards one o'clock, the *Centurion* hoisted her broad pendant and colours, she being then within gun-shot of the enemy, and the commodore perceiving the Spaniards to have neglected clearing their ship till that time, as he saw them throwing overboard cattle and lumber, he gave orders to fire upon them with the chace guns, to disturb them in their work, and prevent them from compleating it, though his general directions had been not to engage before they were within pistol-shot. The galeon returned the fire with two of her stern chace; and the *Centurion* getting her sprit-sail yard fore and aft, that, if necessary, she might be ready for boarding, the Spaniards, in a bravado, rigged their sprit-sail yard fore and aft likewise. Soon after, the *Centurion* came abreast of the enemy within pistol-shot, keeping to the leeward of them, with a view of preventing their putting before the wind and gaining the port of Jalapay, from which they were about seven leagues distant. And now the engagement began in earnest, and for the first half-hour Mr. Anson over-reached the galeon and lay on her bow, where, by the great wideness of his ports, he could traverse almost all his guns upon the enemy, whilst the galeon could only bring a part of hers to bear. Immediately on the commencement of the action, the mats with which the galeon had stuffed her netting took fire and burnt violently, blazing up half as high as the mizen-top. This accident, supposed to be caused by the *Centurion's* wads, threw the enemy into the utmost terror, and also alarmed the commodore, for he feared lest the galeon should be burnt, and lest he himself too might suffer by her driving on board him. However, the Spaniards at last freed themselves from the fire, by cutting away the netting and tumbling the whole mass, which was in flames, into the sea. All this interval the *Centurion* kept her first advantageous position, firing her cannon with great regularity and briskness, whilst at the same time the galeon's decks lay open to her top-men, who having at their first volley driven the Spaniards from their tops, made prodigious havock with their small arms, killing or wounding every officer but one that appeared on the quarter-deck, and wounding in particular the general of the galeon himself. Thus the action proceeded for at least half an hour; but then the *Centurion* lost the superiority arising from her original situation, and was close alongside the galeon, and the enemy continued to fire briskly for near an hour longer; yet even in this posture the commodore's grape-shot swept their decks so effectually, and the number of their slain and wounded became so considerable, that they began to fall into great disorder, especially as the general, who was the life of the action, was no longer capable of exerting himself. Their confusion was visible from on board the commodore, for the ships were so near that some of the Spanish officers were seen running about with much assiduity, to prevent the desertion of their men from their quarters. But all their endeavours were in vain, for after having, as a last effort, fired five or six guns with more judgment than usual, they yielded up the contest, and, the galeon's colours being singed off the ensign staff in the beginning of the engagement, she struck the standard at her main top-gallant mast-head; the person who was employed to perform this office having been in imminent peril of being killed, had not the commodore, who perceived what he was about, given express orders to his people to desist from firing.

Thus was the *Centurion* possessed of this rich prize, amounting in value to near a million and a half of dollars. She was called the *Nostra Signora de Cabadonga*, and was commanded by General Don Jeronimo de Mentero, a Portuguese, who was the most approved officer for skill and courage of any employed in that service. The galeon was much larger than the *Centurion*, and had five hundred and fifty men, and thirty-six guns mounted for action, besides twenty-eight pedreroes in her gunwale, quarters, and tops, each of which carried a four-pound ball. She was very well furnished with small arms, and was particularly provided against boarding, both by her close quarters, and by a strong network of two-inch rope which was laced over her waist, and was defended by half-pikes. She had sixty-seven killed in the action, and eighty-four wounded, whilst the *Centurion* had only two killed, and a lieutenant and sixteen wounded, all of whom but one recovered: of so little consequence are the most destructive arms in untutored and unpractised hands.

The treasure thus taken by the *Centurion* having been, for at least eighteen months, the great object of their hopes, it is impossible to describe the transport on board when, after all their reiterated disappointments, they at last saw their wishes accomplished. But their joy was near being suddenly damped by a most tremendous incident, for no sooner had the galeon struck, than one of the lieutenants coming to Mr. Anson to congratulate him on his prize, whispered him at the same time that the *Centurion* was dangerously on fire near the powder-room. The commodore received this dreadful news without any apparent emotion, and taking care not to alarm his people, gave the necessary orders for extinguishing the fire, which was happily done in a short time, though its appearance at first was extremely terrible. It seems some cartridges had been blown up by accident between decks, and the blast had communicated its flame to a quantity of oakum in the after hatchway, near the after powder-room, where the great smother and smoke of the oakum occasioned the apprehension of a more extended and mischievous conflagration. All hopes too of avoiding its fury by escaping on board the prize had instantly vanished, for at the same moment the galeon fell on board the *Centurion* on the starboard quarter, though she was fortunately cleared without doing or receiving any considerable damage.

The commodore appointed the Manila vessel to be a post ship in his Majesty's service, and gave the command of her to Mr. Saumarez, his first lieutenant, who before night sent on board the *Centurion* all the Spanish prisoners, except such as were thought the most proper to be retained to assist in navigating the galeon. And now the commodore learnt from some of these prisoners that the other ship, which he had kept in the port of Acapulco the preceding year, instead of returning in company with the present prize, as was expected, had set sail from Acapulco alone much sooner than usual, and had, in all probability, got into the port of Manila long before the *Centurion* arrived off Cape Espiritu Santo, so that Mr. Anson, notwithstanding his present success, had great reason to regret his loss of time at Macao, which prevented him from taking two rich prizes instead of one.

The commodore, when the action was ended, resolved to make the best of his way with his prize for the river of Canton, being the meantime fully employed in securing his prisoners, and in removing the treasure from on board the galeon into the *Centurion*. The last of these operations was too important to be postponed, for as the navigation to Canton was thro' seas but little known, and where, from the season of the year, very tempestuous weather might be expected, it was of great consequence that the treasure should be sent on board the *Centurion*, which ship, by the presence of the commander-in-chief, the larger number of her hands, and her other advantages, was doubtless better provided against all the casualties of winds and seas than the galeon. And the securing the prisoners was a matter of still more consequence, as not only the possession of the treasure but the lives of the captors depended thereon. This was indeed an article which gave the commodore much trouble and disquietude, for they were above double the number of his own people, and some of them, when they were brought on board the *Centurion*, and had observed how slenderly she was manned, and the large proportion which the striplings bore to the rest, could not help expressing themselves with great indignation to be thus beaten by a handful of boys. The method which was taken to hinder them from rising was by placing all but the officers and the wounded in the hold, where, to give them as much air as possible, two hatchways were left open; but then (to avoid any danger that might happen whilst the *Centurion's* people should be employed upon deck) there was a square partition of thick planks, made in the shape of a funnel, which enclosed each hatchway on the lower deck, and reached to that directly over it on the upper deck. These funnels served to communicate the air to the hold better than could have been done without them, and, at the same time, added greatly to the security of the ship, for they being seven or eight feet high, it would have been extremely difficult for the Spaniards to have clambered up; and still to augment that difficulty, four swivel guns, loaded with musket-bullets, were planted at the mouth of each funnel, and a centinel with a lighted match was posted there ready to fire into the hold amongst them, in case of any disturbance. Their officers, who amounted to seventeen or eighteen, were all lodged in the first lieutenant's cabin, under a guard of six men; and the general, as he was wounded, lay in the commodore's cabin with a centinel always with him; every prisoner, too, was sufficiently apprised that any violence or disturbance would be punished with instant death. And, that the *Centurion's* people might be at all times prepared, if, notwithstanding these regulations, any tumult should arise, the small arms were constantly kept loaded in a proper place, whilst all the men went armed with cutlasses and pistols; and no officer ever pulled off his cloaths when he slept, or, when he lay down, omitted to have his arms always ready by him.

These measures were obviously necessary, considering the hazards to which the commodore and his people would have been exposed had they been less careful. Indeed, the sufferings of the poor prisoners, though impossible to be alleviated, were much to be commiserated; for the weather was extremely hot, the stench of the hold loathsome beyond all conception, and their allowance of water but just sufficient to keep them alive, it not being practicable to spare them more than at the rate of a pint a day for each, the crew themselves having only an allowance of a pint and a half. All this considered, it was wonderful that not a man of them died during their long confinement, except three of the wounded, who expired the same night they were taken, though it must be confessed that the greatest part of them were strangely metamorphosed by the heat of the hold; for when they were first brought on board, they were sightly robust fellows, but when, after above a month's imprisonment, they were discharged in the river Canton, they were reduced to mere skeletons, and their air and looks corresponded much more to the conception formed of ghosts and spectres than to the figure and appearance of real men.

Thus employed in securing the treasure and the prisoners, the commodore, as hath been said, stood for the river of Canton, and on the 30th of June, at six in the evening, got sight of Cape Delangano, which then bore west ten leagues distant. The next day he made the Bashee Islands, and the wind being so far to the northward that it was difficult to weather them, it was resolved to stand through between Grafton and Monmouth Islands, where the passage seemed to be clear, though in getting thro' the sea had a very dangerous aspect, for it ripled and foamed with all the appearances of being full of breakers, which was still more terrible as it was then night. But the ships got thro' very safe, the prize keeping ahead; and it was found that the agitation of the sea which had alarmed them, had been occasioned only by a strong tide. I must here observe that tho' the Bashee Islands are usually reckoned to be no more than five, yet there are many more lying about them to the westward, which, seeing the channels amongst them are not at all known, makes it adviseable for ships rather to pass to the northward or southward than thro' them; as indeed the commodore proposed to have gone to the northward between them and Formosa, had it been possible for him to have weathered them. From hence the *Centurion* steering the proper course for the river of Canton, she, on the 8th of July, discovered the island of Supata, the wester-most of the Lema Islands, being the double-peaked rock in the islands of Lema, formerly referred to. This island of Supata they made to be a hundred and thirty-nine leagues distant from Grafton's Island, and to bear from it north 82° 37' west. And on the 11th, having taken on board two Chinese pilots, one for the *Centurion*, and the other for the prize, they came to an anchor off the city of Macao.

By this time the particulars of the cargoe of the galeon were well ascertained, and it was found that she had on board 1,313,843 pieces of eight, and 35,682 oz. of virgin silver, besides some cochineal and a few other commodities, which, however, were but of small account in comparison of the specie. And this being the commodore's last prize, it hence appears that all the treasure taken by the *Centurion* was not much short of £400,000 independent of the ships and merchandize, which she either burnt or destroyed, and which, by the most reasonable estimation, could not amount to so little as £600,000 more: so that the whole damage done the enemy by our squadron did doubtless exceed a million sterling. To which, if there be added the great expence of the court of Spain, in fitting out Pizarro, and in paying the additional charges in America, incurred on our account, together with the loss of their men-of-war, the total of all these articles will be a most exorbitant sum, and is the strongest conviction of the utility of this expedition, which, with all its numerous disadvantages, did yet prove so extremely prejudicial to the enemy. I shall only add that there was taken on board the galeon several draughts and journals, from some of which many of the particulars recited in the tenth chapter of the second book are collected. Among the rest there was found a chart of all the ocean between the Philippines and the coast of Mexico, which was what was made use of by the galeon in her own navigation. With this digression I shall end this chapter, and leave the *Centurion* and her prize at anchor off Macao, preparing to enter the river of Canton.

CHAPTER IX

TRANSACTIONS IN THE RIVER OF CANTON

The commodore having taken pilots on board, proceeded with his prize for the river of Canton, and on the 14th of July cast anchor short of the Bocca Tigris, which is a narrow passage forming the mouth of that river. This entrance he proposed to stand through the next day, and to run up as far as Tiger Island, which is a very safe road, secured from all winds. But whilst the *Centurion* and her prize were thus at anchor, a boat with an officer was sent off from the mandarine commanding the forts at Bocca Tigris to examine what the ships were and whence they came. Mr. Anson informed the officer that his own ship was a man-of-war belonging to the King of Great Britain, and that the other in company with him was a prize he had taken, that he was going into Canton river to shelter himself against the hurricanes which were then approaching, and that as soon as the monsoon shifted he should set sail for England. The officer then desired an account of what men, guns, and ammunition were on board, a list of all which he said was to be sent to the government of Canton. But when these articles were repeated to him, particularly upon his being told that there were in the *Centurion* four hundred firelocks, and between three and four hundred barrels of powder, he shrugged up his shoulders and seemed to be terrified with the bare recital, saying that no ships ever came into Canton river armed in that manner; adding that he durst not set down the whole of this force, lest it should too much alarm the regency. After he had finished his enquiries, and was preparing to depart, he desired to leave two custom-house officers behind him, on which the commodore told him that though as a man-of-war he was prohibited from trading, and had nothing to do with customs or duties of any kind, yet for the satisfaction of the Chinese, he would permit two of their people to be left on board, who might themselves be witnesses how punctually he should comply with his instructions. The officer seemed amazed when Mr. Anson mentioned being exempted from all duties, and answered that the emperor's duty must be paid by every ship that came into his ports: and it is supposed that on this occasion private directions were given by him to the Chinese pilot not to carry the commodore through the Bocca Tigris, which makes it necessary more particularly to describe that entrance.

The Bocca Tigris is a narrow passage, little more than musquet-shot over, formed by two points of land, on each of which there is a fort, that on the starboard side being a battery on the water's edge, with eighteen embrasures, but where there were no more than twelve iron cannon mounted, seeming to be four or six-pounders; the fort on the larboard side is a large castle, resembling those old buildings which here in England we often find distinguished by that name; it is situated on a high rock, and did not appear to be furnished with more than eight or ten cannon, none of which were supposed to exceed six-pounders. These are the defences which secure the river of Canton, and which the Chinese (extremely defective in all military skill) have imagined were sufficient to prevent an enemy from forcing his way through.

But it is obvious from the description of these forts that they could have given no obstruction to Mr. Anson's passage, even if they had been well supplied with gunners and stores; and therefore, though the pilot, after the Chinese officer had been on board, refused at first to take charge of the ship till he had leave from the forts, yet as it was necessary to get through without any delay, for fear of the bad weather which was hourly expected, the commodore weighed on the 15th, and ordered the pilot to carry him by the forts, threatening him that if the ship ran aground he would instantly hang him up at the yard-arm. The pilot, awed by these threats, carried the ship through safely, the forts not attempting to dispute the passage. Indeed the poor pilot did not escape the resentment of his countrymen, for when he came on shore he was seized and sent to prison, and was rigorously disciplined with the bamboo. However, he found means to get at Mr. Anson afterwards, to desire of him some recompence for the chastisement he had undergone, and of which he then carried very significant marks about him; Mr. Anson, therefore, in commiseration of his sufferings, gave him such a sum of money as would at any time have enticed a Chinese to have undergone a dozen bastinadings.

Nor was the pilot the only person that suffered on this occasion; for the commodore soon after seeing some royal junks pass by him from Bocca Tigris towards Canton, he learnt, on enquiry, that the mandarine commanding the forts was a prisoner on board them; that he was already turned out, and was now carrying to Canton, where it was expected he would be severely punished for having permitted the ships to pass. Upon the commodore's urging the unreasonableness of this procedure, from the inability of the forts to have done otherwise, and explaining to the Chinese the great superiority his ships would have had over the forts, by the number and size of their guns, the Chinese seemed to acquiesce in his reasoning, and allowed that their forts could not have stopped him; but they still asserted that the mandarine would infallibly suffer for not having done what all his judges were convinced was impossible. To such indefensible absurdities are those obliged to submit who think themselves concerned to support their authority when the necessary force is wanting. But to return.

On the 16th of July the commodore sent his second lieutenant to Canton with a letter for the viceroy, informing him of the reason of the *Centurion's* putting into that port, and that the commodore himself soon proposed to repair to Canton to pay a visit to his excellency. The lieutenant was very civilly received, and was promised that an answer should be sent to the commodore the next day. In the meantime Mr. Anson gave leave to several of the officers of the galeon to go to Canton, they engaging their parole to return in two days. When these prisoners got to Canton, the regency sent for them and examined them, enquiring particularly by what means they came into Mr. Anson's power. It luckily happened that on this occasion the prisoners were honest enough to declare that as the kings of Great Britain and of Spain were at war they had proposed to themselves the taking of the *Centurion*, and had bore down upon her with that view, but that the event had been contrary to their hopes. And being questioned as to their usage on board, they frankly acknowledged that they had been treated by the commodore much better than they believed they should have treated him, had he fallen into their hands. This confession from an enemy had great weight with the Chinese, who till then, tho' they had revered the commodore's military force, had yet suspected his morals, and had considered him rather as a lawless free-booter than as one commissioned by the state for the revenge of public injuries. But they now changed their opinions, and regarded him as a more important person; to which perhaps the vast treasure of his prize might not a little contribute; the acquisition of wealth being a matter greatly adapted to the esteem and reverence of the Chinese nation.

In this examination of the Spanish prisoners, though the Chinese had no reason in the main to doubt of the account which was given them, yet there were two circumstances which appeared to them so singular as to deserve a more ample explanation; one of them was the great disproportion of men between the *Centurion* and the galeon, the other was the humanity with which the people of the galeon were treated after they were taken. The mandarines therefore asked the Spaniards how they came to be overpowered by so inferior a force? and how it happened, since the two nations were at war, that they were not put to death when they fell into the hands of the English? To the first of these enquiries the Spaniards answered that though they had more men than the *Centurion*, yet she being intended solely for war, had a great superiority in the size of her guns, and in many other articles, over the galeon, which was a vessel fitted out principally for traffic: and as to the second question, they told the Chinese that amongst the nations of Europe it was not customary to put to death those who submitted, though they readily owned that the commodore, from the natural bias of his temper, had treated both them and their countrymen, who had formerly been in his power, with very unusual courtesy, much beyond what they could have expected or than was required by the customs established between nations at war with each other. These replies fully satisfied the Chinese, and at the same time wrought very powerfully in the commodore's favour.

On the 20th of July, in the morning, three mandarines, with a great number of boats and a vast retinue, came on board the *Centurion* and delivered to the commodore the Viceroy of Canton's order for a daily supply of provisions, and for pilots to carry the ships up the river as far as the second bar; and at the same time they delivered him a message from the viceroy in answer to the letter sent to Canton. The substance of the message was that the viceroy desired to be excused from receiving the commodore's visit during the then excessive hot weather, because the assembling the mandarines and soldiers necessary to that ceremony would prove extremely inconvenient and fatiguing; but that in September when the weather would be more temperate he should be glad to see both the commodore himself and the English captain of the other ship that was with him. As Mr. Anson knew that an express had been dispatched to the court at Pekin with an account of the *Centurion* and her prize being arrived in the river of Canton, he had no doubt but the principal motive for putting off this visit was that the regency at Canton might gain time to receive the emperor's instructions about their behaviour on this unusual affair.

When the mandarines had delivered their message they began to talk to the commodore about the duties to be paid by his ships, but he immediately told them that he would never submit to any demand of that kind; that as he neither brought any merchandize thither, nor intended to carry any away, he could not be reasonably deemed within the meaning of the emperor's orders, which were doubtless calculated for trading vessels only, adding that no duties were ever demanded of men-of-war by nations accustomed to their reception, and that his master's orders expressly forbade him from paying any acknowledgment for his ships anchoring in any port whatever.

The mandarines being thus cut short on the subject of the duty, they said they had another matter to mention, which was the only remaining one they had in charge; this was a request to the commodore that he would release the prisoners he had taken on board the galeon; for that the Viceroy of Canton apprehended the emperor, his master, might be displeased if he should be informed that persons, who were his allies and carried on a great commerce with his subjects, were under confinement in his dominions. Mr. Anson himself was extremely desirous to get rid of the Spaniards, having on his first arrival sent about an hundred of them to Macao, and those who remained, which were near four hundred more, were, on many accounts, a great incumbrance to him. However, to inhance the favour, he at first raised some difficulties; but permitting himself to be prevailed on, he at last told the mandarines that to show his readiness to oblige the viceroy he would release the prisoners, whenever they, the Chinese, would order boats to fetch them off. This matter being thus adjusted, the mandarines departed; and on the 28th of July, two Chinese junks were sent from Canton to take on board the prisoners and to carry them to Macao. And the commodore, agreeable to his promise, dismissed them all, and directed his purser to allow them eight days' provision for their subsistence during their sailing down the river: since, before they were dispatched, the *Centurion* was arrived at her moorings, above the second bar, where she and her prize proposed to continue till the monsoon shifted.

Though the ships, in consequence of the viceroy's permit, found no difficulty in purchasing provisions for their daily consumption, yet it was impossible that the commodore could proceed to England without laying in a large quantity both of provisions and naval stores for his use during the voyage. The procuring this supply was attended with much perplexity; for there were people at Canton who had undertaken to furnish him with biscuit and whatever else he wanted; and his linguist, towards the middle of September, had assured him from day to day that all was ready and would be sent on board him immediately. But a fortnight being elapsed, and nothing brought, the commodore sent to Canton to enquire more particularly into the reasons of this disappointment: and he had soon the vexation to be informed that the whole was an illusion; that no order had been procured from the viceroy to furnish him with his sea stores, as had been pretended; that there was no biscuit baked, nor any one of the articles in readiness, which had been promised him; nor did it appear that the contractors had taken the least step to comply with their agreement. This was most disagreeable news, and made it suspected that the furnishing the *Centurion* for her return to Great Britain might prove a more troublesome matter than had been hitherto imagined; especially too as the month of September was nearly ended without Mr. Anson's having received any message from the Viceroy of Canton.

And here perhaps it might be expected that a satisfactory account should be given of the motives of the Chinese for this faithless procedure. However, as I have already, in a former chapter, made some kind of conjectures about a similar event, I shall not repeat them again in this place; but shall content myself with observing that after all it may perhaps be impossible for an European, ignorant of the customs and manners of that nation, to be fully apprized of the real incitements to this behaviour. Indeed, thus much may undoubtedly be asserted, that in artifice, falsehood, and an attachment to all kinds of lucre many of the Chinese are difficult to be paralleled by any other people. But then the particular application of these talents, and the manner in which they operate on every emergency, are often beyond the reach of a foreigner's penetration: so that though it may be surely concluded that the Chinese had some interest in thus amusing the commodore, yet it may not be easy to assign the individual views by which they were influenced. And that I may not be thought too severe in ascribing to this nation a fraudulent and selfish turn of temper, so contradictory to the character given of them in the legendary accounts of the Romish missionaries, I shall here mention an extraordinary transaction or two which I conceive will be some kind of confirmation of what I have advanced.

When the commodore lay first at Macao, one of his officers, who had been extremely ill, desired leave of him to go on shore every day on a neighbouring island, imagining that a walk upon the land would contribute greatly to the restoring of his health. The commodore would have dissuaded him from it, suspecting the tricks of the Chinese, but the officer continued importunate, in the end the boat was ordered to carry him thither. The first day he was put on shore he took his exercise and returned without receiving any molestation or even seeing any of the inhabitants; but the second day he was assaulted just after his arrival by a great number of Chinese who had been hoeing rice in the neighbourhood, and who beat him so violently with the handles of their hoes that they soon laid him on the ground incapable of resistance; after which they robbed him, taking from him his sword, the hilt of which was silver, his money, his watch, gold-headed cane, snuff-box, sleeve buttons, and hat, with several other trinkets. In the meantime, the boat's crew, who were at a little distance and had no arms of any kind with them, were incapable of giving him any relief; till at last one of them flew on the fellow who had the sword in his possession, and wresting it out of his hands, drew it, and with it was preparing to fall on the Chinese, some of whom he could not have failed of killing. But the officer, perceiving what he was about, immediately ordered him to desist, thinking it more prudent to submit to the present violence than to embroil his commander in an inextricable squabble with the Chinese Government by the death of their subjects: which calmness in this gentleman was the more meritorious as he was known to be a person of an uncommon spirit and of a somewhat hasty temper. By this means the Chinese speedily recovered the possession of the sword, when they perceived it was prohibited to be made use of against them, and carried off their whole booty unmolested. No sooner were they gone than a Chinese on horseback, very well dressed, and who had the air and appearance of a gentleman, came down to the seaside and, as far as could be understood by his signs, seemed to censure the conduct of his countrymen and to commiserate the officer, being wonderfully officious to assist in getting him on board the boat: but notwithstanding this behaviour, it was shrewdly suspected that he was an accomplice in the theft, and time fully made out the justice of those suspicions.

When the boat returned on board, and the officer reported what had passed to the commodore, he immediately complained of it to the mandarine who attended to see his ship supplied; but the mandarine coolly observed that the boat ought not to have gone on shore, promising, however, that if the thieves could be found they should be punished: though it appeared plain enough by his manner of answering that he would never give himself any trouble in searching them out. However, a considerable time afterwards, when some Chinese boats were selling provisions to the *Centurion*, the person who had wrested the sword from the Chinese came with eagerness to the commodore to assure him that one of the principal thieves was then in a provision boat alongside the ship; and the officer who had been robbed, viewing the fellow on this report, and well remembering his face, orders were immediately given to seize him; and he was accordingly secured on board the ship where strange discoveries were now made.

This thief on his being first apprehended expressed so much fright in his countenance that it was feared he would have died on the spot; the mandarine too who attended the ship had visibly no small share of concern on the occasion. Indeed he had reason enough to be alarmed, since it was soon apparent that he had been privy to the whole robbery; for the commodore declaring that he would not deliver up the thief, but would himself order him to be shot, the mandarine immediately put off the magisterial air, with which he had at first pretended to demand him, and begged his release in the most abject manner. But the commodore seeming to be inflexible, there came on board, in less than two hours' time, five or six of the neighbouring mandarines, who all joined in the same entreaty, and with a view of facilitating their suit, offered a large sum of money for the fellow's liberty. Whilst they were thus soliciting it was discovered that the mandarine, the most active amongst them, and who was thence presumed to be most interested in the event, was the very gentleman who rode up to the officer just after the robbery and who pretended to be so much displeased with the villainy of his countrymen. On further inquiry it was also found that he was the mandarine of the island, and that he had by the authority of his office ordered the peasants to commit that infamous action. This easily accounted for his extraordinary vigilance in the present conjuncture; since, as far as could be collected from the broken hints which were casually thrown out, it seemed that he and his brethren, who were every one privy to the transaction, were terrified with the fear of being called before the tribunal at Canton, where the first article of their punishment would be the stripping them of all they were worth; though their judges (however fond of inflicting a chastisement so lucrative to themselves) were perhaps of as tainted a complexion as the delinquents. Mr. Anson was not displeased to have caught the Chinese in this dilemma; he entertained himself for some time with their perplexity, rejecting their money with scorn, appearing inexorable to their prayers, and giving out that the thief should certainly be shot; but as he then foresaw that he should be forced to take shelter in their ports a second time, when the influence he might hereby acquire over the magistrates would be of great service to him, he at length permitted himself to be persuaded, and as a favour released his prisoner; though not till the mandarine had collected and returned all that had been stolen from the officer, even to the minutest trifle.

But notwithstanding this instance of the good intelligence between the magistrates and criminals, the strong addiction of the Chinese to lucre often prompts them to break through this awful confederacy, and puts them on defrauding the authority that protects them of its proper quota of the pillage. For not long after the above-mentioned transaction (the former mandarine, attendant on the ship, being in the meantime relieved by another) the commodore lost a top-mast from his stern, which, on the most diligent enquiry, could not be traced out. As it was not his own, but had been borrowed at Macao to heave down by, and was not to be replaced in that part of the world, he was extremely desirous to recover it, and published a considerable reward to any who would bring it him again. There were suspicions from the first of its being stolen, which made him conclude a reward was the likeliest method of getting it back. Hereupon, soon after, the mandarine informed him that some of his, the mandarine's, attendants had found the top-mast, desiring the commodore to send his boats to fetch it, which, being done, the mandarine's people received the promised reward. It seems the commodore had told the mandarine that he would make him a present besides on account of the care he had taken in directing it to be searched for; and accordingly Mr. Anson gave a sum of money to his linguist to be delivered to the mandarine; but the linguist knowing that the Chinese had been paid, and ignorant that a further present had been promised, kept the money himself. However, the mandarine fully confiding in Mr. Anson's word, and suspecting the linguist, he took occasion, one morning, to admire the size of the *Centurion's* masts, and thence on a pretended sudden recollection he made a digression to the top-mast which had been lost, and asked Mr. Anson if he had not got it again. Mr. Anson presently perceived the bent of this conversation, and enquired of him if he had not received the money from the linguist, and finding he had not, he offered to pay him upon the spot. But this the mandarine refused, having now somewhat more in view than the sum which had been detained. For the next day the linguist was seized, and was doubtless mulcted of whatever he had gotten in the commodore's service, which was supposed to be little less than two thousand dollars; being besides so severely bastinadoed that it was wonderful he escaped with his life. And when he was upbraided by the commodore (to whom he afterwards came a-begging) with his folly in risquing this severe chastisement, and the loss of all he was worth, for the lucre of fifty dollars, the present of which he defrauded the mandarine, he had no other excuse to make than the strong bias of his nation to dishonesty, replying in his broken jargon, "Chinese man very great rogue truly, but have fashion, no can help."

It were endless to recount all the artifices, extortions, and frauds which were practised on the commodore and his people by this interested race. The method of buying provisions in China being by weight, the tricks the Chinese made use of to augment the weight of what they sold to the *Centurion* were almost incredible. One time a large quantity of fouls and ducks being bought for the ship's store, the greatest part of them presently died. This spread a general alarm on board, it being apprehended that they had been killed by poison; but on examination it appeared that it was only owing to their being crammed with stones and gravel to increase their weight, the quantity thus forced into most of the ducks being found to amount to ten ounces in each. The hogs too, which were bought ready killed of the Chinese butchers, had water injected into them for the same purpose; so that a carcass hung up all night that the water might drain from it, had lost above a stone of its weight. And when, to avoid this cheat, the hogs were bought alive, it was discovered that the Chinese gave them salt to increase their thirst, and having thus excited them to drink great quantities of water, they then took measures to prevent them from discharging it again by urine, and sold the tortured animal in this inflated state. When the commodore first put to sea from Macao, they practised an artifice of another kind; for as the Chinese never scruple eating any food that dies of itself, they contrived by some secret practices that great part of his live sea-store should die in a short time after it was put on board, hoping to make a second profit of the dead carcasses which they expected would be thrown overboard; and two-thirds of the hogs dying before the *Centurion* was out of sight of land, many of the Chinese boats followed her only to pick up the carrion. These instances may serve as a specimen of the manners of this celebrated nation, which is often recommended to the rest of the world as a pattern of all kinds of laudable qualities. But to return.

The commodore, towards the end of September, having found out (as has been said) that those who had contracted to supply him with sea provisions and stores had deceived him, and that the viceroy had not invited him to an interview according to his promise, he saw it would be impossible for him to surmount the difficulties he was under without going to Canton and visiting the viceroy. And therefore, on the 27th of September, he sent a message to the mandarine who attended the *Centurion*, to inform him that he, the commodore, intended, on the 1st of October, to proceed in his boat to Canton: adding that the day after he got there he should notify his arrival to the viceroy, and should desire him to fix a time for his audience. This message being delivered to the mandarine, he returned no other answer than that he would acquaint the viceroy with the commodore's intentions. In the meantime all things were prepared for this expedition: and the boat's crew which Mr. Anson proposed to take with him were cloathed in an uniform dress, resembling that of the watermen on the Thames; they were in number eighteen and a coxswain; they had scarlet jackets and blue silk waistcoats, the whole trimmed with silver buttons, besides silver badges on their jackets and caps. As it was apprehended, and even asserted, that the payment of the customary duties for the *Centurion* and her prize would be demanded by the regency of Canton, and would be insisted on previous to their granting a permission to victual the ship for her future voyage, the commodore, who was resolved never to establish so dishonourable a precedent, took all possible precaution to prevent the Chinese from facilitating the success of their unseasonable pretensions by having him in their power at Canton. And therefore the better to secure his ship and the great treasure on board her against their projects, he appointed his first lieutenant, Mr. Brett, to be captain of the *Centurion* under him, giving him proper instructions for his conduct; directing him particularly, if he, the commodore, should be detained at Canton on account of the duties in dispute, to take out the men from the *Centurion's* prize and to destroy her, and then to proceed down the river through the Bocca Tigris with the *Centurion* alone, and to remain without that entrance till he received further orders from Mr. Anson.

These necessary steps being taken, which were not unknown to the Chinese, it should seem as if their deliberations were in some sort perplexed thereby. It is reasonable to imagine that they were in general very desirous of getting the duties to be paid them; not perhaps solely in consideration of the amount of those dues, but to keep up their reputation for address and subtlety, and to avoid the imputation of receding from claims on which they had already so frequently insisted. However, as they now foresaw that they had no other method of succeeding than by violence, and that even against this the commodore was prepared, they were at last disposed, I conceive, to let the affair drop rather than entangle themselves in an hostile measure which they found would only expose them to the risque of having the whole navigation of their port destroyed without any certain prospect of gaining their favourite point.

But though there is reason to conclude that these were their thoughts at that time, yet they could not depart at once from the evasive conduct to which they had hitherto adhered. Since when the commodore, on the morning of the 1st of October, was preparing to set out for Canton, his linguist came to him from the mandarine who attended the ship, to tell him that a letter had been received from the Viceroy of Canton, desiring the commodore to put off his going thither for two or three days. The reality of this message was not then questioned; but in the afternoon of the same day, another linguist came on board, who with much seeming fright told Mr. Anson that the viceroy had expected him up that day; that the council was assembled, and the troops had been under arms to receive him; and that the viceroy was highly offended at the disappointment, and had sent the commodore's linguist to prison, chained, supposing that the whole had been owing to the linguist's negligence. This plausible tale gave the commodore great concern, and made him apprehend that there was some treachery designed him which he could not yet fathom. And though it afterwards appeared that the whole was a fiction, not one article of it having the least foundation, yet for reasons best known to themselves this falshood was so well supported by the artifices of the Chinese merchants at Canton, that three days afterwards the commodore received a letter signed by all the supercargoes of the English ships then at that place, expressing their great uneasiness about what had happened, and intimating their fears that some insult would be offered to his boat if he came thither before the viceroy was fully satisfied of the mistake. To this letter Mr. Anson replied that he did not believe there had been a mistake; but was persuaded it was a forgery of the Chinese to prevent his visiting the viceroy; that therefore he would certainly come up to Canton on the 13th of October, confident that the Chinese would not dare to offer him any insult, as well knowing he should want neither power nor inclination to make them a proper return.

On the 13th of October, the commodore continuing firm to his resolution, all the supercargoes of the English, Danish, and Swedish ships came on board the *Centurion*, to accompany him to Canton, for which place he set out in his barge the same day, attended by his own boats, and by those of the trading ships, which on this occasion sent their boats to augment his retinue. As he passed by Wampo, where the European vessels lay, he was saluted by all of them but the French, and in the evening he arrived safely at Canton. His reception in that city, and the most material transactions from henceforward, till the expedition was brought to a period by the return of the *Centurion* to Great Britain, shall be the subject of the ensuing chapter.

CHAPTER X

PROCEEDINGS AT THE CITY OF CANTON, AND THE RETURN OF THE "CENTURION" TO ENGLAND

When the commodore arrived at Canton, he was visited by the principal Chinese merchants, who affected to appear very much pleased that he had met with no obstruction in getting thither, and who thence pretended to conclude that the viceroy was satisfied about the former mistake, the reality of which they still insisted on. In the conversation which passed upon this occasion, they took care to insinuate that as soon as the viceroy should be informed that Mr. Anson was at Canton, which they promised should be done the next morning, they were persuaded a time would be immediately appointed for the visit, which was the principal business that had brought the commodore to that city.

The next day the merchants returned to Mr. Anson and told him that the viceroy was then so fully employed in preparing his dispatches for Pekin that there was no getting admittance to him at present, but that they had engaged one of the officers of his court to give them information as soon as he should be at leisure, when they proposed to notify Mr. Anson's arrival and to endeavour to fix the audience. The commodore was already too well acquainted with their artifices not to perceive that this was a falshood, and had he consulted only his own judgment, he would have applied directly to the viceroy by other hands. But the Chinese merchants had so far prepossessed the supercargoes of our ships with chimerical fears that they, the supercargoes, were extremely apprehensive of being embroiled with the government, and of suffering in their interest, if those measures were taken which appeared to Mr. Anson at that time to be the most prudential: and therefore, lest the malice and double dealing of the Chinese might have given rise to some sinister incident, which would be afterwards laid at his door, he resolved to continue passive as long as it should appear that he lost no time by thus suspending his own opinion. In pursuance of this resolution, he proposed to the English that he would engage not to take any immediate step himself for getting admittance to the viceroy, provided the Chinese, who contracted to furnish his provisions, would let him see that his bread was baked, his meat salted, and his storee prepared with the utmost dispatch. But if by the time when all was in readiness to be shipped off, which it was supposed would be in about forty days, the merchants should not have procured the government's permission to send it on board, then the commodore was determined to apply to the viceroy himself. These were the terms Mr. Anson thought proper to offer to quiet the uneasiness of the supercargoes; and, notwithstanding the apparent equity of the conditions, many difficulties and objections were urged; nor would the Chinese agree to the proposal till the commodore had consented to pay for every article he bespoke before it was put in hand. However, at last, the contract being past, it was some satisfaction to the commodore to be certain that his preparations were now going on; and being himself on the spot, he took care to hasten them as much as possible.

During this interval, in which the stores and provisions were getting ready, the merchants continually entertained Mr. Anson with accounts of their various endeavours to procure a licence from the viceroy and their frequent disappointments. This was now a matter of amusement to the commodore, as he was fully satisfied there was not one word of truth in anything they said. But when all was compleated, and wanted only to be shipped, which was about the 24th of November, at which time, too, the N.E. monsoon was set in, he then resolved to demand an audience of the viceroy, as he was persuaded that, without this ceremony, the grant of a permission to take his stores on board would meet with great difficulty. On the 24th of November, therefore, Mr. Anson sent one of his officers to the mandarine who commanded the guard of the principal gate of the city of Canton with a letter directed to the viceroy. When this letter was delivered to the mandarine, he received the officer who brought it very civilly, and took down the contents of it in Chinese, and promised that the viceroy should be immediately acquainted with it; but told the officer it was not necessary he should wait for an answer, because a message would be sent to the commodore himself.

When Mr. Anson first determined to write this letter, he had been under great difficulties about a proper interpreter, as he was well aware that none of the Chinese usually employed as linguists could be relied on, but he at last prevailed with Mr. Flint, an English gentleman belonging to the factory, who spoke Chinese perfectly well, to accompany his officer. This person, who upon that occasion and many others was of singular service to the commodore, had been left at Canton, when a youth, by the late Captain Rigby. The leaving him there to learn the Chinese language was a step taken by that captain merely from his own persuasion of the considerable advantages which the East India Company might one day receive from an English interpreter, and tho' the utility of this measure has greatly exceeded all that was expected from it, yet I have not heard that it has been to this hour imitated: but we imprudently choose, except in this single instance, to carry on the vast transactions of the port of Canton either by the ridiculous jargon of broken English, which some few of the Chinese have learnt, or by the suspected interpretation of the linguists of other nations.

Two days after the sending the above-mentioned letter, a fire broke out in the suburbs of Canton. On the first alarm Mr. Anson went thither with his officers and his boat's crew to aid the Chinese. When he came there, he found that it had begun in a sailor's shed, and that by the slightness of the buildings, and the aukwardness of the Chinese, it was getting head apace. However, he perceived that by pulling down some of the adjacent sheds it might easily be extinguished; and particularly observing that it was then running along a wooden cornice, which blazed fiercely, and would immediately communicate the flame to a great distance, he ordered his people to begin with tearing away the cornice. This was presently attempted, and would have been soon executed, but in the meantime he was told that as there was no mandarine there, who alone has a power to direct on these occasions, the Chinese would make him, the commodore, answerable for whatever should be pulled down by his command. Hereupon Mr. Anson and his attendants desisted, and he sent them to the English factory, to assist in securing the company's treasure and effects, as it was easy to foresee that no distance was a protection against the rage of such a fire, where so little was done to put a stop to it; since all the while the Chinese contented themselves with viewing it, and now and then holding one of their idols near it, which they seemed to expect should check its progress. Indeed, at last, a mandarine came out of the city, attended by four or five hundred firemen. These made some feeble efforts to pull down the neighbouring houses, but by that time the fire had greatly extended itself and was got amongst the merchants' warehouses, and the Chinese firemen, wanting both skill and spirit, were incapable of checking its violence, so that its fury increased upon them, and it was feared the whole city would be destroyed. In this general confusion the viceroy himself came thither, and the commodore was sent to, and was intreated to afford his assistance, being told that he might take any measures he should think most prudent in the present emergency. Upon this message he went thither a second time, carrying with him about forty of his people, who, in the sight of the whole city, exerted themselves after so extraordinary a manner as in that country was altogether without example. For, behaving with the agility and boldness peculiar to sailors, they were rather animated than deterred by the flames and falling buildings amongst which they wrought; whence it was not uncommon to see the most forward of them tumble to the ground on the roofs and amidst the ruins of houses which their own efforts brought down under them. By their resolution and activity the fire was soon extinguished, to the amazement of the Chinese: and it fortunately happened too, that the buildings being all on one floor, and the materials slight, the seamen, notwithstanding their daring behaviour, escaped with no other injuries than some considerable bruises.

The fire, though at last thus luckily extinguished, did great mischief during the time it continued, for it consumed a hundred shops and eleven streets full of warehouses, so that the damage amounted to an immense sum; and one of the Chinese merchants, well known to the English, whose name was

Succoy, was supposed, for his own share, to have lost near two hundred thousand pounds sterling. It raged indeed with unusual violence, for in many of the warehouses there were large quantities of camphire, which greatly added to its fury, and produced a column of exceeding white flame, which blazed up into the air to such a prodigious height that it was distinctly seen on board the *Centurion*, though she was at least thirty miles distant.

Whilst the commodore and his people were labouring at the fire, and the terror of its becoming general still possessed the whole city, several of the most considerable Chinese merchants came to Mr. Anson to desire that he would let each of them have one of his soldiers (for such they stiled his boat's crew, from the uniformity of their dress) to guard their warehouses and dwelling-houses, which, from the known dishonesty of the populace, they feared would be pillaged in the tumult. Mr. Anson granted them this request, and all the men that he thus furnished behaved much to the satisfaction of the merchants, who afterwards highly applauded their great diligence and fidelity.

By this means, the resolution of the English in mastering the fire, and their trusty and prudent conduct where they were employed as safeguards, was the general subject of conversation amongst the Chinese. And the next morning many of the principal inhabitants waited on the commodore to thank him for his assistance, frankly owning to him that he had preserved their city from being totally consumed, as they could never have extinguished the fire of themselves. Soon after, too, a message came to the commodore from the viceroy, appointing the 30th of November for his audience, which sudden resolution of the viceroy, in a matter that had been so long agitated in vain, was also owing to the signal services performed by Mr. Anson and his people at the fire; of which the viceroy himself had been in some measure an eye-witness.

The fixing this business of the audience was on every account a circumstance with which Mr. Anson was much pleased, since he was satisfied the Chinese Government would not have determined this point without having agreed among themselves to give up their pretensions to the duties they claimed, and to grant him all he could reasonably ask. For, as they well knew the commodore's sentiments, it would have been a piece of imprudence, not consistent with their refined cunning, to have admitted him to an audience only to have contested with him. Being therefore himself perfectly easy about the result of this visit, he made the necessary preparations against the day, and engaged Mr. Flint, whom I have mentioned before, to act as interpreter in the conference: and Mr. Flint, in this affair as in all others, acquitted himself much to the commodore's satisfaction, repeating with great boldness, and doubtless with exactness, whatever was given him in charge, a part which no Chinese linguist would have performed with any tolerable fidelity.

At ten o'clock in the morning, on the day appointed, a mandarine came to the commodore to let him know that the viceroy was prepared, and expected him, on which the commodore and his retinue immediately set out. As soon as he entered the outer gate of the city, he found a guard of two hundred soldiers ready to receive him; these attended him to the great parade before the emperor's palace, where the viceroy then resided. In this parade, a body of troops, to the number of ten thousand, were drawn up under arms, who made a very fine appearance, they being all of them new cloathed for this ceremony. Mr. Anson, with his retinue, having passed through the middle of them, he was then conducted to the great hall of audience, where he found the viceroy seated under a rich canopy in the emperor's chair of state, with all his council of mandarines attending. Here there was a vacant seat prepared for the commodore, in which he was placed on his arrival. He was ranked the third in order from the viceroy, there being above him only the two chiefs of the law and of the treasury, who in the Chinese Government have precedence of all military officers. When the commodore was seated, he addressed himself to the viceroy by his interpreter, and began with reciting the various methods he had formerly taken to get an audience; adding that he imputed the delays he had met with to the insincerity of those he had employed, and that he had therefore no other means left than to send, as he had done, his own officer with a letter to the gate. On the mention of this the viceroy interrupted the interpreter, and bid him assure Mr. Anson that the first knowledge they had of his being at Canton was from that letter. Mr. Anson then proceeded, and told him that the subjects of the King of Great Britain trading to China had complained to him, the commodore, of the vexatious impositions both of the merchants and inferior custom-house officers, to which they were frequently necessitated to submit, by reason of the difficulty of getting access to the mandarines, who alone could grant them redress. That it was his, Mr. Anson's, duty, as an officer of the King of Great Britain, to lay before the viceroy these grievances of the British subjects, which he hoped the viceroy would take into consideration, and would give orders that hereafter there should be no just reason for complaint. Here Mr. Anson paused, and waited some time in expectation of an answer, but nothing being said, he asked his interpreter if he was certain the viceroy understood what he had urged; the interpreter told him he was certain it was understood, but he believed no reply would be made to it. Mr. Anson then represented to the viceroy the case of the ship *Haslingfield*, which, having been dismasted on the coast of China, had arrived in the river of Canton but a few days before. The people on board this vessel had been great sufferers by the fire; the captain in particular had all his goods burnt, and had lost besides, in the confusion, a chest of treasure of four thousand five hundred tahel, which was supposed to be stolen by the Chinese boatmen. Mr. Anson therefore desired that the captain might have the assistance of the government, as it was apprehended the money could never be recovered without the interposition of the mandarines. And to this request the viceroy made answer that, in settling the emperor's customs for that ship, some abatement should be made in consideration of her losses.

And now the commodore having dispatched the business with which the officers of the East India Company had entrusted him, he entered on his own affairs, acquainting the viceroy that the proper season was already set in for returning to Europe, and that he wanted only a licence to ship off his provisions and stores, which were all ready; and that as soon as this should be granted him, and he should have gotten his necessaries on board, he intended to leave the river of Canton and to make the best of his way for England. The viceroy replied to this that the licence should be immediately issued, and that everything should be ordered on board the following day. And finding that Mr. Anson had nothing farther to insist on, the viceroy continued the conversation for some time, acknowledging in very civil terms how much the Chinese were obliged to him for his signal services at the fire, and owning that he had saved the city from being destroyed: then observing that the *Centurion* had been a good while on their coast, he closed his discourse by wishing the commodore a prosperous voyage to Europe, after which the commodore, thanking him for his civility and assistance, took his leave.

As soon as the commodore was out of the hall of audience, he was much pressed to go into a neighbouring apartment, where there was an entertainment provided; but finding, on enquiry, that the viceroy himself was not to be present, he declined the invitation and departed, attended in the same manner as at his arrival, only on his leaving the city he was saluted with three guns, which are as many as in that country are ever fired on any ceremony. Thus the commodore, to his great joy, at last finished this troublesome affair, which, for the preceding four months, had given him much disquietude. Indeed he was highly pleased with procuring a licence for the shipping off his stores and provisions, as thereby he was enabled to return to Great Britain with the first of the monsoons, and to prevent all intelligence of his being expected: but this, though a very important point, was not the circumstance which gave him the greatest satisfaction, for he was more particularly attentive to the authentic precedent established on this occasion, by which his Majesty's ships of war are for the future exempted from all demands of duty in any of the ports of China.

In pursuance of the promises of the viceroy, the provisions were begun to be sent on board the day succeeding the audience, and four days after, the commodore embarked at Canton for the *Centurion*. And now all the preparations for putting to sea were pursued with so much vigilance, and were so soon compleated, that the 7th of December the *Centurion* and her prize unmoored and stood down the river, passing through the Bocca Tigris on the 10th. On this occasion I must observe that the Chinese had taken care to man the two forts on each side of that passage with as many men as they could well contain, the greatest part of them armed with pikes and matchlock musquets. These garrisons affected to shew themselves as much as possible to the ships, and were doubtless intended to induce Mr. Anson to think more reverently than he had hitherto done of the Chinese military power. For this purpose they were equipped with extraordinary parade, having a great number of colours exposed to view; and on the castle in particular there was laid considerable heaps of large stones, and a soldier of unusual size, dressed in very sightly armour, stalked about on the parapet with a battle-ax in his hand, endeavouring to put on as important and martial an air as possible, though some of the observers on board the *Centurion* shrewdly suspected, from the appearance of his armour, that instead of steel it was composed only of a particular kind of glittering paper.

The *Centurion* and her prize being now without the river of Canton, and consequently upon the point of leaving the Chinese jurisdiction, I beg leave, before I quit all mention of the Chinese affairs, to subjoin a few remarks on the disposition and genius of that celebrated people. And though it may be supposed that observations made at Canton only, a place situated in a corner of the empire, are very imperfect materials on which to found any general conclusions, yet as those who have had opportunities of examining the inner parts of the country have been evidently influenced by very ridiculous prepossessions, and as the transactions of Mr. Anson with the regency of Canton were of an uncommon nature, in which many circumstances occurred different perhaps from any which have happened before, I hope the following reflections, many of them drawn from these incidents, will not be altogether unacceptable to the reader.

That the Chinese are a very ingenious and industrious people is sufficiently evinced from the great number of curious manufactures which are established amongst them, and which are eagerly sought for by the most distant nations; but though skill in the handicraft art seems to be the most valuable qualification of this people, yet their talents therein are but of a second-rate kind, for they are much outdone by the Japanese in those manufactures which are common to both countries, and they are in numerous instances incapable of rivalling the mechanic dexterity of the Europeans. Indeed, their principal excellency seems to be imitation, and they accordingly labour under that poverty of genius which constantly attends all servile imitators. This is most conspicuous in works which require great truth and accuracy, as in clocks, watches, fire-arms, etc., for in all these, though they can copy the different parts, and can form some resemblance of the whole, yet they never could arrive at such a justness in their fabric as was necessary to produce the desired effect. If we pass from those employed in manufactures to artists of a superior class, as painters, statuaries, etc., in these matters they seem to be still more defective; their painters, though very numerous and in great esteem, rarely succeeding in the drawing or colouring of human figures, or in the grouping of large compositions; and though in flowers and birds their performances are much more admired, yet even in these some part of the merit is rather to be imputed to the native brightness and excellency of the colours than to the skill of the painter, since it is very unusual to see the light and shade justly and naturally handled, or to find that ease and grace in the drawing which are to be met with in the works of European artists. In short, there is a stiffness and minuteness in most of the Chinese productions which are extremely displeasing: and it may perhaps be truly asserted that these defects in their arts are entirely owing to the peculiar turn of the people, amongst whom nothing great or spirited is to be met with.

If we next examine the Chinese literature (taking our accounts from the writers who have endeavoured to represent it in the most favourable light), we shall find that on this head their obstinacy and absurdity are most wonderful; since though, for many ages, they have been surrounded by nations to whom the use of letters was familiar, yet they, the Chinese alone, have hitherto neglected to avail themselves of that almost divine invention, and have continued to adhere to the rude and inartificial method of representing words by arbitrary mark—a method which necessarily renders the number of their characters too great for human memory to manage, makes writing to be an art that requires prodigious application, and in which no man can be otherwise than partially skilled; whilst all reading and understanding of what is written is attended with infinite obscurity and confusion, as the connexion between these marks and the words they represent cannot be retained in books, but must be delivered down from age to age by oral tradition—and how uncertain this must prove in such a complicated subject is sufficiently obvious to those who have attended to the variation which all verbal relations undergo when they are transmitted thro' three or four hands only. Hence it is easy to conclude that the history and inventions of past ages recorded by these perplexed symbols must frequently prove unintelligible, and consequently the learning and boasted antiquity of the nation must, in numerous instances, be extremely problematical.

However, we are told by many of the missionaries that tho' the skill of the Chinese in science is confessedly much inferior to that of the Europeans, yet the morality and justice taught and practised by them are most exemplary: so that from the description given by some of these good fathers, one should be induced to believe that the whole empire was a well-governed affectionate family, where the only contests were who should exert the most humanity and social virtue. But our preceding relation of the behaviour of the magistrates, merchants, and tradesmen at Canton sufficiently refutes these Jesuitical fictions. Beside, as to their theories of morality, if we may judge from the specimens exhibited in the works of the missionaries, we shall find them frequently employed in recommending ridiculous attachment to certain frivolous points, instead of discussing the proper criterion of human actions, and regulating the general conduct of mankind to one another on reasonable and equitable principles. Indeed, the only pretension of the Chinese to a more refined morality than their neighbours is founded not on their integrity or beneficence, but solely on the affected evenness of their demeanor, and their constant attention to suppress all symptoms of passion and violence. But it must be considered that hypocrisy and fraud are often not less mischievous to the general interests of mankind than impetuosity and vehemence of temper: since these, though usually liable to the imputation of imprudence, do not exclude sincerity, benevolence, resolution, nor many other laudable qualities. And perhaps, if this matter was examined to the bottom, it would appear that the calm and patient turn of the Chinese, on which they so much value themselves, and which distinguishes the nation from all others, is in reality the source of the most exceptionable part of their character; for it has been often observed by those who have attended to the nature of mankind, that it is difficult to curb the more robust and violent passions without augmenting, at the same time, the force of the selfish ones. So that the timidity, dissimulation, and dishonesty of the Chinese may, in some sort, be owing to the composure and external decency so universally prevailing in that empire.

Thus much for the general disposition of the people: but I cannot dismiss this subject without adding a few words about the Chinese Government, that too having been the subject of boundless panegyric. And on this head I must observe that the favourable accounts often given of their prudent regulations for the administration of their domestic affairs are sufficiently confuted by their transactions with Mr. Anson, as we have seen that their magistrates are corrupt, their people thievish, and their tribunals venal and abounding with artifice. Nor is the constitution of the empire, or the general orders of the state, less liable to exception, since that form of government which does not in the first place provide for the security of the public against the enterprizes of foreign powers is certainly a most defective institution: and yet this populous, this rich and extensive country, so pompously celebrated for its refined wisdom and policy, was conquered about an age since by a handful of Tartars; and even now, through the cowardice of the inhabitants, and the want of proper military regulations, it continues exposed not only to the attempts of any potent state, but to the ravages of every petty invader. I have already observed, on occasion of the commodore's disputes with the Chinese, that the *Centurion* alone was an overmatch for all the naval power of that empire. This perhaps may appear an extraordinary position, but it is unquestionable, for I have examined two of the vessels made use of by the Chinese. The first of these is a junk of about a hundred and twenty tuns burthen, and was what the *Centurion* hove down by; these are most used in the great rivers, tho' they sometimes serve for small coasting voyages. The other junk is about two hundred and eighty tuns burthen, and is of the same form with those in which they trade to Cochinchina, Manila, Batavia, and Japan, tho' some of their trading vessels are of a much larger size; its head is perfectly flat, and when the vessel is deep laden, the second or third plank of this flat surface is oft-times under water. The masts, sails, and rigging of these vessels are ruder than the built, for their masts are made of trees, no otherwise fashioned than by barking them and lopping off their branches. Each mast has only two shrouds of twisted rattan, which are often both shifted to the weather side; and the halyard, when the yard is up, serves instead of a third shroud. The sails are of mat, strengthened every three feet by an horizontal rib of bamboo; they run upon the mast with hoops, and when they are lowered down they fold upon the deck. These traders carry no cannon, and it appears from this whole description that they are utterly incapable of resisting any European armed vessel. Nor is the state provided with ships of considerable force, or of a better fabric, to protect their merchantmen: for at Canton, where doubtless their principal naval power is stationed, we saw no more than four men-of-war junks, of about three hundred tuns burthen, being of the make already described, and mounted only with eight or ten guns, the largest of which did not exceed a four-pounder. This may suffice to give an idea of the defenceless state of the Chinese Empire. But it is time to return to the commodore, whom I left with his two ships without the Bocca Tigris, and who, on the 12th of December, anchored before the town of Macao.

Whilst the ships lay there, the merchants of Macao finished their purchase of the galeon, for which they refused to give more than 6000 dollars: this was greatly short of her value, but the impatience of the commodore to get to sea, to which the merchants were no strangers, prompted them to insist on these unequal terms. Mr. Anson had learnt enough from the English at Canton to conjecture that the war with Spain was still continued, and that probably the French might engage in the assistance of Spain before he could arrive in Great Britain; and therefore, knowing that no intelligence could come to Europe of the prize he had taken and the treasure he had on board till the return of the merchantmen from Canton, he was resolved to make all possible expedition in getting back, that he might be himself the first messenger of his own good fortune, and might thereby prevent the enemy from forming any projects to intercept him. For these reasons, he, to avoid all delay, accepted of the sum offered for the galeon, and she being delivered to the merchants the 15th of December 1743, the *Centurion* the same day got under sail on her return to England. On the 3d of January she came to anchor at Prince's Island in the Streights of Sunda, and continued there wooding and watering till the 8th, when she weighed and stood for the Cape of Good Hope, where, on the 11th of March, she anchored in Table Bay.

The Cape of Good Hope is situated in a temperate climate, where the excesses of heat and cold are rarely known, and the Dutch inhabitants, who are numerous, and who here retain their native industry, have stocked it with prodigious plenty of all sorts of fruits and provision, most of which, either from the equality of the seasons, or the peculiarity of the soil, are more delicious in their kind than can be met with elsewhere: so that by these, and by the excellent water which abounds there, this settlement is the best provided of any in the known world for the refreshment of seamen after long voyages. Here the commodore continued till the beginning of April, highly delighted with the place, which, by its extraordinary accommodations, the healthiness of its air, and the picturesque appearance of the country, the whole enlivened too by the addition of a civilized colony, was not disgraced on a comparison with the vallies of Juan Fernandes and lawns of Tinian. During his stay he entered about forty new men, and having, by the 3d of April 1744, compleated his water and provision, he, on that day, weighed and put to sea. The 19th of April they saw the island of St. Helena, which, however, they did not touch at, but stood on their way, and arriving in soundings about the beginning of June, they, on the 10th of that month, spoke with an English ship bound for Philadelphia, from whom they received the first intelligence of a French war. By the 12th of June they got sight of the Lizard, and the 15th, in the evening, to their infinite joy, they came safe to an anchor at Spithead. But that the signal perils which had so often threatned them in the preceding part of the enterprize might pursue them to the very last, Mr. Anson learnt on his arrival that there was a French fleet of considerable force cruising in the chops of the Channel, which, from the account of their position, he found the *Centurion* had ran through, and had been all the time concealed by a fog. Thus was this expedition finished, when it had lasted three years and nine months, after having, by its event, strongly evinced this important truth, that though prudence, intrepidity, and perseverance united are not exempted from the blows of adverse fortune, yet in a long series of transactions they usually rise superior to its power, and in the end rarely fail of proving successful.